ELEMENTARY MATHEMATICS:
Its Structure
and Concepts

Elementary Mathematics: Its Structure and Concepts

SECOND EDITION

Margaret F. Willerding

Professor of Mathematics
San Diego State College

John Wiley & Sons, Inc.
New York   London   Sydney   Toronto

Library of Congress Catalog Card Number: 71-100328

SBN 471 94665 6

Printed in the United States of America

3 4 5 6 7 8 9 10

To Herman, with deep affection

Preface

This revision of *Elementary Mathematics: Its Structure and Concepts* is based upon classroom experiences of many people at many colleges and universities since it first appeared in 1966. Although much of the original text has been rewritten and new material has been added, the underlying philosophy and purpose of the book remain the same. The fundamental purpose is to present a modern treatment of the topics of elementary mathematics to meet the needs of those students preparing to teach in the elementary schools.

The book is recommended for use as a text in a two-semester sequence of three units each semester. With certain omissions, as suggested in the instructor's manual, the book may be used for a one-semester course of four units, a two-quarter course of three units each quarter, a one-semester course of three units, or a one-quarter course of five units.

The first edition of *Elementary Mathematics: Its Structure and Concepts* evolved over a period of five years during which I taught the subject matter many times to both pre-service and in-service teachers. In selecting the material presented in the first edition I relied heavily upon my experience working with these pre-service and in-service teachers, and on the recommendations made by national committees and groups concerned with teaching mathematics in the elementary schools.

In preparing this revision I not only relied on the above mentioned groups and my experiences using the first edition with many classes, but upon my study of contempory textbooks for grades 1–8 and the suggestions from instructors from numerous colleges and universities using the first edition. However, the book reflects my own opinion as to content and presentation and does not follow any particular textbook series or outline of any particular committee or study group.

Although the revision gives a more formal treatment of the subject matter, its style and presentation is still designed for the student with little mathematical maturity and background. This simplicity of presentation has been deliberate. I have found that even though more and more students have studied elementary and high school mathematics from a curriculum emphasizing the structure and concepts of mathematics, many still enter colleges and universities unprepared for a completely rigorous treatment of the material presented here.

The chief changes from the first edition are as follows:

(1) Numerous illustrative examples are presented throughout the book to clarify the use of mathematical ideas.

(2) All proofs of theorems are given in greater detail to help the student develop an insight into the concept of deductive proof.

(3) The number of exercises in all chapters have been substantially increased.

(4) In Chapter 2, on logic, the sections on proof have been expanded and clarified.

(5) The chapter entitled Topics from Geometry has been lengthened to include the subject of geometric solids.

(6) Chapters 9 and 10, entitled Fractions and Non-Negative Rational Numbers respectively, have been completely revised and rewritten. Chapter 9 establishes the relationship between the concept of a number pair and a fraction. In this chapter we take the point of view that a fraction is a symbol for a number pair. Working with sets of equivalent fractions is an important part of this chapter and is a preparation for the introduction of non-negative rational numbers presented in Chapter 10.

(7) Having introduced fractions in Chapter 9, Chapter 10 introduces the concept of non-negative rational number. Just as we introduced the cardinal number concept from sets, we introduce the concept of non-negative rational number from sets of equivalent fractions.

(8) Chapter 15, Measurement of Geometric Figures, is completely new. This chapter includes measurement of line segments, perimeters and areas of polygons, circumference and area of the circle, and volumes of some solids.

(9) Chapter 16, Probability, is also new. This chapter is included because many of the contemporary elementary school mathematics texts contain topics from probability theory. Keeping this in mind, I felt that the elementary teacher needs to be introduced to some simple probability theory before he begins his teaching career. This chapter depends heavily upon set theory presented in Chapter 1, and may be presented as soon as Chapter 1 has been completed.

The answers to all odd-numbered exercises are included in the text. A separate teachers manual giving answers to even-numbered exercises, suggested material for review and testing purposes, a time schedule and recommended topics for emphasis is available.

I wish to thank all those persons who in one way or another, either consciously or unconsciously, have helped me write this book. To my colleagues, Dr. Marguerite Brydegaard, Dr. James Inskeep, Dr. Richard Madden, and Mrs. Alma Marosz, who have used this book in their classes, I am sincerely grateful. I wish to thank Professor Roy Dubisch and Professor Marjorie Pickering who reviewed the manuscript and who made valuable suggestions for improvements in presentation. In particular, I would like to thank Mrs. Ruth A. Hayward for her suggestions for making certain sections of the book more readable and for her help in proof-reading the manuscript.

MARGARET F. WILLERDING

San Diego, California
January, 1970

Contents

1.1. What Is a Set?

Every time we use the term "set" we are expressing a mathematical concept that is useful in all of mathematics. A **set** is simply a collection of things or objects. Familiar examples of sets are:

> A set of dishes
> A flock of geese (another way of saying a set of geese)
> A herd of cattle (another way of saying a set of cattle)
> An army (another way of saying a set of soldiers).

The objects in a set are called the **elements** or **members** of a set. In a set of dishes the individual cups, saucers, plates, etc., are the elements of the set. The individual members of a set are said to **belong to** or to be **contained in** the set.

DEFINITION 1.1.   *The objects in a set are called the **elements** or*

members of the set.

A mathematical set is **well-defined.** By this we mean that there is a method or rule whereby set membership or nonmembership can be determined. The sets we shall consider in this book are all well-defined,

Sets

that is, they are described so clearly that membership is unambiguous. Some examples of well-defined sets are:

The set of days of the week
The set of months of the year
The set of even natural numbers
The set of primary colors.

In each of the above sets no question arises as to whether or not an element belongs to a particular set.

1.2. Set Notation

A set may be specified by listing its elements. Thus the set pictured in Figure 1.1 consists of a dog, a cat, a squirrel, and a pig.

Figure 1.1

To denote this set we may use the symbol

$$\{\text{dog, cat, squirrel, pig}\}$$

The members of the set are enclosed in **braces,** { }, with a comma separating them. This symbol is read: "The set whose elements are dog, cat, squirrel, and pig." When each member of a set is tabulated, as in the set above, we say that we have **listed** its members.

If a set has a great many elements, we often abbreviate when listing its members. For example, we may denote the set of letters of the English alphabet by

$$\{a, b, c, \ldots, z\}$$

The three dots indicate that the letters d through y inclusive are also members of the set but have not been listed.

If we wish to list the members of a non-ending set, for example, the set of numbers used in counting, 1, 2, 3, 4, etc., called the set of **counting numbers** or **natural numbers,** we use three dots and write

$$\{1, 2, 3, 4, \ldots\}$$

A set may also be specified by giving some characteristic that each element must have to belong to the set. Let us consider the set

$$\{a, b, c, d\}$$

To belong to this set an element must be one of the first four letters of the English alphabet. We may use the following notation, called **set-builder notation,** to denote this set:

$$\{x \mid x \text{ is one of the first four letters of the English alphabet}\}$$

We read this symbol: "The set of all x such that x is one of the first four letters of the English alphabet." The vertical line in the set-builder notation is read "such that." The symbol x used in the set-builder notation is called a variable. A **variable** is a symbol representing an unspecified element of a given set, called the **replacement set** of the variable.

We customarily use capital letters such as A, B, C, and so forth, as names of sets. Thus we may write

$$A = \{a, b, c, d\}$$

We read this: "A equals the set whose elements are a, b, c, and d." Note that the symbol $=$ is read "equals" and means, in this case, that A and $\{a, b, c, d\}$ are names for the same thing.

Symbols to denote individual members of a set are usually lower case letters of the alphabet, a, b, c, d, and so forth. We use the symbol $\in$ to mean "is a member of" or "belongs to" a set. Thus if

$$B = \{p, q, r, s\}$$

then

$$p \in B$$

We read this: "p is a member of set B." We use the symbol $\notin$ to mean "is not a member of." Hence

$$a \notin B$$

EXAMPLE i.   List the elements of the set of the seasons of the year.
Solution: {summer, winter, fall, spring}.

EXAMPLE ii.   List the elements of the set

$$\{x \mid x \text{ is a month of the year with exactly 30 days}\}.$$

Solution: {April, June, September, November}.

EXAMPLE iii.   Use set-builder notation to denote the set of days of the week whose names begin with the letter "s."
Solution: $\{x \mid x$ is a day of the week whose name begins with "s"$\}$.

1.3   The Empty Set.

It is useful to have a set that contains no elements. Consider the set consisting of all the four-dollar bills in circulation in the United States. This set contains no elements. A set that contains no elements is called the **empty** or **null set.** The empty set is represented by the symbol ϕ or the symbol $\{\ \ \}$. Some examples of the empty set are:

The set of all rectangles with five sides
The set of icebergs in the Sahara desert
The set of coconut palms native to Alaska.

DEFINITION 1.2.   *The* **empty set,** *denoted by ϕ or $\{\ \ \}$, is the set that contains no elements.*

1.4 Equality of Sets

Two sets with exactly the same elements are called **equal** sets. Sets

$$C = \{a, b, c, d\} \quad \text{and} \quad D = \{b, a, d, c\}$$

are equal because every element of C is an element of D and every element of D is an element of C. We write

$$C = D$$

to indicate that set C and set D are equal.

We observe that the order in which the elements of a set are listed is immaterial. Just as a committee consisting of Albert, Brown, Cohen, and Davis is the same committee as one consisting of Brown, Albert, Davis, and Cohen, so a set having elements a, b, c, and d is the same as a set having elements b, a, d, and c.

Now let us consider the two sets:

$$A = \{\text{April, June, September, November}\}$$
$$B = \{x \mid x \text{ is a month with exactly 30 days}\}$$

Although we have denoted these two sets in different ways, they have exactly the same elements and hence are equal, that is, $A = B$.

DEFINITION 1.3. *Two sets, A and B, are defined to be* **equal,** *denoted by $A = B$, if they have exactly the same elements.*

Notice that:

(1) Every set is equal to itself; that is, $A = A$ for every set A.
(2) If set A is equal to set B, then set B is equal to set A; that is, if $A = B$, then $B = A$.
(3) If set A is equal to set B and set B is equal to set C, then set A is equal to set C; that is, if $A = B$ and $B = C$, then $A = C$.

1.5. One-to-One Correspondence

If the elements of two sets can be matched in some manner such that one element of each set is associated with a single element of the other, then the elements of the sets are said to be in **one-to-one correspondence.** The

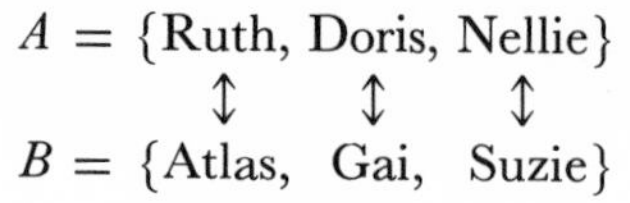

$$A = \{\text{Ruth, Doris, Nellie}\}$$
$$\updownarrow \qquad \updownarrow \qquad \updownarrow$$
$$B = \{\text{Atlas, Gai, Suzie}\}$$

Figure 1.2

elements of sets A and B in Figure 1.2 are matched and a one-to-one correspondence is exhibited.

Other pairings that set up a one-to-one correspondence between the elements of sets A and B above are shown in Figure 1.3. If the elements of two sets can be put into one-to-one correspondence in one way, they can be put into a one-to-one correspondence in at least one other way unless the sets have only one element.

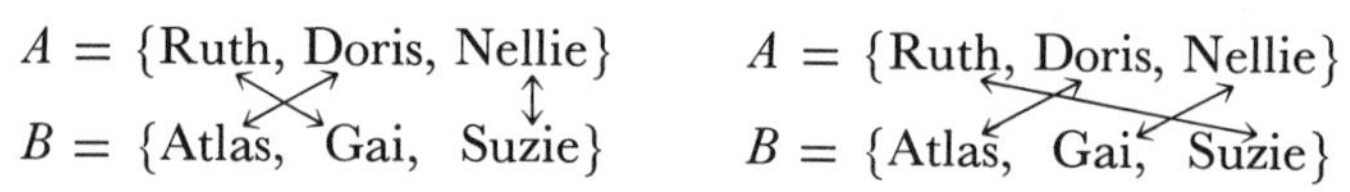

$$A = \{\text{Ruth, Doris, Nellie}\} \qquad A = \{\text{Ruth, Doris, Nellie}\}$$
$$B = \{\text{Atlas, Gai, Suzie}\} \qquad B = \{\text{Atlas, Gai, Suzie}\}$$

Figure 1.3

Instances of one-to-one correspondence between the elements of sets are easy to observe every day. For example, if a professor enters his class and finds that every student has a chair and there are no empty chairs, he knows immediately that there is a one-to-one correspondence between the students in the class and the chairs in the room.

Two sets whose elements can be put into one-to-one correspondence are called **equivalent sets.** This equivalence relationship between sets is not the same as the equality relation defined in Definition 1.3. Equal sets are equivalent, but equivalent sets are not necessarily equal. For example, consider the sets below:

$$P = \{a, b, c, d, e\}$$
$$Q = \{h, o, r, s, e\}$$
$$R = \{s, h, o, r, e\}$$

Set Q is equal to set R because the two sets have exactly the same elements. Set Q is also equivalent to set R since the elements of the two sets can be put into a one-to-one correspondence. Set Q is also equivalent to set P, but these two sets are not equal. Why?

We use the symbol $\sim$ (read: is equivalent to) to denote the equivalence of two sets. Thus sets P and Q above are equivalent (even though they are not equal) and we write $P \sim Q$.

DEFINITION 1.4. *Two sets, A and B, are* **equivalent,** *denoted by $A \sim B$, if the elements of each set may be put into a one-to-one correspondence with the elements of the other.*

We observe that:

(1) Every set is equivalent to itself; that is $A \sim A$.
(2) If set A is equivalent to set B, then set B is equivalent to set A; that is if $A \sim B$ then $B \sim A$.
(3) If set A is equivalent to set B and set B is equivalent to set C, then set A is equivalent to set C; that is if $A \sim B$ and $B \sim C$ then $A \sim C$.

Questions involving "more" and "fewer" can be resolved by a one-to-one pairing of elements of two sets that have a limited number of elements and by observing the existence of unpaired elements in one of the sets when the elements of the other are exhausted. For example, if we compare sets S and T where

$$S = \{a, b, c\}$$
$$T = \{x, y, z, w\}$$

it is apparent that they are not equivalent. If we begin pairing the elements of S with the elements of T, the elements of S are exhausted before those of T and we say that S has **fewer** elements than T or that T has **more** elements than S.

Exercise 1.1

List the elements of the following sets using braces. (Ex. 1–12)
1. The set of the months of the year whose names begin with A.
2. The set of months of the year with exactly 31 days.
3. The set of states of the United States whose names begin with the letter N.
4. The set of states of the United States whose names begin with the letter A.
5. The set of states of the United States that border Mexico.
6. The set of letters of the English alphabet.

7. The set of days of the week.
8. The set of months of the calendar year.
9. The set of days of the week whose names begin with the letter T.
10. The set of months of the year whose names begin with the letter J.
11. The set of the first five days of the week.
12. The set of natural numbers greater than 6.

Use set-builder notation to denote the following sets. (Ex. 13–17)
13. {Washington, Oregon, California, Alaska, Hawaii}
14. {Sunday, Saturday}
15. {January, February, March, April, May, June}
16. $\{w, x, y, z\}$
17. {Hawaii, Alaska}
18. What are the individual objects in a set called?
19. Which of the following are examples of the empty set?
    a. The set of days of the week whose names begin with the letter D.
    b. The set of men six feet tall.
    c. The set of triangles with five sides.
    d. The set of natural numbers greater than 25.
    e. The set of U. S. astronauts that have landed on the planet Pluto.
    f. $\{x \mid x$ is a rectangle with three sides$\}$
    g. $\{x \mid x$ is a natural number$\}$
    h. $\{x \mid x$ is a month of the calendar year with exactly 45 days$\}$
    i. $\{x \mid x$ is a day of the week with 48 hours$\}$
20. Which of the following sets are equivalent to $\{a, b, c, d, e\}$?
    a. $\{p, e, a, c, h\}$
    b. $\{0, 1, 2, 3, 4, 5\}$
    c. $\{x \mid x$ is a season of the year$\}$
    d. $\{x \mid x$ is the number of toes on the two feet of a human being$\}$
21. What symbols are used to denote the empty set?

Put the members of each of the following pairs of sets into a one-to-one correspondence in two ways. (Ex. 22–25)
22. $\{0, 5, 10, 15\}$; $\{1, 2, 3, 4\}$
23. $\{x, y, z\}$; $\{a, b, c\}$
24. $\{a, b\}$; $\{0, K\}$
25. {Mary, Jane, Bee, Ann}; {Joe, Charles, Walter, Oscar}
26. Put the members of set G and set H into a one-to-one correspondence in three different ways.
$$G = \{3, 4, 5, 6, 7\} \qquad H = \{a, b, c, d, e\}$$

27. Which of the following sets are equal to each other?

$A = \{a, b, c, d\}$ $\qquad$ $B = \{9, h, i, j\}$
$C = \{a, c, e, f\}$ $\qquad$ $D = \{d, c, a, b\}$
$E = \{4, 8, 16, 9\}$ $\qquad$ $F = \{16, 9, 4, 8\}$

28. Which of the following sets are equivalent to each other?

$A = \{1, 2, 3, 4\}$ $\qquad$ $B = \{2, 3, 4, 5, 6\}$
$C = \{2, 3, 4, 5\}$ $\qquad$ $D = \{a, b, c, d, e\}$
$E = \{4, 8, 15\}$ $\qquad$ $F = \{9, 12\}$

29. Which of the following sets are equivalent to $\{\alpha, \beta, \gamma, \delta, \epsilon, \lambda, \sigma\}$?

a. $\{a, b, c, d, e, f, g\}$
b. $\{x \mid x$ is a day of the week$\}$
c. $\{0, 1, 2, 3, 4, 5, 6, 7\}$
d. $\{x \mid x$ is a month of the calendar year$\}$

30. Replace the symbol * with $\in$ or $\notin$ to make true statements given
$A = \{1, 2, 3, 4\}$, $B = \{5, 6, 7, 8\}$, and $C = \{3, 6, 9, 12\}$

a. $1 * A$ $\qquad$ d. $1 * B$ $\qquad$ g. $1 * C$
b. $6 * A$ $\qquad$ e. $6 * B$ $\qquad$ h. $6 * C$
c. $3 * A$ $\qquad$ f. $3 * B$ $\qquad$ i. $3 * C$

31. Which of the following pairs of sets are equal? Which are equivalent? Which are neither equal nor equivalent?

a. $\{a, b, c\}$; $\{c, b, a\}$
b. $\emptyset$; $\{0\}$
c. $\{2, 4, 6, 8\}$; $\{1, 2, 3, 7\}$
d. $\{1, 3, 5, 7, 9, 11\}$; $\{a, b, c, d\}$
e. $\{3, 6, 9, 12, 15\}$; $\{2, 4, 6, 8\}$

32. If $A \sim B$, is A necessarily equal to B?

33. If $A \sim B$ and $B \sim C$, what relation exists between A and C?

34. Define equality of two sets.

35. Define equivalence of two sets.

1.6. Subsets

Any set B all of whose elements are also elements of set A is called a **subset** of A. For example, if

$$A = \{1, 2, 3, 4, 5\}$$

and

$$B = \{2, 4\}$$

since every element of B is also an element of A, B is a subset of A. To denote that B is a subset of A we write

$$B \subseteq A$$

We read this symbol, "B is a subset of A." Notice that the set $\{4, 5, 3, 2, 1\}$ is a subset of A since all of its elements are elements of A. However, this set is equal to A since it has exactly the same elements as A. In fact, *every set is a subset of itself. The empty set is also a subset of every set.*

DEFINITION 1.5. *Set B is a* **subset** *of set A, denoted by $B \subseteq A$, if every element of B is an element of A.*

According to Definition 1.5:

(1) Every set is a subset of itself; $A \subseteq A$.
(2) The empty set is a subset of every set: $\phi \subseteq A$.
(3) If set A is a subset of set B and set B is a subset of set C, then set A is a subset of set C; if $A \subseteq B$ and $B \subseteq C$, then $A \subseteq C$.

All of the subsets of a set A are called **proper** subsets of A except A itself which is called an **improper** subset of A. We use the symbol $\subset$ to denote a proper subset of a set. Note that the symbol $\subseteq$ denotes *any* subset of a given set, whether proper or improper, but the symbol $\subset$ denotes *only* a proper subset.

Observe that if $A \subseteq B$ and $B \subseteq A$, then $A = B$. Suppose $A \subseteq B$ and $B \subseteq A$ and $A \neq B$. Since $A \neq B$, A must contain some element that is not an element of B. This is impossible since $A \subseteq B$. Also B must contain some element that is not an element of A. This is also impossible since $B \subseteq A$. Hence A and B have exactly the same elements and are equal.

EXAMPLE i. Is it true that

$$\{x \mid x \text{ is a dog}\} \subseteq \{x \mid x \text{ is an animal}\}?$$

Solution: Since all dogs are animals, the set of all dogs is a subset of the set of all animals, and the given statement is true.

EXAMPLE ii. Let $A = \{1, 3, 5, 7\}$, $B = \{3\}$, $C = \{3, 5, 7\}$, $D = \phi$, and $E = \{3, 5, 7, 9, 11\}$.
(a) Which of the given sets are subsets of A?
(b) Which of the given sets are proper subsets of A?

Solution: (a) A, B, C, and D.
            (b) B, C, and D.

1.7. Number of Subsets

A given set may have many subsets. How many? Perhaps we can determine this number by considering a few specific cases. Let us consider the set

$$A_1 = \{1\}.$$

(The symbol A_1 is read "A sub-one." The "1" is called a subscript.) This set has two subsets:

$$\{1\}, \phi.$$

Next let us consider the set

$$A_2 = \{1, 2\}.$$

Its subsets are:

$$\{1, 2\}, \phi, \{1\}, \{2\}.$$

We see that a set of two elements has four subsets or two times as many as a set with one element.

A set of three elements

$$A_3 = \{1, 2, 3\}$$

has eight subsets:

$$\{1, 2, 3\}, \phi, \{1\}, \{2\}, \{3\}, \{1, 2\}, \{1, 3\}, \{2, 3\}.$$

We see that a set of three elements has twice as many subsets as a set with two elements.

Examine Table 1.1.

Although we shall not prove it here, the above examples should intuitively convince us that if a set has n elements then it has 2^n (that is, the product of n factors each of which is 2) subsets. Thus a set of eight elements has $2^8 = 256$ subsets; a set containing ten elements has $2^{10} = 1024$ subsets.

EXAMPLE i. How may subsets does $\{s, o, m, e\}$ have?
Solution: Since the given set has four elements, it has

$$2^4 = 2 \times 2 \times 2 \times 2 = 16$$

subsets.

Table 1.1

Set	Subsets	No. of Subsets
$\{1\}$	$\{1\}\phi$	$2 = 2^1$
$\{1, 2\}$	$\{1, 2\}\{1\}\{2\}\phi$	$4 = 2 \times 2 = 2^2$
$\{1, 2, 3\}$	$\{1, 2, 3\}\{1\}\{2\}\{3\}\phi$	$8 = 2 \times 2 \times 2 = 2^{3*}$
	$\{1, 2\}\{1, 3\}\{2, 3\}$	
$\{1, 2, 3, 4\}$	$\{1, 2, 3, 4\}\{1\}\{2\}\{3\}\{4\}$	$16 = 2 \times 2 \times 2 \times 2 = 2^4$
	$\{1, 2\}\{1, 3\}\{1, 4\}\{2, 3\}\{2, 4\}$	
	$\{3, 4\}\{1, 2, 3\}\{1, 2, 4\}$	
	$\{1, 3, 4\}\{2, 3, 4\}\phi$	
$\{1, 2, 3, 4, 5\}$	$\{1, 2, 3, 4, 5\}\{1\}\{2\}\{3\}\{4\}$	$32 = 2 \times 2 \times 2 \times 2 \times 2 = 2^5$
	$\{5\}\{1, 2\}\{1, 3\}\{1, 4\}\{1, 5\}$	
	$\{2, 3\}\{2, 4\}\{2, 5\}\{3, 4\}\{3, 5\}$	
	$\{4, 5\}\{1, 2, 3\}\{1, 2, 4\}$	
	$\{1, 2, 5\}\{1, 3, 4\}\{1, 4, 5\}\{1, 3, 5\}$	
	$\{2, 3, 4\}\{2, 3, 5\}\{3, 4, 5\}\{2, 4, 5\}$	
	$\{1, 2, 3, 4\}\{1, 2, 3, 5\}\{1, 2, 4, 5\}$	
	$\{1, 3, 4, 5\}\{2, 3, 4, 5\}\phi$	

* The superscript, 3 in this case, is called an exponent. Exponents will be discussed in Chapter 4.

EXAMPLE ii. How many subsets does the set given below have?

$$\{x \mid x \text{ is a day of the week}\}$$

Solution: This set is the set of all x such that x is a day of the week. Since there are seven days in a week this set has seven elements and

$$2^7 = 2 \times 2 \times 2 \times 2 \times 2 \times 2 \times 2 = 128$$

subsets.

EXAMPLE iii. How many subsets does the set given below have?

$$\{x \mid x \text{ is a state of the United States}\}$$

Solution: This set is the set of all the states of the United States.
This set has 50 elements since there are 50 states in the United States.
This set has 2^{50}* subsets.

* When the exponent is large, as in 2^{50}, and it is laborious to compute a standard numeral for the exponential form, we usually leave the result in exponential form, that is, in the form 2^{50}.

Exercise 1.2

1. List all of the subsets of $\{11, 12, 13\}$.
2. List all of the subsets of $\{$red, blue, green, yellow$\}$ that have three elements.
3. List all of the subsets of $\{$Marilyn, Sharon, Joan, Ruth$\}$ that have two elements.
4. For each set in the left column below, choose the sets from the right column which are subsets of it.

   a. $\{d, i, n\}$ (1) ϕ

   b. Set of letters of the word "dream"   (2) $\{e, d\}$

   c. $\{1, 2, 3, 4, 5\}$ (3) $\{2, 4\}$

   d. $\{2, 4, 6, 8, 10, 12\}$ (4) $\{d, m\}$

   e. $\{5, 10, 15, 20\}$ (5) $\{10\}$

   f. The set of numbers named on a clock face   (6) $\{m, a, d, e\}$

   g. The set of letters in the word "reading"   (7) $\{d, n\}$

                                                         (8) $\{10, 12\}$

5. Let $B = \{0, 2, 4, 6, 8, 10\}$
   a. Give an improper subset of B.
   b. An odd number is a number that has a remainder of 1 when divided by 2. Thus 17 is an odd number because 17 divided by 2 gives a quotient of 8 and a remainder of 1. Give the subset of B that contains all the odd numbers which are elements of B.
   c. A multiple of 3 is a number that has a remainder of 0 when divided by 3. Thus 21 is a multiple of 3 because 21 divided by 3 gives a quotient of 7 and a remainder of 0. Give the subset of B that contains all of the elements of B that are multiples of 3.
   d. Give a subset of B that contains all of the elements of B greater than 7.

6. Given $A = \{1, 2, 3, 4, 5, 6, 7, 8, 9\}$
   a. An even number is a number that has a remainder of 0 when divided by 2. Thus 12 is an even number because 12 divided by 2 gives a quotient of 6 and a remainder of 0. Give the subset of A that contains all the even numbers that are elements of A.
   b. Give the subset of A that contains no elements.
   c. $B = \{9, 8, 7, 6, 5, 4, 3, 2, 1\}$. Is B a subset of A?
   d. Is B in part c a proper or improper subset of A?

e.   Is set B in part c equal to set A?

f.   Is set B in part c equivalent to set A?

7.   Give all of the subsets of B when

a. $B = \{1, 2, 3, 4\}$

b. $B = \{a, b, c\}$

c. $B = \{5, 7, 9, 11, 13\}$

8.   Give the number of subsets of B when

a. $B = \{☆, \text{①}\}$

b. $B = \{□, △, ○, ◁\}$

c. $B = \{5, 7, 9, 11, 13\}$

d. $B = \{1, 2, 3, \ldots, 300\}$

9.   Given $A = \{a, b, c, d, e, f, g, h\}$, $B = \{a, c, e\}$, and $C = \{a, c, e, f, g, h\}$, which of the following are true?

a. $A \subset C$ c. $B \subseteq A$

b. $C \subset A$ d. $B \subseteq C$

10. Which of the following sets are subsets of the set of letters in the word *matching*?

a. $\{t, a, m\}$

b. $\{m, a, t, c, h\}$

c. $\{t, h, i, n, g\}$

d. ϕ

e. $\{t, a, m, i, n, h\}$

11. Give the number of subsets for each of the following sets. Give the answers in exponential form, as 2^7.

a. $\{x \mid x$ is a member of the United States Senate$\}$

b. $\{x \mid x$ is a letter of the English alphabet$\}$

c. $\{x \mid x$ is a season of the year$\}$

d. $\{x \mid x$ is a month with exactly 31 days$\}$

e. $\{x \mid x$ is a state of the United States admitted to the union after 1940$\}$

12. If $A \subseteq B$, $B \subseteq C$, $C \subseteq D$, and $D \subseteq E$, what relation does A have to E?

13. The only proper subset of set B is ϕ. How many elements does set B have?

14. In the freshman class, all students are preparing to take the mathematics placement test. Is the set of freshmen preparing for the placement test a proper or an improper set of the freshman class?

15. How many different basketball teams (each a subset of five elements)

can be formed from a set of eight students? (Disregard the actual position held by the players.)

16. Is the defensive team a proper or an improper subset of the set of players on the varsity football team?

17. Let $U = \{x \mid x$ is a state of the United States and x is east of the Mississippi River$\}$. Which of the following are subsets of U?

    a. $\{x \mid x$ is a state of the United States that borders the Atlantic Ocean$\}$

    b. $\{x \mid x$ is a state of the United States that borders Canada$\}$

    c. $\{x \mid x$ is a New England State$\}$

    d. $\{x \mid x$ is a state north of the Mason Dixon Line$\}$

    e. $\{x \mid x$ is a state bordering the Mississippi River$\}$

1.8. Universal Set

It is useful when discussing sets to have the members of a given set come from some specified "population." For example, if we are talking about sets of committees of a club, we must know the general population or membership of the club whom we consider to be eligible members of the committees. The specified club from which we are forming our sets (committees in this case) is called the **universal set** or, simply, the **universe.** We denote the universal set by the capital letter U.

In any particular discussion involving sets, every set in the discussion is a subset of the universe. Of course, as the sets under discussion change, so does the universe.

EXAMPLE i. What is a suitable universe for the following sets?

    {Babe Ruth, Grover Cleveland Alexander}
    {Stan Musial, Joe DiMaggio}
    {Mickey Mantle, Roger Maris}

Solution: Since the elements in each of the given sets are baseball players, a suitable universe is the set of baseball players.

EXAMPLE ii. What is a suitable universe for the following sets?

    {San Francisco, Los Angeles, San Diego}
    $\{x \mid x$ is a city of the state of New York$\}$
    {Miami, Atlanta, Dallas, New Orleans}

Solution: Since each element in the sets above is a city of the United States, a suitable universe is

$$U = \{x \mid x \text{ is a city of the United States}\}$$

1.9. Venn Diagrams

There is a schematic representation used to depict set concepts. These representations are called **Venn diagrams** or **Euler diagrams.** In a Venn diagram a rectangular region usually represents the universal set. The elements of U are represented as points inside the rectangle. The size of the rectangular region has nothing to do with the number of elements of U. Venn diagrams are used to picture relationships between sets rather than their comparative sizes. A subset of U is usually represented by a circular region inside the rectangular region representing the universe.

Figure 1.4 shows two subsets, A and B, of a universe U that have no elements in common. Two sets which have no elements in common are called **disjoint sets.**

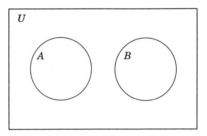

Figure 1.4

Sets which may be represented by the diagram in Figure 1.4 are:

i. $U = \{x \mid x \text{ is an animal}\}$; $A = \{x \mid x \text{ is a dog}\}$;
$B = \{x \mid x \text{ is a cat}\}$.

ii. $U = \{x \mid x \text{ is a human being}\}$; $A = \{x \mid x \text{ is a female}\}$;
$B = \{x \mid x \text{ is a male}\}$.

iii. $U = \{x \mid x \text{ is a college student}\}$; $A = \{x \mid x \text{ is a freshman}\}$;
$B = \{x \mid x \text{ is a senior}\}$.

Figure 1.5 shows two subsets D and E of U with $E \subseteq D$.

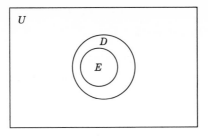

Figure 1.5

Sets which may be represented by the diagram in Figure 1.5 are:

iv. $U = \{x \mid x$ is an animal$\}$; $E = \{x \mid x$ is a poodle$\}$;
$D = \{x \mid x$ is a dog$\}$.

v. $U = \{x \mid x$ is a human being$\}$;
$E = \{x \mid x$ is a Senator of the U.S.A.$\}$;
$D = \{x \mid x$ is over 21 years of age$\}$.

vi. $U = \{x \mid x$ is a college student$\}$;
$E = \{x \mid x$ is a member of the men's glee club$\}$;
$D = \{x \mid x$ is a male$\}$.

Figure 1.6 shows two set, P and Q, which have elements in common and which are subsets of U. Sets which have elements in common are sometimes called **overlapping sets.**

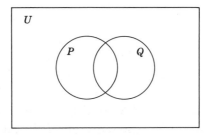

Figure 1.6

Sets which may be represented by the diagram in Figure 1.6 are:

vii. $U = \{x \mid x$ is an animal$\}$; $P = \{x \mid x$ is a poodle$\}$;
$Q = \{x \mid x$ is a white dog$\}$.

viii. $U = \{x \mid x$ is a human being$\}$; $P = \{x \mid x$ has blonde hair$\}$;
$Q = \{x \mid x$ is 6 feet tall$\}$.

ix. $U = \{x \mid x$ is a college student$\}$; $P = \{x \mid x$ is an honor student$\}$; $Q = \{x \mid x$ is a senior$\}$.

When more than two sets are related, their relationship may also be shown in a Venn diagram. Examples of such relationships are shown in Figure 1.7.

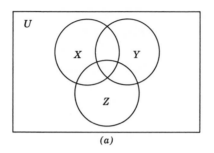

(a)

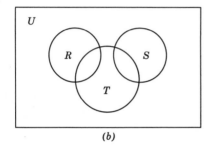 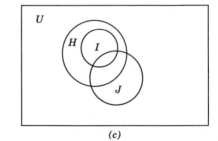

(b)   (c)

Figure 1.7

Figure 1.7a shows a Venn diagram representing three overlapping sets, X, Y, and Z. Each set represented contains some elements in both of the other sets. Figure 1.7b represents three sets, R, S, and T. Sets R and T are overlapping, that is, they have elements in common. Sets S and T are also overlapping sets. Sets R and S are disjoint, that is, they have no elements in common. Figure 1.7c represents sets H, I, and J, where I is a subset of H, and J overlaps both H and I.

Exercise 1.3

1.   What is a suitable universal set for each of the following pairs of sets?
    a. $A = \{$horse, dog, cat, lion$\}$; $B = \{$poodle, boxer, terrier$\}$.

b. $A = \{$lily, rose, jonquil$\}$; $B = \{$cherry, apple, peach, lemon$\}$.

c. $A = \{$football, basketball, soccer$\}$; $B = \{$poker, bridge, rummy$\}$.

2. Let $U = \{x \mid x$ is a college student$\}$. Use Venn diagrams to illustrate the relation of the following subsets of U.

a. $A = \{x \mid x$ is taking Mathematics courses$\}$

$B = \{x \mid x$ is taking Education courses$\}$

b. $A = \{x \mid x$ is on the football team$\}$

$B = \{x \mid x$ is on the basketball team$\}$

c. $A = \{x \mid x$ is over twenty-one years of age$\}$

$B = \{x \mid x$ is under twenty years of age$\}$

d. $A = \{x \mid x$ is a male$\}$

$B = \{x \mid x$ plays in the band$\}$

$C = \{x \mid x$ plays in the orchestra$\}$

3. Draw a Venn diagram illustrating the relationships between the various sets. In each case U is the universe and A and B are subsets of U.

a. $U = \{x \mid x$ is a boat anchored at San Diego Yacht Club$\}$

$A = \{x \mid x$ is a sailboat$\}$

$B = \{x \mid x$ is a boat over 25 feet in length$\}$

b. $U = \{x \mid x$ is a counting number$\} = \{1, 2, 3, 4, \ldots\}$

$A = \{x \mid x$ is less than 25$\}$

$B = \{x \mid x$ is greater than 10$\}$

c. $U = \{x \mid x$ is a four-sided polygon$\}$

$A = \{x \mid x$ is a rectangle$\}$

$B = \{x \mid x$ is a square$\}$

4. What is a suitable universal set for each of the following sets?

a. The set of honor students at New York University.

b. The set of Chevrolet cars.

c. The set of coffee cups.

d. The set of ski sweaters.

5. Draw a Venn diagram to show the relationship between the following sets, all subsets of the set of university students.

A is the set of freshmen. D is the set of blonde female students.

B is the set of seniors. E is the set of honor students.

C is the set of male students. F is the set of freshmen women.

6. Draw a Venn diagram which illustrates the proper relationship between the following sets, which are subsets of the set of all animals.

A is the set of all four-legged animals.

B is the set of all two-legged animals.

C is the set of all cows.

D is the set of all male human beings.

7. Draw a Venn diagram which illustrates the proper relationship between the following sets which are subsets of the set of countries of the world.

A is the set of all European countries.

B is the set of all North American countries.

C is the set of all countries which are members of the United Nations.

D is the set of all communist countries.

8. Draw a Venn diagram which illustrates the proper relationship between the following sets which are subsets of the set of all living Americans.

A is the set of all members of the House of Representatives.

B is the set of all persons who are registered voters in the state of Montana.

C is the set of all persons over twenty-one years of age.

D is the set of all persons who live in a New England state.

9. Draw a Venn diagram which illustrates the proper relationship between the following sets which are all subsets of the set of college students.

A is the set of members of the marching band.

B is the set of all juniors.

C is the set of all students with a B average.

D is the set of all seniors.

1.10. Number and Numeral

Number is a mathematical abstraction. There is a distinction between a number and the symbol used to represent it. **Numerals** are names for numbers. Thus the symbol "4" is a numeral used to name the number four. The same idea is used when we write the name "Ruth." The symbol names a girl, but is not the girl herself.

In ordinary language there is usually little difficulty in distinguishing between the object and its name or symbol. In writing, since it is sometimes more difficult to make the distinction, we use quotation marks (either single or double) to enclose the symbol which names an object. For example:

(1) Ruth has four letters.

(2) "Ruth" has four letters.

Statement (1) means that a girl, Ruth, has received four letters in the mail.

Statement (2) means that the word "Ruth" contains four letters of the alphabet.

In mathematics we handle comparable situations similarly:

(3) Write "4" on the board.
(4) The sum of 2 and 3 is 5.
(5) He said 3, but he wrote "8" on the board.

Statement (3) means that the numeral 4, which is a name for the number four, is to be written on the board. Statement (4) means that the sum of numbers 2 and 3 is the number 5. Statement (5) means that someone said "three", meaning the number three, and wrote the numeral 8 on the board.

In this book, when the context is clear, quotation marks will be omitted. The important thing to remember is that symbols like 4, 5, and 25 are *names* for numbers and not numbers themselves. These symbols are numerals.

1.11. Cardinality

Let us consider the following sets:

$$A = \{a, b\}$$
$$B = \{\text{Nellie, Ruth}\}$$
$$C = \{\text{Gai, Suzie}\}$$
$$D = \{\text{Nhu, Atlas}\}$$
$$E = \{1, 2\}$$

Each of the above sets is equivalent to any other since a one-to-one correspondence can be established between the elements of any two of the sets. Let us imagine all possible sets that are equivalent to set E; the common characteristic they all have is that they are equivalent to each other.

Using this idea we can classify sets into sets of equivalent sets. Any set that has a limited number of elements belongs to only one set of equivalent sets. Formally, we see that every set in a particular class of sets is equivalent to every other set in this class. Informally, we say that any two sets in a particular class have the same number of elements.

We use this idea of classification of sets to define the abstract concept of number. The common property of all sets equivalent to set E above, for example, is called the **number property** or the **cardinality** of the set, in this case 2. We denote the cardinality of E by the symbol $n(E)$ and write $n(E) = 2$. Similarly

$$n(A) = n(B) = n(C) = n(D) = 2$$

The cardinality of

$$S = \{1, 2, 3, 4, 5, 6\}$$

is 6 and we write $n(S) = 6$.

We see that we are using the concept of equivalent sets to define an abstract idea—the concept of cardinal number. For any set A with a limited number of elements $n(A)$ denotes the number property or cardinality of A and is called a **cardinal number.** The empty set, ϕ, is assigned the cardinal number zero, that is, $n(\phi) = 0$.

EXAMPLE i.   What is the cardinality of $A = \{a, b, c, d, e, f\}$?
Solution: $n(A) = 6$
EXAMPLE ii.   What is the cardinality of $P = \{1, 2, 3, \ldots, 300\}$?
Solution: We see that the elements of P are the first three hundred natural numbers, hence

$$n(P) = 300$$

1.12.   Finite and Infinite Sets

Some sets have an unlimited number of elements. For example, the set, N, of natural numbers

$$N = \{1, 2, 3, 4, 5, \ldots\}$$

is an unending set. We say that the set of natural numbers is an **infinite set.** The set of cardinal numbers

$$\{0, 1, 2, 3, \ldots\}$$

is also an infinite set. Other examples of infinite sets are:

The set of even cardinal numbers: $\{0, 2, 4, 5, \ldots\}$
The set of odd cardinal numbers: $\{1, 3, 5, 7, \ldots\}$
The set of nonzero multiples of 5: $\{5, 10, 15, 20, \ldots\}$

A set that is not an infinite set is called a **finite** set. A set is a finite set if the elements of the set can be arranged in some order and counted and such a counting terminates. In other words, a finite set has a limited number of elements.

A set may contain a fantastic number of elements and still be a finite set. For example, the set of persons who live in the United States at a particular moment is a finite set because the members of the set can be counted and the counting will ultimately come to an end. Some examples of finite sets are:

The set of men who have been president of the United States
The set of all grains of sand on the beach at Coney Island at any particular moment
The set of cardinal numbers less than 1,000,000,000.

Let us consider the set C of cardinal numbers and the set E of even cardinal numbers.

$$C = \{0, 1, 2, 3, 4, \ldots\}$$
$$E = \{0, 2, 4, 6, 8, \ldots\}$$

Both of these sets are infinite sets. Set E is a non-empty proper subset of set C since every element of E is an element of C. Let us set up a one-to-one correspondence between the elements of these two sets:

$$
\begin{array}{cccccc}
0 & 1 & 2 & 3 & 4 & 5 \quad \cdots \\
\updownarrow & \updownarrow & \updownarrow & \updownarrow & \updownarrow & \updownarrow \\
0 & 2 & 4 & 6 & 8 & 10 \quad \cdots
\end{array}
$$

The rule for correspondence is $n \leftrightarrow 2n$, where n is a cardinal number. For example, $0 \leftrightarrow 2 \cdot 0 = 0$; $1 \leftrightarrow 2 \cdot 1 = 2$; $2 \leftrightarrow 2 \cdot 2 = 4$; $\ldots$, $20 \leftrightarrow 2 \cdot 20 = 40$; and so forth.

According to our definition of equivalence, $C \sim E$. But $E \subset C$. We see then that an *infinite* set may be put into one-to-one correspondence with a non-empty proper subset of itself. This is never possible with finite sets.

Exercise 1.4

1. Given $A = \{1, 2, 3, 4\}$, $B = \{1, 3, 5, 7, 9\}$, and $C = \{5, 10, 15\}$, find
   a. $n(A)$
   b. $n(B)$
   c. $n(C)$

2. What is the cardinality of $\{x \mid x$ is a natural number less than 12$\}$?
3. What is the number property of $\{x \mid x$ is a natural number less than 5 and greater than 2$\}$?
4. What is the number property of $\{x \mid x$ is a natural number greater than 8 and less than 25$\}$?
5. Which of the following sets are infinite sets? Which are finite sets?
   a. The set of ships in the United States Navy.
   b. The set of natural numbers.
   c. The set of all people living on the planet Earth at this moment.
   d. The set of all vowels in the English alphabet.
   e. The set of grains of sand on Jones Beach at this moment.
   f. The set of all cities in Europe.
   g. The set of all countries in Africa.
   h. The set of monkeys in the St. Louis Zoo.
   i. The set of cardinal numbers greater than 4000.
6. Which of the following are infinite sets? Which are finite sets?
   a. $\{x \mid x$ is a cardinal number less than 1,000,000,000$\}$
   b. $\{x \mid x$ is a natural number greater than 1,000,000,000$\}$
   c. $\{x \mid x$ is a college student enrolled in a Junior College in California$\}$
   d. $\{x \mid x$ is a city in Pennsylvania$\}$
   e. $\{x \mid x$ is a name listed in the New York City telephone directory$\}$
   f. $\{x \mid x$ is a cardinal number less than 45$\}$
7. Give the cardinality of each of the following sets.
   a. $\{1, 5, 7, 8, 11, 2, 9\}$
   b. $\{h, o, r, s, e\}$
   c. $\{1, 2, 3, \ldots, 465\}$
   d. The set of states of the United States
   e. The set of Senators in the United States Congress
   f. $\{x \mid x$ is a cardinal number less than 25$\}$
   g. $\{x \mid x$ is a cardinal number greater than 6 and less than 30$\}$
8. What is the cardinality of $\{x \mid x$ is a cardinal number less than 8$\}$?
9. What is the cardinality of $\{3, 7, 11, 15, \ldots, 39\}$
10. Show that the set of natural numbers can be put into one-to-one correspondence with the set of odd natural numbers.
11. Show that the set of natural numbers can be put into a one-to-one correspondence with the non-zero multiples of 3.
12. Explain why the following statement may be used as a definition of an infinite set: An infinite set is a set whose elements can be put into

one-to-one correspondence with the elements of one of its non-empty proper subsets.

1.13. Union of Sets

Suppose we consider the following sets:

$$A = \{a, b, c\}$$
$$B = \{k, m, r, s\}$$

Let us combine the elements of sets A and B to form a new set. This new set consists of all the elements belonging to A or to B or to both A and B and is called the **union** of A and B. We write

$$A \cup B = \{a, b, c, k, m, r, s\}$$

We read the symbol $A \cup B$: "A union B."

DEFINITION 1.6 *If A and B are two sets, the **union** of A and B, denoted by $A \cup B$, is the set of elements belonging to A or to B or to both A and B.*

This process of forming the union of two sets is called an **operation** on sets.

Since the operation of forming the union of two sets is a set, we say that the universe is **closed** under the operation of union.

Consider the sets:

$$P = \{a, b, c, d\}$$
$$Q = \{a, c, d, e, f\}$$
$$R = \{d, e, a, f\}$$
$$S = \{i\}$$

Let us form $P \cup Q$. We see that

$$P \cup Q = \{a, b, c, d, e, f\}$$

Notice that $P \cup Q$ is not written $\{a, b, c, d, a, c, d, e, f\}$ since there is no point in naming a, c, and d, which are elements of both sets, twice.

Now let us form $P \cup S$ and $S \cup P$:

$$P \cup S = \{a, b, c, d, i\}$$
$$S \cup P = \{i, a, b, c, d\}$$

We see that $P \cup S = S \cup P$. This is true in general. If A and B are any two sets

$$A \cup B = B \cup A$$

The order in which the sets are considered in forming the union is of no consequence. When the order in which we operate on two things does not affect the result, we say that the operation is **commutative.** The operation of forming the union of two sets is a **commutative operation.**

Not all operations are commutative. For example, the operation of putting on a pair of socks and a pair of shoes is not a commutative operation. The result of putting on a pair of shoes followed by putting on a pair of socks is quite different from the result of putting on a pair of socks followed by putting on a pair of shoes.

The Venn diagram in Figure 1.8 shows, by shading, the region representing $A \cup B$.

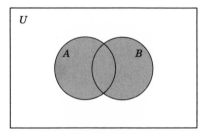

Figure 1.8

Let us see if we can find the union of three sets, A, B, and C in that order. Since we can find the union of only two sets at a time, to find the union of A, B, and C, in that order, we could find the union of sets A and B, that is $A \cup B$, and then find the union of this new set with C. We indicate this order of operation as

$$(A \cup B) \cup C$$

We could also have found the union of B and C, that is $B \cup C$, and then found the union of A and this new set, $B \cup C$. We indicate this order of operation by

$$A \cup (B \cup C)$$

We see that the parentheses indicate which operation is performed first.

Let us consider the sets:

$$K = \{1, 2, 3\}$$
$$L = \{2, 4\}$$
$$M = \{1, 3, 5\}$$

Then

$$(K \cup L) \cup M = \{1, 2, 3, 4\} \cup \{1, 3, 5\} = \{1, 2, 3, 4, 5\}$$
$$K \cup (L \cup M) = \{1, 2, 3\} \cup \{2, 4, 1, 3, 5\} = \{1, 2, 3, 4, 5\}$$

Notice that $(K \cup L) \cup M = K \cup (L \cup M)$. This is true in general, if A, B, and C are any three sets, then

$$(A \cup B) \cup C = A \cup (B \cup C)$$

When the grouping of three operations does not change the result of an operation we say that the operation is **associative.** We see that the operation of forming the union of sets is associative. Since $(A \cup B) \cup C = A \cup (B \cup C)$, we usually omit the parentheses and simply write $A \cup B \cup C$.

The Venn diagram in Figure 1.9 shows, by shading, the region representing $A \cup B \cup C$.

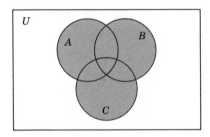

Figure 1.9

EXAMPLE i. Given $A = \{1, 3, 5, 7, 9\}$ and $B = \{3, 6, 9, 12\}$.
List the elements in $A \cup B$.
Solution: $A \cup B = \{1, 3, 5, 6, 7, 9, 12\}$

EXAMPLE ii. Given $P = \{a, b, c, d, e\}$ and $B = \{b, e, a, d\}$
List the elements in $P \cup B$.
Solution: $P \cup B = \{a, b, c, d, e\} = P$.

Notice in Example ii that set B is a subset of set P and that $P \cup B = P$. This is true in general, if $B \subseteq A$, then $A \cup B = A$.

EXAMPLE iii. Let $A = \{1, 2, 5, 8, 9\}$, $B = \{3, 5, 7, 9\}$, and $C = \{1, 5, 10, 15\}$. Find $A \cup B \cup C$.

Solution: $(A \cup B) \cup C = \{1, 2, 3, 5, 7, 8, 9\} \cup \{1, 5, 10, 15\}$
$$= \{1, 2, 3, 5, 7, 8, 9, 10, 15\}.$$

1.14. Intersection of Sets

Another operation on sets is the intersection. Suppose that

$$A = \{o, r, a, n, g, e\}$$
and
$$B = \{r, o, s, e\}$$

Sets A and B have some common elements. The set of those elements which are common to both A and B is called the **intersection** of sets A and B. We write

$$A \cap B = \{o, r, e\}$$

We read the symbol $A \cap B$: "A intersection B."

DEFINITION 1.7. *If A and B are two sets, the* **intersection** *of A and B, denoted by $A \cap B$, is the set of elements belonging to both A and B.*

In Figure 1.10, the region representing $A \cap B$ is shown, by shading, in a Venn diagram.

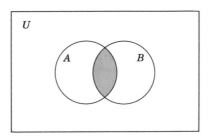

Figure 1.10

If
$$P = \{a, b, c, d\}$$
and
$$Q = \{r, s, t\}$$

what is $P \cap Q$? Since P and Q have no elements in common,

$$P \cap Q = \phi.$$

If A and B are two disjoint sets, their intersection, $A \cap B$, is the empty set ϕ.

Suppose we form $A \cap B$ and $B \cap A$. In each case we get exactly the same set. Thus $A \cap B = B \cap A$, and the operation of intersection is commutative.

Although we find the intersection of only two sets at a time, we can find the intersection of three sets in the same way that we found the union of three sets. We recall that the symbol $(A \cap B) \cap C$ means that we find the intersection of sets A and B and then find the intersection of this new set, $A \cap B$, and C. The symbol $A \cap (B \cap C)$ means that we find the intersection of sets B and C, and then find the intersection of A and this new set $B \cap C$.

Let

$$A = \{1, 2, 3, 4, 5, 6, 7, 8\}$$
$$B = \{2, 4, 6, 8, 10\}$$
$$C = \{3, 9, 6\}$$

Then

$$(A \cap B) \cap C = \{2, 4, 6, 8\} \cap \{3, 9, 6\}$$
$$= \{6\}$$

and

$$A \cap (B \cap C) = \{1, 2, 3, 4, 5, 6, 7, 8\} \cap \{6\}$$
$$= \{6\}$$

We see that $(A \cap B) \cap C = A \cap (B \cap C)$.

This is true in general: if A, B, and C are any three sets

$$(A \cap B) \cap C = A \cap (B \cap C).$$

The operation of intersection is associative. Since $(A \cap B) \cap C = A \cap (B \cap C)$, we usually omit the parentheses and simply write $A \cap B \cap C$.

In Figure 1.11a we see a Venn diagram with the region representing $A \cap B$ shaded; in Figure 1.11b we see a Venn diagram with the region representing $(A \cap B) \cap C$ shaded.

In Figure 1.12a the region representing $B \cap C$ is shaded; in Figure 1.12b the region representing $A \cap (B \cap C)$ is shaded.

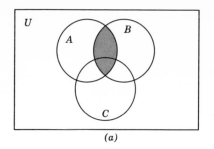
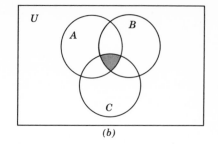

(a)   (b)

Figure 1.11

Figures 1.11 and 1.12 should convince us that in general $(A \cap B) \cap C = A \cap (B \cap C)$.

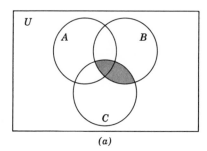
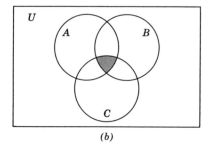

(a)   (b)

Figure 1.12

EXAMPLE i. Given $A = \{1, 2, 3, 4, 5, 6, 7\}$ and $B = \{3, 6, 9, 12\}$. Find $A \cap B$.
Solution: $A \cap B = \{3, 6\}$

EXAMPLE ii. Given $P = \{a, b, c\}$ and $Q = \{x, y, z\}$. Find $P \cap Q$.
Solution: Since P and Q have no elements in common their intersection is the empty set:

$$P \cap Q = \phi$$

EXAMPLE iii. Given $A = \{1, 2, 5, 7, 9\}$; $B = \{2, 3, 6, 7\}$; and $C = \{1, 3, 6, 8\}$. Find $A \cap B \cap C$.
Solution: $(A \cap B) \cap C = \{2, 7\} \cap \{1, 3, 6, 8\}$
$$= \phi$$

1.15. Complement of a Set

Let A be a subset of a universe U. The set of all elements of U that are not elements of A is called the **complement** of A. We use the symbol A' to denote the complement of A.

DEFINITION 1.8.   *If A is a subset of U, the **complement** of A, denoted by A', is the set of elements of U that are not elements of A.*

EXAMPLE i.   Let U be the set of all people and A the set of all men. What is the complement of A?
Solution: The complement of A is the set of all elements of U that are not elements of A. Hence A' is the set of all women.

EXAMPLE ii.   Let $U = \{1, 2, 3, 4, 5, 6, 7, 8, 9, 10\}$, $A = \{2, 4, 6, 8, 10\}$ and $B = \{3, 6, 9\}$. Find (a) A'; (b) B'; (c) $(A \cup B)'$.
Solution:

(a) $A' = \{1, 3, 5, 7, 9\}$
(b) $B' = \{1, 2, 4, 5, 7, 8, 10\}$
(c) $A \cup B = \{2, 3, 4, 6, 8, 9, 10\}$. The set $(A \cup B)'$, that is the complement of $A \cup B$, is the set of all elements of U not in $A \cup B$. Hence

$$(A \cup B)' = \{1, 5, 7\}.$$

Notice that

$$\mathbf{A \cup A' = U}$$
$$\mathbf{A \cap A' = \phi}$$
$$\mathbf{U' = \phi}$$
$$\mathbf{\phi' = U.}$$

EXAMPLE iii.   Let $U = \{x \mid x \text{ is a human being}\}$ and $A = \{x \mid x \text{ is over 21 years of age}\}$. What is A'?
Solution: $A' = \{x \mid x \text{ is twenty-one years of age or younger}\}$.

Exercise 1.5

1.   Given $A = \{1, 2, 3, 4\}$, $B = \{3, 6, 9, 14\}$, and $C = \{2, 4, 6, 8\}$ all subsets of the set of cardinal numbers. Find
   a. $n(A)$ c. $n(A \cap B)$ e. $n(A \cap C)$
   b. $n(A \cup B)$ d. $n(A \cup C)$ f. $n(A \cap B \cap C)$

2. Given $A = \{1, 2, 3, 4, 5, 6, 7, 8\}$, $B = \{3, 6, 9, 12, 15\}$ and $C = \{4, 8, 12, 16, 18, 20\}$. Find

   a. $A \cup B$ d. $A \cap C$ g. $A \cup B \cup C$

   b. $A \cap B$ e. $B \cup C$ h. $A \cup (B \cap C)$

   c. $A \cup C$ f. $B \cap C$ i. $A \cap (B \cup C)$

3. What does $P \cap (Q \cup R)$ equal if:

   a. P and $(Q \cup R)$ are disjoint sets?

   b. $P \subset R$ and $R \subset Q$?

   c. $Q \cup R = \phi$?

   d. $Q = R$?

4. Let $U = \{a, b, c, d, e\}$ and $C = \{a, e\}$. Suppose that both A and B are nonempty sets. Find A in each of the following cases.

   a. $A \cup B = U, A \cap B = \phi$ and $B = \{a\}$

   b. $A \subset C$ and $A \cup B = \{d, e\}$

   c. $A \cap B = \{c\}, A \cup B = \{b, c, d\}$ and $B \cup C = \{a, b, c, e\}$

5. What is $A \cup B$ if

   a. $B \subset A$?   c. $A = \phi$?

   b. $A \subset B$?   d. $A = B$?

6. What is $A \cap B$ if

   a. $B \subset A$?   c. $A = \phi$?

   b. $A \subset B$?   d. $A = B$?

7. Given: $U = \{x \mid x$ is an elementary school teacher$\}$

          $A = \{x \mid x$ is an elementary school teacher who has visited Canada$\}$

          $B = \{x \mid x$ is an elementary school teacher who has visited Mexico$\}$

Describe in words the following sets.

   a. $A \cup B$ c. A'

   b. $A \cap B$ d. B'

8. Given: $U = \{x \mid x$ is a city in California$\}$

          $A = \{$Los Angeles, San Francisco, San Diego$\}$

          $B = \{$Fresno, Chico, San Diego, San Bernardino$\}$

          $C = \{$Eureka, Sacramento, Los Angeles$\}$

          $D = \{$San Diego, Cardiff, Oceanside, Riverside$\}$

List the members of

   a. $A \cup B$ c. $A \cap B$ e. $A \cap B \cap C$

   b. $A \cap C$ d. $(A \cap B) \cup D$ f. $(A \cup B) \cap (C \cup D)$

9. Give the set represented by each of the following.
   a. $U \cup A$ e. $A \cap \phi$
   b. $U \cup \phi$ f. $U \cap A$
   c. $A \cup \phi$ g. $U \cap \phi$
   d. $A \cup B$ if $B \subseteq A$ h. $A \cap B$ if $B \subseteq A$.
10. Given $U = \{x \mid x$ is a human being$\}$, find the complement of each of the following.
    a. $\{x \mid x$ is a female$\}$
    b. $\{x \mid x$ is less than 18 years old$\}$
    c. $\{x \mid x$ is married$\}$
    d. $\{x \mid x$ earns more than \$10,000 annually$\}$
11. Given $U = \{1, 2, 3, 4, 5, 6, 7, 8, 9, 10\}$, $A = \{5, 7, 9\}$, $B = \{2, 4, 6\}$, and $C = \{1, 4, 7, 10\}$. Find each of the following.
    a. $A \cup B$ f. $(A \cup B)'$
    b. $A \cap B$ g. $(A \cap B)'$
    c. A' h. $A \cap C$
    d. B' i. $(A \cap C)'$
    e. C' j. $(A' \cap B')'$
12. Given $A = \{2, 4, 5, 6\}$, $B = \{1, 2, 3, 4, 5, 6\}$, $C = \{1, 3\}$, $D = \{1, 2, 3, 5\}$, and $E = \{4, 6\}$. Which of the following are true statements?
    a. $A \cup C = B$ e. $C \subset A$
    b. $A \cup B = C$ f. $E \subset E$
    c. $C \subset B$ g. $\phi \subset B$
    d. $D \subset B$ h. $A \subseteq A$
13. The following sets were used as classifications on a personality profile:
    $A = \{x \mid x$ is disliked$\}$ $D = \{x \mid x$ is antisocial$\}$
    $B = \{x \mid x$ is aggressive$\}$ $E = \{x \mid x$ is emotional$\}$
    $C = \{x \mid x$ is an introvert$\}$ $F = \{x \mid x$ is an extrovert$\}$
    The following persons took the test and were given the classification after his or her name. Describe the personality of each according to the results of the test.
    a. Evelyn: $C \cap D$
    b. George: $E \cap C$
    c. Grace: $C \cap D \cap E$
    d. Doris: $A \cap B \cap E$
    e. Clarence: $F \cap A \cap B$

1.16. Ordered Pairs

In everyday language when we speak of a "pair" we simply mean two objects, usually distinct, although similar. In mathematics, an ordered pair is a pair of objects, not necessarily different, which are considered in a particular order. If we consider a pair of objects a and b in that order, we designate this **ordered pair** by the symbol (a, b). We call a in the ordered pair (a, b) the **first component;** we call b the **second component.**

Examples of ordered pairs are (cat, dog), (5, 10), (knife, fork), (x, y). The first component in the ordered pair (cat, dog) is cat, the second component is dog. The two components of an ordered pair may be the same; for example, the elements of the ordered pair $(9, 9)$.

We now define equality of two ordered pairs.

DEFINITION 1.9. *Two ordered pairs (a, b) and (c, d) are defined to be* **equal** *if and only if $a = c$ and $b = d$.*

The phrase "If and only if" is a way of combining two statements at once. The statement in Definition 1.9 is equivalent to the two statements: "If $a = c$ and $b = d$, the ordered pairs (a, b) and (c, d) are equal." and "If $(a, b) = (c, d)$, then $a = c$ and $b = d$."

From the above definition we see that (a, b) equals (b, a) if and only if $a = b$.

1.17. Cartesian Products

Suppose the cafeteria offers the following choices of meat: roast beef and pork chops; and the following choice of vegetables: potatoes, corn, asparagus, and beets. On the special plate lunch you may have a choice of meat and one vegetable. How many choices do you have? The choices are:

roast beef and potatoes	pork chops and potatoes
roast beef and corn	pork chops and corn
roast beef and asparagus	pork chops and asparagus
roast beef and beets	pork chops and beets.

We see from the list above that there are eight choices. In this example we have two sets:

$$M = \{\text{roast beef, pork chops}\}$$
$$V = \{\text{potatoes, corn, asparagus, beets}\}$$

All combinations of meat and vegetable, in that order, form the set of all possible ordered pairs in which the first component of the ordered pair is a member of set M and the second component of the ordered pair is a member of the set V. This set of ordered pairs is called the **Cartesian product** of M and V and denoted by $M \times V$ (read: M cross V). If we let B represent roast beef, P, pork chops, p, potatoes, c, corn, a, asparagus, and b, beets, then sets M and V may be represented by

$$M = \{B, P\}$$
$$V = \{p, c, a, b\}$$

and

$$M \times V = \{(B, p), (B, c), (B, a), (B, b), (P, p), (P, c), (P, a), (P, b)\}$$

DEFINITION 1.9. *The* **Cartesian product*** *of two sets A and B, denoted by $A \times B$, is the set of all ordered pairs whose first components are elements of A and whose second components are elements of B.*

In a Cartesian product we have a set whose elements are not simple objects, but ordered pairs. From the two sets

$$A = \{1, 2, 3\}$$
$$B = \{4, 5\}$$

we can form four Cartesian products as follows:

$A \times B = \{(1, 4), (1, 5), (2, 4), (2, 5), (3, 4), (3, 5)\}$
$B \times A = \{(4, 1), (4, 2), (4, 3), (5, 1), (5, 2), (5, 3)\}$
$A \times A = \{(1, 1), (1, 2), (1, 3), (2, 1), (2, 2), (2, 3), (3, 1), (3, 2), (3, 3)\}$
$B \times B = \{(4, 4), (4, 5), (5, 4), (5, 5)\}$

$A \times B$ is the set of all ordered pairs whose first components are elements of A and whose second components are elements of B. $B \times A$ is the set of all ordered pairs whose first components are elements of B and whose second components are elements of A. $A \times A$ is the set of all ordered pairs both of whose components are elements of A. $B \times B$ is the set of all ordered pairs both of whose components are elements of B.

* The Cartesian product is sometimes called the product set.

By comparing the elements of $A \times B$ with those of $B \times A$, we see that $A \times B \neq B \times A$, when $A \neq B$, (the symbol $\neq$ means "is not equal to"), hence the formation of Cartesian products is not commutative. Note that although $A \times B \neq B \times A$, $n(A \times B) = n(B \times A)$.

We also notice that set A has three elements, set B has two elements and that sets $A \times B$ and $B \times A$ have $2 \times 3 = 6$ elements. Set A has three elements and $A \times A$ has $3 \times 3 = 3^2 = 9$ elements; set B has two elements and $B \times B$ has $2 \times 2 = 2^2 = 4$ elements. This is true in general, if A has m elements and B has n elements, $A \times B$ and $B \times A$ have $m \times n$ elements, $A \times A$ has $m \times m = m^2$ elements, and $B \times B$ has $n \times n = n^2$ elements.

The Cartesian product $A \times B$ of two sets A and B may be represented pictorially using a grid called a **lattice**. (Figure 1.13). On squared paper we draw two perpendicular lines called **axes**. Along the horizontal axis we label equally spaced points with elements of the first set A. Along the vertical axis we label equally spaced points with the elements of the second set B. Through the labeled points on the horizontal axis we draw lines parallel to the vertical axis; through the labeled points on the vertical axis we draw lines parallel to the horizontal axis. The points of intersection of these lines represent the ordered pairs of the Cartesian product of the two sets. Thus each point, called a **lattice point,** represents an ordered pair

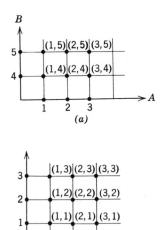

Figure 1.13

and the totality of points of intersection represent the Cartesian product. Figure 1.13, represents the Cartesian product $A \times B$ of sets $A = \{1, 2, 3\}$ and $B = \{4, 5\}$. When we are picturing $A \times A$, the two axes have exactly the same labels, (Figure 1.13b).

Exercise 1.6

1. Given $A = \{c, a, b\}$ and $B = \{9, 7\}$, list the elements of
   a. $A \times B$ c. $A \times A$
   b. $B \times A$ d. $B \times B$
2. Given $P = \{x, y, z\}$ and $Q = \{t, a, p\}$, list the elements of
   a. $P \times Q$ c. $P \times P$
   b. $Q \times P$ d $Q \times Q$
3. Which of the following pairs of ordered pairs are equal?
   a. $(a, b); (b, a)$ d. $(9, 0); (5 + 4, 5 - 5)$
   b. $(3, 2); (1 + 2, 1 + 1)$ e. $(17, 4); (4, 17)$
   c. $(7, 6); (7, 3 + 2)$ f. $(3 + 0, 4 \times 0), (3, 0)$
4. Given $C = \{$Ford, Rambler$\}$ and $M = \{$coupe, sedan, convertible$\}$.
   a. List the elements of $C \times M$
   b. List the elements of $M \times C$
   c. What is $n(M \times C)$?
   d. What is $n(C \times M)$?
   e. Is $n(C \times M)$ equal to $n(M \times C)$?
5. List the ordered pairs of $A \times B$ for each of the following pairs of sets.
   a. $A = \{a\}; B = \{1\}$
   b. $A = \{1, 2, 3, 4\}; B = \{p, y\}$
   c. $A = \{$red, blue, green$\}; B = \{3, 4\}$
   d. $A = \{m, w, f\}; B = \{1, 2\}$
6. List all the ordered pairs in $A \times A$ if $A = \{d, o, g\}$.
7. Graph $A \times A$ found in Problem 6 on a lattice.
8. If set A has 3 elements and set B has 4 elements, how many elements are in
   a. $A \times B$?   c. $B \times B$?
   b. $A \times A$?   d. $B \times A$?
9. If set A has k elements and set B has 5 elements how many elements are in $A \times B$?

10. If set A has p elements and set B has q elements, how many elements are in $B \times A$?

11. Let $F = \{x \mid x \text{ is an area}\}$ and $S = \{x \mid x \text{ is a salesman}\}$. If $n(F) = 15$ and $n(S) = 12$, in how many ways can the salesmen be assigned to the areas?

12. Let $S = \{x \mid x \text{ is a pair of slacks}\}$ and $J = \{x \mid x \text{ is a jacket}\}$. If $n(S) = 5$ and $n(J) = 3$, how many different outfits can he put together?

13. Given $A = \{1, 2, 3, 4\}$.
    a. Form $A \times A$
    b. Graph $A \times A$ on a lattice
    c. What is the subset of $A \times A$ satisfying the condition that the first component of every ordered pair is less than 3 and the second component is 2?

14. Given $P = \{0, 1, 2, 3, 4, 5, 6, 7, 8, 9\}$
    a. Form $P \times P$
    b. Graph $P \times P$ on a lattice
    c. What is the subset of $P \times P$ satisfying the condition that the sum of the two components of the ordered pair is 9.

15. Given $B = \{1, 2, 3, 4, 5\}$
    a. Form $B \times B$
    b. Graph $B \times B$ on a lattice.
    c. What is the subset of $B \times B$ satisfying the condition that the first component of each ordered pair is twice the second?

1.18. Combinations of Set Operations

Combinations of the operations of union and intersection offer many possibilities for the formation of new sets from given sets. Venn diagrams are useful aids for visualizing the resulting sets when set operations are combined.

Let us consider

$$A \cap (B \cup C)$$

In Figure 1.14, we have used vertical line segments to show the region representing $B \cup C$ and horizontal line segments to show the region representing A.

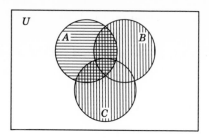

Figure 1.14

We are interested in the intersection of sets A and $B \cup C$. The intersection of two sets is the set of elements common to the two given sets. Hence the region in the diagram that is cross hatched, ▦, is the region representing $A \cap (B \cup C)$. This region is shown by shading in Figure 1.15.

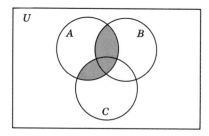

Figure 1.15

Now let us consider

$$(A \cap B) \cup (A \cap C)$$

In Figure 1.16 we have used horizontal line segments to show the region representing $A \cap B$ and vertical line segments to show the region representing $A \cap C$.

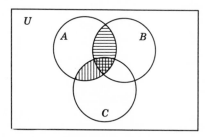

Figure 1.16

We are interested in the union of $A \cap B$ and $A \cap C$. The union of two sets is the set of elements in one or the other or both of the given sets. Hence $(A \cap B) \cup (A \cap C)$ is represented by the region which has either vertical or horizontal lines shading it. This region is shaded in Figure 1.17.

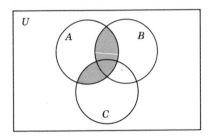

Figure 1.17

Notice that in Figures 1.15 and 1.17, the same area is shaded in both cases. This leads us to state that

$$A \cap (B \cup C) = (A \cap B) \cup (A \cap C)$$

This is known as the **distributive property of intersection over union.**
It can also be shown that

$$A \cup (B \cap C) = (A \cup B) \cap (A \cup C).$$

This is called the **distributive property of union over intersection.** Figure 1.18a shows (by shading) the region representing $A \cup (B \cap C)$. Figure 1.18b shows (by shading) the region representing $(A \cup B) \cap (A \cup C)$. In both cases the same area is shaded.

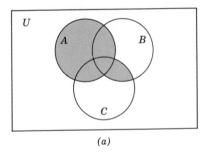
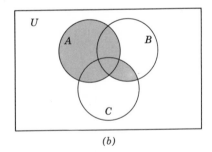

(a)   (b)

Figure 1.18

EXAMPLE i. Show on a Venn diagram the region representing the set

$$(A \cap B)'$$

Solution:

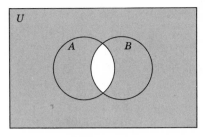

Figure 1.19

EXAMPLE ii. Show on a Venn diagram the region representing the set

$$A' \cap B'$$

Solution:

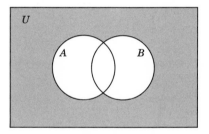

Figure 1.20

Exercise 1.7

1. Make a sequence of Venn diagrams to illustrate that the following are true statements.
   a. $(A \cup B) \cup C = A \cup (B \cup C)$
   b. $(A \cup B) \cap (A \cup C) = A \cup (B \cap C)$
   c. $(A \cap B) \cap C = A \cap (B \cap C)$
   d. $(A \cap B) \cup (A \cap C) = A \cap (B \cup C)$
2. Let $A = \{1, 2, 3, 4, 5, 6, 7\}$, $B = \{1, 3, 5, 7, 9, 11\}$, and
   $C = \{2, 4, 6, 8\}$
   Find
   a. $(A \cap B) \cup C$

b. $(B \cup C) \cap (A \cup C)$
c. $(A \cap B) \cup (C \cap \phi)$

3. Let $A = \{0, 1, 2, 3, 4, 5, 6, 7, 8, 9\}$, $B = \{0, 2, 4, 6\}$, $C = \{3, 6, 9\}$, $D = \{1, 3, 5, 7, 9, 11, 13\}$, and $E = \{4, 8, 12, 16\}$. Find
   a. $(A \cup B) \cup (C \cap D)$ d. $(E \cap D) \cup (B \cap C)$
   b. $(A \cap C) \cup (A \cap B)$ e. $E \cap B \cap C$
   c. $A \cap (B \cup C)$ f. $E \cap (D \cup A)$

4. Let $U = \{1, 2, 3, 4, 5, 6, 7, 8, 9\}$, $A = \{1, 3, 5, 7, 9\}$, $B = \{2, 4, 6, 8\}$, $C = \{3, 6, 9\}$, and $D = \{2, 3, 5, 7\}$. Find
   a. $B' \cup A'$ d. $(A \cap C)' \cup (B \cap D)'$
   b. $(B \cap D) \cup (A \cap C)$ e. $(A')'$
   c. $(A \cap D)'$ f. $(A \cup B \cup C)' \cap D$

5. Draw Venn diagrams showing by shading the region representing each of the following sets.
   a. $(A \cap B) \cup (B \cup C)$ c. $(A \cup B)'$
   b. A' d. $A' \cup B'$

2.1. Undefined Words

The first requirement for understanding any subject, be it mathematics or tennis, is to know the meanings of the words that are used. We acquire early in life the habit of looking up the definitions of unfamiliar words in the dictionary. A little experience using the dictionary convinces us that some words must be undefined. Without knowing the meaning of some words our use of the dictionary leads us around in a circle. For example, suppose we look up the verb "to desire." We may find that as we look up each definition given, we are ultimately led back to "to desire." (Figure 2.1)

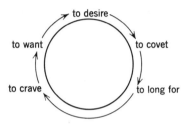

Figure 2.1

Logic

II

We must make a beginning somewhere, and in mathematics we do this by choosing a small number of words which we accept as **undefined.** The choice of these undefined words depends upon the subject, as we shall see when we study geometry (Chapter 8).

With these basic undefined words, we now define other words in terms of them. Of course, we must assume that we have some ordinary nontechnical English words at our disposal. For example, if we accept *point, line,* and *betweenness* as our undefined terms we define a line segment as the set of all points between two given points and including these two points.

2.2. Statements

With our undefined and defined words we form sentences called **propositions** or **statements.** These statements must have **truth value,** that is, they must be either true or false, but not both. The following sentences are statements:

Abraham Lincoln was president in 1862.
Chicago is west of New York City.
A cow purrs.
Snakes can fly.

The following sentences are not statements by our definition because, as they stand, they have no truth value, that is, we cannot tell whether or not they are true or false.

He is the best student in the class.
That city has a population of 40,000.
This number is a prime number.

In a deductive system such as mathematics we accept as true a set of statements. These statements that are assumed to be true are called **axioms** or **postulates.** Once our axioms are stated, we use them together with a process called **logical reasoning** to prove other statements called **theorems.**

2.3. Compound Statements

There are several ways in which we combine two statements to form a third statement. Those that we shall discuss are (1) conjunction, (2) disjunction, (3) implication, and (4) negation.

In a **conjunction** we combine two given statements by placing an "and" between them. Thus the conjunction of the two statements:

(a) It is raining
(b) The streets are flooded

is "It is raining and the streets are flooded."

We define a conjunction to be true if both of the statements, called **components,** are true and false otherwise. If "It is raining" is true and "The streets are flooded" is true, then the conjunction "It is raining and the streets are flooded" is true. If "It is raining" is true, and "The streets are flooded" is false, the conjunction is false. If "It is raining" is false and "The streets are flooded" is true, the conjunction is false. If both "It is raining" and "The streets are flooded" are false, the conjunction is false.

We usually denote statements by lower case letters of the alphabet such as p, q, r, and s. If p and q are letters which represent statements used to form a conjunction, we represent the conjunction by

$$p \wedge q$$

We read this symbol "p and q." Table 2.1 shows the truth values of the conjunction $p \wedge q$. Such a table is called a **truth table.** In the table T means true and F means false.

Table 2.1

p	q	$p \wedge q$
T	T	T
T	F	F
F	T	F
F	F	F

EXAMPLE i. Give the truth value of each of the following conjunctions.
1) Ice is cold and sugar is sweet.
2) Cats are animals and people are rodents.

3) New York City is north of Miami and Miami is west of Chicago.

4) Easter is in November and Christmas is in June.

Solution:

1) True because both components are true.

2) False because the component "People are rodents" is false.

3) False because the component "Miami is west of Chicago" is false.

4) False because both components are false.

In a **disjunction** we combine two statements by placing "or" between them. The disjunction formed using statements (a) and (b) above is "It is raining or the streets are flooded." A disjunction is defined to be true if one or the other or both of its components are true, and false otherwise. If "It is raining" is true and "The streets are flooded" is true, the disjunction "It is raining or the streets are flooded" is true. If either "It is raining" or "The streets are flooded" is true, the disjunction is true. If both "It is raining" and "The streets are flooded is false" the disjunction is false.

If p and q represent the components used to form a disjunction, we represent the disjunction by

$$p \vee q$$

We read this symbol "p or q." Table 2.2 shows the truth value of the disjunction $p \vee q$.

Table 2.2

p	q	$p \vee q$
T	T	T
T	F	T
F	T	T
F	F	F

Note that the connective "or" when used in a disjunction is used in the **inclusive** sense, that is "or" means p or q or both p and q. In everyday language we often use "or" in the **exclusive** sense; that is, "or" means p or q but not both. For example, we say, "I shall get a grade of A or a grade of B in this course." This means that a grade of A or a grade of B will be received, but not both. Hereafter, the word "or" will *always* be used in this book in the inclusive sense.

EXAMPLE ii.   Give the truth value of the following disjunctions.

   (a) Ice is cold or sugar is sweet.
   (b) Cats are animals or people are rodents.
   (c) New York is north of Miami or Chicago is on the Pacific Ocean.
   (d) $2 \times 2 = 6$ or $3 + 3 = 8$.

Solution:

   (a) True because both components are true.
   (b) True because the component "cats are animals" is true.
   (c) True because the component "New York is north of Miami" is true.
   (d) False because both components are false.

2.4.   Implication

A very important statement in mathematics is the **implication.** It takes the form of an "if-then" sentence. Examples of implications are:

   If it is raining, then the streets are flooded.
   If $2 + 2 = 4$, then 5 is greater than 2.
   If a polygon has three sides, then it is a triangle.

From any two statements we can form two implications. From "It is raining" and "The streets are flooded" we can form the implications:

   (a) If it is raining, then the streets are flooded.
   (b) If the streets are flooded, then it is raining.

In implication (a), the statement "It is raining" is called the **antecedent** and the statement "The streets are flooded" is called the **conclusion.** In implication (b) the statement "The streets are flooded" is the **antecedent** and the statement "It is raining" is the **conclusion.** The two implications that can be formed from two given statements are related as we shall see later.

If p and q are symbols denoting two statements, we use the symbol

$$p \rightarrow q$$

to denote the implication "If p then q." The symbol $p \rightarrow q$ is also read "p implies q." Table 2.3 shows the truth values of the implication $p \rightarrow q$.

The first line in the table is easily accepted. In fact, this may be precisely our interpretation of the implication $p \rightarrow q$: "If p is true, then q

Table 2.3

p	q	$p \rightarrow q$
T	T	T
T	F	F
F	T	T
F	F	T

is true." The second line is intuitively easy to accept: From the understanding that "If p is true then q is true" it must be false to have p to be true and q to be false. The third and fourth lines of the table are less familiar and more difficult to accept.

We shall however accept the truth table as the definition of the truth values of the implication $p \rightarrow q$. That is, we define the implication $p \rightarrow q$ as true unless the antecedent p is true and the conclusion q is false, in which case it is false. Note that an implication is always true if the conclusion is true.

In ordinary speech, it is customary for the antecedent and the conclusion of an implication to be related as "If I work hard, then I shall pass this course." It is generally meaningless to combine two apparently unrelated statements as "If $3 \times 3 = 9$, then this class is a snap." In mathematical logic, we shall not hesitate to combine unrelated statements in an implication.

EXAMPLE i. Give the truth values of the following implications.
    (a) If dogs have wings, then cows have six legs.
    (b) If $3 + 8 = 10$, then $3 \times 4 = 12$.
    (c) If cats purr, then all men are 6 feet tall.
    (d) If $3 \times 4 = 12$, then $3 \times 5 = 15$.
Solution:
    (a) This implication is true because both the antecedent and the conclusion are false.
    (b) The implication is true because the conclusion is true.
    (c) This implication is false, because the conclusion is false and the antecedent is true.
    (d) This implication is true because both the antecedent and the conclusion are true.

EXAMPLE ii.   What are the antecedent and the conclusion of the following implication?

> I usually get tipsy if I drink too much wine.

Solution: In "if-then" form this implication is:

> If I drink too much wine, then I usually get tipsy.

The antecedent is "I drink too much wine" and the conclusion is "I usually get tipsy."

2.5.   Negation

It is convenient to have a symbol to indicate that a statement is false. This is called **negation.** If p is the statement "Cows have six legs," its negation, denoted by $\sim p$, asserts that the statement "Cows have six legs" is false. It does not assert "Cows have four legs." Cows may have two legs or one hundred legs or no legs at all. "Cows do not have six legs" or "It is false that cows have six legs" is all that we can conclude from $\sim p$.

Clearly p and $\sim p$ have opposite truth values. This is shown in Table 2.4.

Table 2.4

p	$\sim p$
T	F
F	T

Exercise 2.1

1.   Which of the following are statements? Give the truth value of each statement.
   a.   Black coffee contains no calories.
   b.   General Eisenhower was president of the United States.
   c.   Her birthday is February 29.
   d.   June is the sixth month of the year.

e. The year 1935 was a presidental election year.

f. All cats eat lettuce.

g. That number is the sum of 6 and 7.

h. He is twenty-one years old.

i. All cows have lavender eyes.

j. Poodles are people.

2. Form the (1) conjunction and (2) disjunction of the following pairs of statements.

a. The sky is blue. The grass is green.

b. Today is Monday. Yesterday was Sunday.

c. The train is late. We shall be late to work.

d. All birds have wings. All dogs are intelligent.

3. Form the implication $p \rightarrow q$ of the statements in Problem 2. Use the first statement as the antecedent and the second statement as the conclusion.

4. Form the negation of the following statements.

a. It is raining.

b. Dogs are loyal friends.

c. People work hard.

d. Water is wet.

e. Ice is cold.

5. Give the antecedent and the conclusion in each of the following implications.

a. If I go skiing, I have fun.

b. If he cooperates with his boss, he will probably get a promotion.

c. A number is even if it is divisible by two.

d. You were born in leap year if you were born in 1952.

e. If Charles likes Elaine and Elaine likes Charles, then miracles do happen.

6. Choose an ordinary nontechnical word and build a circular definition from this word back to itself. Use a standard dictionary for your definitions. Do not put simple connectives such as "the", "and", "is", and so forth, in your chain.

7. Use the following symbolic representations:

p: Chicago is the Windy City.

q: New York is the Empire State.

Write each of the following in symbolic notation.

a. Chicago is not the Windy City.

b. Chicago is the Windy City and New York is the Empire State.

c. If New York is the Empire State, then Chicago is not the Windy City.

d. Chicago is the Windy City or New York is the Empire State.

8. Given p: All rectangles are quadrilaterals.

   q: All squares are rectangles.

   Write the English statements for the following symbolic notations.

   a. $p \wedge q$ d. $p \to q$

   b. $p \vee q$ e. $q \to p$

   c. $(\sim p) \wedge (\sim q)$ f. $(\sim q) \to p$

9. Give the truth value of each of the following.

   a. Birds have wings and dogs have four legs.

   b. If $3 \times 6 = 9$, then $7 + 7 = 14$.

   c. If a triangle is a square, then a cow is a bird.

   d. If turnips are carrots, then carrots are money.

   e. If Hawaii borders the Atlantic Ocean, then New York is Russian territory.

   f. Cats eat meat and meat grows on trees.

   g. It is raining or it is not raining.

   h. The legal voting age is 16 or monkeys are people.

   i. April comes after March if Christmas is in June.

   j. Three fours are five if two threes are one.

   k. Today is Friday or today is not Friday.

10. Write the following in symbolic notation.

    a. Jane likes Charles. (statement p)

    b. Charles likes Jane. (statement q)

    c. Jane and Charles like each other.

    d. If Charles likes Jane, then Jane likes Charles.

    e. Jane and Charles dislike each other.

11. Write the following in symbolic notation. Let p be "Modern music is noise." and q be "The Tricky Trio is good."

    a. Either modern music is noise or the Tricky Trio is not good.

    b. If the Tricky Trio is good, then modern music is noise.

    c. Modern music is noise and the Tricky Trio is not good.

    d. The Tricky Trio is good if modern music is not noise.

12. Let p be "Taxes are high" and q be "Property values are rising." Give an English statement for each of the following.

    a. $p \wedge q$ d. $\sim(p \wedge q)$

b. $p \rightarrow q$ e. $\sim[(\sim p) \vee q]$

c. $(\sim p) \rightarrow (\sim q)$ f. $p \wedge (\sim q)$

2.6. Equivalence Relations

We are familiar with many kinds of relations. For example:

Tom *is older than* Larry.
Mary *is the sister of* Alice.
New York *is east of* Chicago.

"Is older than," "is the sister of" and "is east of" are **relations.** These relations are **binary relations** because they relate two elements of a set. Some binary relations of mathematics are described by such phrases as "is less than," "is equal to," "is greater than," "is a factor of," and "is a multiple of."

Symbolically we shall denote a relation with the capital letter R. If the relation R relates two elements a and b, in that order, where a and b are elements of set S, we shall express this by the symbol aRb. We read this symbol "a is related to b." For example, "Tom is older than Larry" may be written aRb, where a stands for Tom, b stands for Larry, and R for the relation "is older than."

If we have a given relation on a set S we require that for every a and b in S, either aRb is true or $a\cancel{R}b$ is true where $a\cancel{R}b$ is the symbol for the negation of aRb.

Suppose R is a relation on S and a, b, and c are elements of S. The relation R is said to be an **equivalence relation** if it satisfies the following properties:

Reflexive Property. aRa for all a in S.
Symmetric Property. aRb implies bRa for all a and b in S.
Transitive Property. aRb and bRc implies aRc for all a, b, and c in S.

Let us consider the relation "lives in the same house as" defined on the set of all people. This relation is certainly reflexive since aRa is true, that is every person lives in the same house as himself. This relation is also symmetric since aRb implies bRa. That is, if a lives in the same house as b it is certainly true that b lives in the same house as a. This relation is also transitive since aRb and bRc together imply aRc. That is, if a lives in the same house as b and b lives in the same house as c, then it is true that a lives in the same house as c. Since the relation "lives in the same house as" defined on the set of all people is reflexive, symmetric, and transitive, it is an equivalence relation.

EXAMPLE i. Is the relation "has the same first name as" defined on the set of all people an equivalence relation?
Solution: aRa: a has the same first name as a (himself).

If aRb then bRa: If a has the same first name as b, then it is true that b has the same first name as a.

If aRb and bRc, then aRc: If a has the same first name as b and b has the same first name as c, then it is true that a has the same first name as c.

Since the relation "has the same first name as" defined on the set of all people is reflexive, symmetric, and transitive, it is an equivalence relation.

EXAMPLE ii. Is the relation "is west of" defined on the set of all cities in the United States an equivalence relation?
Solution: a$\not\!R$a: a is not west of a (itself)

If aRb then $b\not\!Ra$: If a is west of b, then b is not west of a.

If aRb and bRc then aRc: If a is west of b and b is west of c then it is true that a is west of c.

This relation is not an equivalence relation because it is neither reflexive nor symmetric.

EXAMPLE iii. Is the relation "weighs within 5 pounds of" defined on the set of all people an equivalence relation?
Solution: aRa: a weighs within 5 pounds of a.

If aRb then bRa: If a weighs within 5 pounds of b then it is true that b weighs within 5 pounds of a.

If aRb and bRc, then it may happen that aRc: If a weighs within 5 pounds of b and b weighs within 5 pounds of c it is not necessarily true that a weighs within 5 pounds of c. For example, if a weighs 110 pounds, b weighs 113 pounds, and c weighs 117 pounds, a weighs within 5 pounds of b, and b weighs within 5 pounds of c, but a does not weigh within 5 pounds of c.

One of the important equivalence relations of mathematics is the equality relation. If R is the relation "is equal to" defined on the set of all numbers, then

aRa: $a = a$

If aRb, then bRa: If $a = b$, then $b = a$

If aRb and bRc, then aRc: If $a = b$ and $b = c$, then $a = c$.

2.7. Tautology

A **tautology** is a statement which is true regardless of the truth or falsity of the components used to form it. For example, the statement

$$\sim[p \wedge (\sim p)]$$

is a tautology. This is demonstrated in the truth table (Table 2.5) below.

Table 2.5

p	$\sim p$	$p \wedge (\sim p)$	$\sim[p \wedge (\sim p)]$
T	F	F	T
F	T	F	T

In making the truth table to show that $\sim[p \wedge (\sim p)]$ is a tautology, we write the headings

$$p, \sim p, p \wedge (\sim p), \sim[p \wedge (\sim p)]$$

Under p we enter its possible truth values, T and F, and carry each line across to the right. If p is true, $\sim p$ is false, $p \wedge (\sim p)$ is false, and the negation of $p \wedge (\sim p)$, that is, $\sim[p \wedge (\sim p)]$, is true. Upon completion of the table we see that $\sim[p \wedge (\sim p)]$ is true in all cases, and is a tautology. The statement $\sim[p \wedge (\sim p)]$ is called the **Law of Contradiction.**

The Law of Contradiction states that a statement and its negation cannot both be true, that is, a statement cannot be both true and false at the same time. For example, if p is the statement "John is here," the statement "It is false that John is here and John is not here" is true regardless of whether the statement p is true or false.

Now let us show that the statement

$$p \vee (\sim p),$$

called the **Law of the Excluded Middle,** is a tautology. Again we form a truth table as shown in Table 2.6. Since $p \vee (\sim p)$ is true regardless of the truth or falsity of p, it is a tautology.

Table 2.6

p	$\sim p$	$p \vee (\sim p)$
T	F	T
F	T	T

The Law of the Excluded Middle says that a statement or its negation is true. That is, a statement is either true or false. For example, if p is the statement "John is here," the disjunction "John is here or John is not here" is true regardless of whether p is true or false.

We shall now construct a truth table to show that the statement

$$[(p \rightarrow q) \wedge (q \rightarrow r)] \rightarrow (p \rightarrow r),$$

called the **Law of Syllogism,** is a tautology.

Table 2.7 verifies that the Law of Syllogisms is a tautology.

Table 2.7

p	q	r	$p \rightarrow q$	$q \rightarrow r$	$(p \rightarrow q) \wedge (q \rightarrow r)$	$p \rightarrow r$	$[(p \rightarrow q) \wedge (q \rightarrow r)] \rightarrow (p \rightarrow r)$
T	T	T	T	T	T	T	T
T	T	F	T	F	F	F	T
T	F	T	F	T	F	T	T
T	F	F	F	T	F	F	T
F	T	T	T	T	T	T	T
F	T	F	T	F	F	T	T
F	F	T	T	T	T	T	T
F	F	F	T	T	T	T	T

To construct Table 2.7 we begin with three columns headed p, q, and r since three statements are used to form the compound statement. There are eight lines in the table since each of the two truth values, T and F, of each statement can be combined with the two truth values of each of the other statements making $2 \times 2 \times 2 = 8$ combinations in all. On the basis of the truth values of p, q, and r, we find the truth values of $p \rightarrow q$, $q \rightarrow r$,

$(p \rightarrow q) \land (q \rightarrow r)$, $p \rightarrow r$, and $[(p \rightarrow q) \land (q \rightarrow r)] \rightarrow (p \rightarrow r)$ as shown in Table 2.7. Since $[(p \rightarrow q) \land (q \rightarrow r)] \rightarrow (p \rightarrow r)$ is true regardless of the truth or falsity of p, q, and r, it is a tautology.

The Law of Syllogism is also called the **Chain Rule.** The use of successive applications of the Law of Syllogisms permits a chain of implications of any desired length. For example, if $p \rightarrow q$ and $q \rightarrow r$ and $r \rightarrow s$, then $p \rightarrow s$.

Anyone who has taken a course in plane geometry is familiar with the application of the Chain Rule. In using a direct proof to prove a theorem, we start with a hypothesis, establish a sequence of implications, and end with the desired conclusion. Each step in the proof of the theorem is justified in terms of axioms, definitions, previously proved theorems, or principles of logic.

Using truth tables we can derive a series of tautologies known as the **rules of inference.** Some of the important rules of inference were derived above.

Exercise 2.2

1. Construct a truth table for $p \rightarrow (p \lor r)$
2. Construct a truth table for $(p \land q) \rightarrow p$.
3. Construct a truth table for $p \rightarrow (q \lor r)$
4. Construct a truth table for $[p \land (\sim p)] \rightarrow q$
5. Construct a truth table for $[p \lor (\sim q)] \land r$
6. Construct a truth table for $[\sim(p \lor q)] \lor (p \lor q)$
7. Construct a truth table for $(p \rightarrow q) \rightarrow [(\sim q) \rightarrow (\sim p)]$
8. Construct a truth table for $(q \land r) \rightarrow (q \lor r)$
9. Construct a truth table for $(p \land q) \lor r$
10. Construct a truth table for $\sim(p \lor q) \rightarrow [(\sim p) \land (\sim q)]$

Using truth tables decide whether or not the following are tautologies. (Ex. 11-20)

11. $(p \land q) \rightarrow p$
12. $(p \lor q) \rightarrow p$
13. $[(p \rightarrow q) \land (r \rightarrow s) \land (p \lor r)] \rightarrow (q \lor s)$
14. $(p \lor q) \rightarrow \{\sim[(\sim p) \land (\sim q)]\}$
15. $q \rightarrow p$
16. $(\sim p) \rightarrow (\sim q)$

17. $(\sim p) \to (\sim p)$
18. $p \to (\sim q)$
19. $(\sim p) \to q$
20. $(p \to q) \to \{\sim[p \wedge (\sim q)]\}$
21. State in words and symbols:
    a. The Law of Contradiction
    b. The Law of the Excluded Middle
    c. The Law of Syllogisms

Determine whether or not each of the following relations on the given sets are (a) reflexive; (b) symmetric; (c) transitive. (Ex. 22–32).

22. "Is intelligent as" on the set of all people.
23. "Is the cousin of" on the set of all people.
24. "Is greater than" on the set of all counting numbers, $\{1, 2, 3, \ldots\}$
25. "Is the brother of" on the set of all males.
26. "Is the brother of" on the set of all people.
27. "Is the same age as" on the set of all children.
28. "Is the same height as" on the set of all sixth graders.
29. "Lives within one mile of" on the set of residents of New Orleans.
30. "Has the same color hair as" on the set of all women.
31. "Drives the same kind of car as" for the set of all college students.
32. "Weighs within $\frac{1}{2}$ pound of" on the set of all dogs.
33. Which of the relations in Problems 22–32 are equivalence relations?

2.8. Derived Implications

From the implication $p \to q$, we can form several related implications which may or may not be true if the given implication is true. The most important are:

Converse: $q \to p$
Inverse: $(\sim p) \to (\sim q)$
Contrapositive: $(\sim q) \to (\sim p)$

Comparing the truth values for $p \to q$ and $q \to p$ (Table 2.8) we see that when a given implication is true, its converse may be true or false.

For example, the implication given below is true, but its converse is false:

Table 2.8

p	q	$p \rightarrow q$	$q \rightarrow p$
T	T	T	T
T	F	F	T
F	T	T	F
F	F	T	T

Implication: If a quadrilateral is a square, then it is a rectangle.
Converse: If a quadrilateral is a rectangle, then it is a square.

On the other hand, both the following implication and its converse are true.

Implication: If it is ten o'clock C.S.T., it is eleven o'clock E.S.T.
Converse: If it is eleven o'clock E.S.T., it is ten o'clock C.S.T.

Now let us construct a truth table and investigate whether the inverse is true when a given implication is true, and false when it is false.

Since the last two columns in Table 2.9 differ, the inverse of a true implication may be true or false.

Table 2.9

p	q	$\sim p$	$\sim q$	$p \rightarrow q$	$(\sim p) \rightarrow (\sim q)$
T	T	F	F	T	T
T	F	F	T	F	T
F	T	T	F	T	F
F	F	T	T	T	T

Now let us construct a truth table (Table 2.10) and investigate whether the contrapositive of a true implication is true.

Table 2.10

p	q	$\sim p$	$\sim q$	$p \rightarrow q$	$(\sim q) \rightarrow (\sim p)$
T	T	F	F	T	T
T	F	F	T	F	F
F	T	T	F	T	T
F	F	T	T	T	T

Since the last two columns of Table 2.10 are identical, we assert that an implication and its contrapositive are simultaneously true or false, that is they are logically equivalent. Likewise, it can be shown that the inverse and the converse of a given implication are logically equivalent. Any two statements that are simultaneously true or false are called **logically equivalent** or simply **equivalent** statements.

We use the fact that an implication and its contrapositive are logically equivalent statements in two ways:

(1) If we know that an implication is true, we can infer that its contrapositive is true and vice versa.

(2) Proving or disproving the contrapositive of an implication is equivalent to proving or disproving the implication.

EXAMPLE i. Show that p and $[\sim(\sim p)]$ are logically equivalent.
Solution: We construct a truth table

p	$\sim p$	$\sim(\sim p)$
T	F	T
F	T	F

Since the first and third columns of the table are exactly alike, p and $[\sim(\sim p)]$ are logically equivalent.

EXAMPLE ii. Show that $p \wedge (\sim q)$ and $\sim[(\sim p) \vee q]$ are logically equivalent.
Solution: We construct a truth table.

p	q	$\sim p$	$\sim q$	$p \wedge (\sim q)$	$(\sim p) \vee q$	$\sim[(\sim p) \vee q]$
T	T	F	F	F	T	F
T	F	F	T	T	F	T
F	T	T	F	F	T	F
F	F	T	T	F	T	F

Since the fifth and seventh columns in the table are exactly alike, the two given statements are logically equivalent

2.9. Laws of Substitution and Detachment

Two more laws of logic used in logical reasoning are the Law of Detachment and the Law of Substitution.

The **Law of Substitution** states that we may substitute at any point in the deductive process one statement for an equivalent statement. Thus we may at any point in an argument substitute the contrapositive, $(\sim q) \rightarrow (\sim p)$, for $p \rightarrow q$ because they are logically equivalent statements.

The **Law of Detachment** (often called **Modus Ponens**) states that if p is true and $p \rightarrow q$ is true, then q is true. For example, suppose we know that the statement "Perkins has a 3.0 grade point average" and the implication "If Perkins has a 3.0 grade point average, then he will graduate with honors" are true statements. The Law of Detachment assures us that "Perkins will graduate with honors."

The validity of the Law of Detachment follows immediately from Table 2.11.

Table 2.11

p	q	$p \rightarrow q$
T	T	T
T	F	F
F	T	T
F	F	T

Since $p \rightarrow q$ is given true, we are given the conditions in lines 1, 3, or 4 of Table 2.11. Since p is also given true, we base our conclusions about q on lines 1 or 2. Since line 1 is the only line that satisfies both conditions, i.e., that p and $p \rightarrow q$ are true, we see that q must be true.

Exercise 2.3

Write the (a) converse; (b) inverse; and (c) contrapositive of each of the following implications. (Ex. 1–10)

1. If Alice is Tim's sister, then Tim is Alice's brother.
2. If 9 divides 27, then 3 divides 27.
3. If the alarm does not go off, Jack oversleeps.

4. If Christmas falls on Wednesday, then New Year's Day falls on Wednesday.
5. If Jay was born on February 29, then he was born in a leap year.
6. If a quadrilateral is a square, then it is a rectangle.
7. If set A is equal to set B, then set B is equal to set A.
8. If wages get higher, then taxes rise.
9. If a polygon has five congruent sides, then it is a regular pentagon.
10. If the White Rabbit wears white gloves, then he is going to a party.

Construct truth tables to show that the following pairs of statements are logically equivalent. (Ex. 11–15)

11. $\sim(p \wedge q)$ and $(\sim p) \vee (\sim q)$
12. $\sim(p \to q)$ and $p \wedge (\sim q)$
13. $(p \to q) \to [(\sim p) \vee q]$
14. $(\sim p) \wedge (\sim q)$ and $\sim(p \vee q)$
15. $\sim[(\sim p) \vee q]$ and $p \wedge (\sim q)$
16. Use the Law of Detachment and draw a conclusion from the following.
    a. If it is 5 o'clock E.S.T. in New York City, it is 2 o'clock P.S.T. in San Francisco. It is 5 o'clock E.S.T. in New York City.
    b. If Jerry lives in Los Angeles, then he lives in California. Jerry lives in Los Angeles.
    c. If everyone in this class fails, then the instructor is not a good teacher or the students are stupid. Everyone in this class fails.
    d. If all rats are finks, then all mice are jerks. All rats are finks.
    e. All girls are bunnies, if Kerry is dumb. Kerry is dumb.

2.10. Proof

Mathematics is a deductive science. It is based on a method of reasoning called **deduction.** Each branch of mathematics such as arithmetic, algebra, and geometry has been developed and organized according to the rules of logic. A branch of mathematics which is organized logically is called a **logical system** or a **deductive system.**

The essentials of a deductive system are:

(1) Undefined words.
(2) Words defined in terms of the undefined words.

(3) An initial set of statements about the undefined and defined words which are assumed to be true. These statements are called **axioms** or **postulates.**

(4) The laws of logic (also called the laws of inference).

We now have what we need to prove statements about the undefined and defined words of our system. These statements which we prove to be true using the laws of logic are called **theorems.** We will not discuss how mathematicians conjecture theorems that they wish to prove true. This process of guessing statements about the terms in the system (and it is educated guessing!) requires a high degree of intellectual ability and mathematical experience. We are interested here in the process of proving that certain conjectured statements are true theorems.

Proofs fall into two classes: (1) direct and (2) indirect. In a direct proof a chain of implications is arranged from the given statements to the desired conclusion. In an indirect proof we assume that the statement which we wish to prove true is false, that is, we assume that its negation is true. We then use the laws of logic to reach a contradiction. When we reach this contradiction we know that our assumption is false; hence, the desired conclusion is true.

There is no given technique for proving a theorem. Cleverness, innate ability, and experience are helpful. We learn a great deal about the technique of proving theorems by studying proofs that others have made and by understanding the types of proof in a deductive system.

2.11.   Direct Proof

Most deductive proofs consist of more than one step. Suppose we are given that statement h is true and we wish to prove the theorem $h \rightarrow c$. Suppose that we are able to show that $h \rightarrow c_1$—some conclusion other than c, the conclusion we desire. Suppose also that we can show $c_1 \rightarrow c_2$, another conclusion different from c. Suppose we can continue establishing implications until we finally reach an implication whose conclusion is the desired conclusion c. We now have a chain of implications which we have shown to be true:

$$h \rightarrow c_1$$
$$c_1 \rightarrow c_2$$

$$c_2 \rightarrow c_3$$
$$c_3 \rightarrow c$$

We can now assert that $h \rightarrow c$ by the law of syllogisms.

Direct proof consists of starting with a hypothesis, establishing a chain of implications and ending with the desired conclusion. Each individual step in the chain of implications must be justified in terms of axioms, definitions, previously proved statements, and the laws of inference.

Study the following examples. In each case the theorem is proved by means of a direct proof.

THEOREM 2.1. **If p, q $\rightarrow$ ($\sim$p) and ($\sim$q) $\rightarrow$ s are true, then s is true.**
Proof: We desire a chain of implications from p to s. The statement $q \rightarrow (\sim p)$ involves the statement $(\sim p)$ rather than p which is given to be true. Fortunately, we may substitute the contrapositive of $q \rightarrow (\sim p)$ and have as true

$$p \rightarrow (\sim q).$$

We now have

p	true by hypothesis
$p \rightarrow (\sim q)$	contrapositive of a true statement
$(\sim q) \rightarrow s$	true by hypotheses

We now conclude that $p \rightarrow s$ is true by the law of syllogisms. Since $p \rightarrow s$ is true and p is true, s is true by the Law of Detachment.

THEOREM 2.2. **Given: Either George or William is the leader. George is not the leader.**

Prove: William is the leader.

Proof: Let p be the statement "William is the leader" and q be the statement "George is the leader". We have given:

$$p \vee q$$
$$\sim q$$

We are asked to prove that p is true.
Since $(\sim q)$ is true, q is false. Why? Since $p \vee q$ is true and q is false, p must be true by the definition of disjunction.

THEOREM 2.3. **Given: If this is an interesting course, then it is worth**

taking. Either the grading is fair or the course is not worth taking. The grading is not fair.

Prove: This is not an interesting course.

Proof: Let us use the following symbolic representation:

    p: This is an interesting course.

    q: This course is worth taking.

    r: The grading is fair.

The hypotheses are:

    $p \rightarrow q$

    $r \vee (\sim q)$

    $\sim r$

We are asked to prove $(\sim p)$.

Since $(\sim r)$ is true, r is false. Since r is false and $r \vee (\sim q)$ is true, $(\sim q)$ is true by the definition of disjunction. Since $p \rightarrow q$ is true, its contrapositive, $(\sim q) \rightarrow (\sim p)$ is true. We now have $(\sim q) \rightarrow (\sim p)$ and $(\sim q)$ true, hence $(\sim p)$ is true by the law of detachment.

2.12. Indirect Proof

The **indirect method** of proof is often called proof by contradiction. By contradiction we mean a statement of the form "p and not p" which is false because a statement cannot be true and false at the same time.

For example, suppose a boy told his mother that he had run all the way home from baseball practice so that he would not be late for dinner. His mother said, "No, you did not! If you had run all the way you would be out of breath. You are not out of breath. Therefore, you did not run all the way home." The mother is using an indirect proof.

The essence of indirect proof consists of negating the desired conclusion and arriving at some contradiction. This method of proof relies on the fact that if $(\sim p)$ is false, p is true. Hence to prove p is true, we attempt to show that $(\sim p)$ is false. The best way to do this is to show that $(\sim p)$ is not consistent with the given hypotheses. In other words, we add $(\sim p)$ to the list of statements given to be true and show that with this statement added we have a contradiction When this contradiction is reached we know our assumption is not true—that is, $(\sim p)$ is false and p is true.

Study the following indirect proofs.

THEOREM 2.4.   Given: p, q → (~p) and (~q) → r.
Prove: r

Proof: Assume $(\sim r)$ is true. Since $(\sim q) \to r$ is true, the contrapositive $(\sim r) \to q$ is true. We now have true:

$$p$$
$$(\sim r)$$
$$(\sim r) \to q$$
$$q \to (\sim p).$$

Since $(\sim r) \to q$ and $q \to (\sim p)$ are true, $(\sim r) \to (\sim p)$ is true by the law of syllogisms. Since $(\sim r)$ is true by assumption and $(\sim r) \to (\sim p)$ is true, $(\sim p)$ is true by the law of detachment. But we now have a contradiction: both p and $(\sim p)$ true. Hence our assumption that $(\sim r)$ is true is false, and hence r is true.

THEOREM 2.5.   Given: Carl dates Alice. Carl dates Betty. If Carl does not date Alice, he spends all of his allowance on skin diving equipment. If Carl dates Betty, he does not spend all of his allowance on skin diving epuipment.
Prove: Carl does not spend all of his allowance on skin diving equipment.

Proof: Let us use the following symbolic notation:

p: Carl dates Alice.
q: Carl dates Betty.
r: Carl spends all of his allowance on skin diving equipment.

We are given:

$$p$$
$$q$$
$$(\sim p) \to r$$
$$q \to (\sim r)$$

We are asked to prove $(\sim r)$.

Let us assume that $(\sim r)$ is false. Then r is true. But $q \to (\sim r)$ is true and q is true, hence $(\sim r)$ is true by the law of detachment. We now have a contradiction: $(\sim r)$ both true and false. Therefore our assumption is false and $(\sim r)$ is true.

THEOREM 2.6. Given: This book is well written. If this book is difficult to read, then it is not well written. If this book is not difficult to read, then I enjoy reading it.

Prove: I enjoy reading this book.

Proof: Let us use the following symbolic notation:

p: This book is well written.
q: This book is difficult to read.
s: I enjoy reading this book.

The hypotheses are:

$$p$$
$$q \to (\sim p)$$
$$(\sim q) \to s$$

We want to prove s.

Let us assume that $(\sim s)$ is true. Since $(\sim q) \to s$ is true, the contrapositive $(\sim s) \to q$ is true. We now have a chain of implications that are true:

$$(\sim s) \to q$$
$$q \to (\sim p)$$

We conclude, by the law of syllogisms, that $(\sim s) \to (\sim p)$ is true. Now $(\sim s) \to (\sim p)$ is true and $(\sim s)$ is true, hence $(\sim p)$ is true by the law of detachment. But this leads to a contradiction because by hypotheses p is true. Hence our assumption is false and s is true.

EXERCISE 2.4

Prove the following.

1. Given: p, q, $(\sim p) \to s$ and $q \to (\sim s)$
    Prove: $\sim s$
2. Given: p and $[\sim(p \to q)]$
    Prove: $(\sim q)$
3. Given: $p \lor q$ and $(\sim p)$
    Prove: q
4. Given: $p \to q$ and $q \to (\sim p)$
    Prove: $(\sim p)$

---

5. Given: $p \rightarrow q$, $q \rightarrow t$, $t \rightarrow r$ and p
   Prove: r

6. Given: If John graduates with honors, his father gives him a trip to Europe. If John's father gives him a trip to Europe, he will visit Germany. If John does not visit France, he will not visit Germany. John graduates with honors.
   Prove: John visits France.

7. Given: If $ABCD$ is a square, then it is a rectangle. $ABCD$ is a square or a rhombus. $ABCD$ is not a rhombus.
   Prove: $ABCD$ is a rectangle.

8. Given: If a number is divisible by 4, then it is divisible by 2. If a number is divisible by 2, then it is even. If a number is even then it has a factor 2. This number does not have a factor 2.
   Prove: This number in not divisible by 4.

9. Given: Geometric figure $ABCD$ is a rectangle or a parallelogram. If $ABCD$ is a rectangle, then it is a parallelogram. If $ABCD$ is a parallelogram, then it is a quadrilateral.
   Prove: $ABCD$ is a parallelogram or a quadrilateral.

10. Given: If Gus doesn't date Suzie, he dates Pat. When Gus dates Pat, Suzie is unhappy. Suzie is not unhappy.
    Prove: Gus dates Suzie.

11. Given: If Mary is telling the truth, John is a true friend. If Helen is not reliable, John is not a true friend.
    Prove: If Mary is telling the truth then Helen is reliable.

12. Given: If the dormouse sleeps and the hatter spills his tea, then the March Hare laughs and Alice is amazed. Alice is not amazed. The hatter spills his tea.
    Prove: The dormouse does not sleep.

13. Given: If the white rabbit wears white gloves, then the Queen of Hearts is giving a party. If the Queen of Hearts is giving a party, then the duchess is angry. The duchess is not angry.
    Prove: The white rabbit is not wearing white gloves.

14. Given: If William goes to Las Vegas, he gambles. If William gambles, he loses money. If William does not play roulette, he does not lose money. William goes to Las Vegas.
    Prove: William plays roulette.

15. Given: If Gene advocates peace, then Dick advocates war. If Dick advocates war, then Carrie does not marry him.

    Prove: If Carrie marries Dick, Gene does not advocate peace.

16. Given: If I have a cat, then I don't have a dog. If I don't have a cat then I dislike animals. I have a dog.

    Prove: I dislike animals.

3.1. The Whole Numbers

The set of whole numbers contains the set, N, of **counting numbers** or **natural numbers**

$$N = \{1, 2, 3, 4, \ldots\}$$

and zero. That is, the set, W, of **whole numbers** is

$$W = \{0, 1, 2, 3, 4, \ldots\}$$

The whole number zero is defined as the cardinal number of the empty set:

$$n(\phi) = 0.$$

The whole number one is defined as the cardinal number of all the sets equivalent to a set with a single element. Thus

$$n(\{1\}) = 1.$$

The System of
Whole Numbers

The whole number two is defined as the cardinal number of all the sets equivalent to a set with a pair of elements. Thus

$$n(\{1, 2\}) = 2.$$

The other whole numbers are defined in a similar fashion:

$$n(\{1, 2, 3\}) = 3$$
$$n(\{1, 2, 3, 4\}) = 4$$
$$n(\{1, 2, 3, 4, 5\}) = 5, \text{ and so forth.}$$

We see that whole numbers are cardinal numbers. We see also that the set of whole numbers is an infinite set.

We take as a fundamental property of whole numbers that each whole number has an immediate **successor.** The successor of 0 is 1; the successor of 1 is 2; the successor of 2 is 3; and so forth. The successor of any whole number k is denoted by $k + 1$. We see, then, that the whole numbers present themselves in a natural order:

$$0, 1, 2, 3, \ldots$$

We start with zero and follow it by its successor 1; we take 1 and follow it by its successor 2; and so forth.

We now state two more fundamental properties of whole numbers: (1) no two different whole numbers have the same successor; and (2) there is no whole number whose successor is 0.

3.2. The Equality Relation

The **equality relation,** denoted by $=$, defined on a set of numbers is one of the most important relations in mathematics. The equality relation is an **equivalence relation** (see Chapter 2, Section 2.6). That is, for all numbers a, b, and c, the equality relation is:

E-1. Reflexive. For all a, $a = a$.
E-2. Symmetric. For all a and b, if $a = b$, then $b = a$.
E-3. Transitive. For all a, b, and c, if $a = b$ and $b = c$, then $a = c$.

In addition to the above properties, the equality relation has the following additional properties:

E-4. Addition Property. For all a, b, and c, if $a = b$, then $a + c = b + c$ and $c + a = c + b$.

E-5. Subtraction Property. For all a, b, and c, if $a = b$, then $a - c = b - c$.

E-6. Multiplication Property. For all a, b, and c, if $a = b$, then $ac = bc$ and $ca = cb$.

E-7. Division Property. For all a, b, and c, $c \neq 0$, if $a = b$, then $a \div c = b \div c$.

E-8. Substitution Property. For all a and b, if $a = b$, then in any statement involving a, b may be substituted for a without changing the truth (or falsity) of the statement.

The addition property of equality (E-4) assures us that we may add the same number to both sides (called **members**) of an equality. Thus if n is a number and $n - 6 = 8$, then, by the addition property, $(n - 6) + 6 = 8 + 6$.

The subtraction property of equality (E-5) assures us that we may subtract the same number from both members of an equality. Thus, if n is a number, and $n + 7 = 12$, by the subtraction property, $(n + 7) - 7 = 12 - 7$.

The multiplication property of equality (E-6) assures us that we may multiply both members of an equality by the same number. Thus if n is a number and $n = 7$, then by the multiplication property of equality $6n = 6 \times 7$.

The division property of equality (E-7) assures us that if n is a number and if $5n = 15$, then $5n \div 5 = 15 \div 5$.

The substitution property of equality (E-8) assures us that if p and q are equal and if $q + 6 = 25$, then $p + 6 = 25$.

3.3. Order Property of Whole Numbers

The whole numbers present themselves in a definite order: 0, 1, 2, 3, ... When the whole numbers are listed in order we see that 3 comes before 6, 6 comes before 10, and 10 comes before 15. We now define the order relation, "is less than" on the set W of whole numbers.

DEFINITION 3.1. *If $n(A) = a$ and $n(B) = b$ are two whole numbers, we say that a is less than b, denoted by $a < b$, provided A is equivalent to a proper subset of B.*
If $a < b$ we say that b is greater than a and write $b > a$.

Let $A = \{a, b, c, d\}$ and $B = \{k, m, n, p, s, t\}$. Then $n(A) = 4$ and $n(B) = 6$. Since A is equivalent to a proper subset of B (for example $A \sim C$ where $C = \{k, m, n, p\}$ and $C \subset B$), we say that 4 is less than 6 and write $4 < 6$. We can also say that 6 is greater than 4 and write $6 > 4$.

When the whole numbers are named in order

$$0, 1, 2, 3, \ldots$$

each number in the sequence is less than any number that comes later in the sequence. Thus $4 < 9$, $9 < 17$, $17 < 26$, and so forth.

If we choose an ordered pair of whole numbers, (a, b), then exactly one of the following statements is true:

$$a < b$$
$$a = b$$
$$a > b.$$

This is called the **trichotomy property.**

In addition to the trichotomy property, the order relation "is less than" defined on the set of whole numbers has the following additional properties:

O-1. **Transitive Property.** For all whole numbers a, b, and c, if $a < b$ and $b < c$, then $a < c$.

O-2. **Addition Property.** For all whole numbers a, b, and c, if $a < b$, then $a + c < b + c$ and $c + a < c + b$.

O-3. **Subtraction Property.** For all whole numbers a, b, and c, if $a < b$, $c < a$ and $c < b$, then $a - c < b - c$.

O-4. **Multiplication Property.** For all whole numbers a, b, and c, if $a < b$ and $c \neq 0$, then $ac < bc$ and $ca < cb$.

O-5. **Division Property.** For all whole numbers a, b, and c, if $ac < bc$, $c \neq 0$, then $a < b$.

The above properties are also true when the symbol $<$ is replaced by the symbol $>$.

Between any two whole numbers there is a definite number of whole numbers. Thus between 6 and 9 there are two whole numbers, 7 and 8. Between 25 and 31, there are five whole numbers, 26, 27, 28, 29, and 30; between 5 and 6 there are no whole numbers. Between any two whole numbers a and b, $a < b$, there are $(b - a) - 1$ whole numbers.

Example i. How many whole numbers are between 16 and 25?
Solution: Between 16 and 25 there are eight whole numbers, 17, 18, 19, 20,

21, 22, 23, and 24. Using the formula given above we have with $b = 25$ and $a = 16$:

$$(b - a) - 1 = (25 - 16) - 1$$
$$= 9 - 1 = 8$$

EXAMPLE ii. How many whole numbers are between 185 and 298?

Solution: Using the formula we have with $a = 185$ and $b = 298$:

$$(b - a) - 1 = (298 - 185) - 1$$
$$= 113 - 1 = 112.$$

There are 112 whole numbers between 185 and 298.

3.4. The Number Line

The number line gives a model for visualizing the order of the whole numbers. The whole numbers may be represented on a line as shown in Figure 3.1. A line is drawn and a sequence of equally spaced points are marked on it. A point is chosen to correspond to zero and labeled "0." The points to the right of the point labeled "0" are labeled successively "1," "2," "3," "4," and so on. Thus we have a matching of the whole numbers with the equally spaced points on the line. Each point is called the **graph** of the whole number to which it corresponds and each number is called the **coordinate** of the point to which it corresponds.

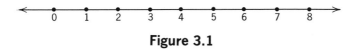

Figure 3.1

Observation of the number line shows the following:

(1) To each whole number there corresponds one and only one point on the number line.
(2) A whole number a is less than a whole number b if the point corresponding to a is to the left of the point corresponding to b.
(3) A whole number a is greater than a whole number b if the point corresponding to a is to the right of the point corresponding to b.
(4) There is no greatest whole number; the sequence of equally spaced points chosen to correspond to the whole numbers continues indefinitely to the right.
(5) Zero is the least whole number.

3.5. Binary Operations

Addition and multiplication of whole numbers are **binary operations.** The operation of addition, denoted by $+$, associates with each ordered pair (a, b) of whole numbers, a unique whole number, $a + b$. Thus the operation of addition associates with the ordered pair $(4, 3)$, the unique whole number $4 + 3$.

The operation of multiplication of whole numbers, denoted by $\times$ or a raised dot,* associates with each ordered pair (a, b) of whole numbers a unique whole number $a \times b$ (also denoted by $a \cdot b$ or ab).

With the ordered pair	The operation	Associates the unique number		That can be renamed
(2, 3)	$\xrightarrow{+}$	2 + 3	=	5
(6, 7)	$\xrightarrow{+}$	6 + 7	=	13
(5, 6)	$\xrightarrow{\times}$	5 × 6	=	30
(2, 3)	$\xrightarrow{\times}$	2 × 3	=	6

The term "binary" refers to the two numbers in the ordered pair. (The prefix "bi" denotes two as in the words "bicycle" and "bilateral.")

We now give a formal definition of a binary operation.

DEFINITION 3.2. *A **binary operation,** denoted by *, on a set S associates with each ordered pair (a, b) of elements of S a uniquely determined element denoted by a * b.*

When we say that the element $a * b$ is uniquely determined we mean that the binary operation * associates with each ordered pair (a, b) one and only one element $a * b$.

The binary operations of addition, subtraction, multiplication, and division are called the **basic operations** of arithmetic. Addition and multiplication are called the **primary operations;** subtraction and division are called **secondary operations.**

* Multiplication may also be indicated by placing two letters which represent numbers adjacent to each other. For example $a \times b$ may be denoted by ab. Similarly, 2×3 may be denoted by $(2)(3)$. Note that we do not indicate 2×3 by 23 since this symbol means 2 tens + 3 ones as we shall see in Chapter 4.

EXAMPLE i. A binary operation * on the set of whole numbers is described as taking the larger number of the components of an ordered pair. Thus

$$a * b = a \text{ if } a > b$$
$$a * b = b \text{ if } a < b.$$

Find 6 * 9.

Solution: $6 * 9 = 9$

EXAMPLE ii. A binary operation * on the set of whole numbers is described as multiplying the first component of the ordered pair by 4 and adding the second component to the product. Thus

$$a * b = 4a + b$$

For example,

$$2 * 3 = (4 \times 2) + 3 = 8 + 3 = 11$$

Find 6 * 7

Solution:
$$6 * 7 = (4 \times 6) + 7$$
$$= 24 + 7$$
$$= 31$$

Exercise 3.1

1. List the whole number elements of each of the following sets in their natural order.
   a. {3, 7, 5, 1, 9}
   b. {34, 95, 12, 16, 56, 11}
   c. {44, 11, 66, 88, 22, 101}
   d. {4, 7, 0, 18, 93, 45, 6}
   e. {101, 404, 12, 7, 0, 16, 666}
2. Draw a number line. Mark the graph of 7 by X; the graph of 3 by Y; the graph of 1 by Z; the graph of 9 by T; and the graph of 12 by K.
3. On the number line below what is the coordinate of the point marked by
   a. X? b. Y? c. Z?

4. Replace each * with $<$, $=$ or $>$ to make true statements.
   a. 15 * 26
   b. 4 * 0
   c. $(3 \times 3) * (5 \times 3)$
   d. $(7 + 8) * (7 + 6)$
   e. $(4 + 3) * (3 + 4)$
   f. $(5 \times 0) * (0 \times 9)$
   g. $[7 \times (3 + 6)] * (7 \times 9)$
   h. $[6 \times (4 + 5)] * [6 \times (3 + 4)]$

5. How many whole numbers are between each of the following pairs of whole numbers?
   a. 16 and 27
   b. 37 and 92
   c. 0 and 18
   d. 118 and 119
   e. 426 and 727
   f. 399 and 1000

6. Consider the ordered pairs below and the number associated with each ordered pair. What binary operation of arithmetic, that is, addition, subtraction, multiplication, or division, has been performed on the components of the ordered pairs to give the designated results?

Ordered pair	Result of binary operation
a. (7, 9)	16
b. (18, 4)	14
c. (27, 9)	3
d. (576, 9)	64
e. (4, 28)	32
f. (186, 97)	89
g. (17, 25)	425

7. A binary operation * on the set of whole numbers is described as doubling the first number of the ordered pair and adding the second to the product. Thus $a * b = (2a) + b; 5 * 3 = (2)(5) + 3 = 13$. Find
   a. 2 * 6
   b. 3 * 9
   c. 5 * 4
   d. 8 * 4
   e. 6 * 2
   f. 4 * 5

8. A binary operation * on the set of whole numbers is described as taking the smaller number of the ordered pair and multiplying it by 3. Thus $2 * 5 = (3)(2) = 6; 6 * 1 = (3)(1) = 3$. Find
   a. 8 * 5
   b. 2 * 3
   c. 3 * 2
   d. 1 * 8
   e. 11 * 6
   f. 100 * 200

9. A binary operation * on the set of whole numbers is described as taking four times the first number of the ordered pair and adding six times the second number to the product. Thus $a * b = 4a + 6b$; $1 * 2 = (4)(1) + (6)(2) = 4 + 12 = 16$. Find

a. $1 * 3$ c. $5 * 2$ e. $9 * 10$
b. $7 * 7$ d. $0 * 0$ f. $11 * 2$

10. What is the least whole number?
11. Define a binary operation.
12. What are the primary operations of arithmetic?
13. What are the secondary operations of arithmetic?
14. What is the Trichotomy Principle?
15. What is another name for the set of counting numbers?
16. What is the greatest whole number?
17. What are the four basic operations of arithmetic?
18. A binary operation * on the set of whole numbers is defined as the remainder when the product of the two numbers is divided by 3. Find
    a. $(4 * 7) * 2$ c. $(7 * 8) * 9$
    b. $(2 * 4) * 5$ d. $(5 * 8) * 11$
19. A binary operation * on the set of whole numbers is defined as the remainder when the product of two numbers is divided by 6. Find
    a. $(8 * 4) * 9$ c. $(5 * 10) * 7$
    b. $(6 * 8) * 7$ d. $(3 * 12) * 15$
20. A binary operation * on the set of whole numbers is defined as the remainder when the product of two numbers is divided by 7. Find
    a. $(4 * 5) * 8$ c. $(9 * 11) * 14$
    b. $(7 * 6) * 12$ d. $(15 * 25) * 30$
21. What properties of the equality relation are illustrated by the following?
    a. If $a = 3$ then $3 = a$.
    b. If $a - 2 = 5$, then $(a - 2) + 2 = 7$.
    c. If $k + n = 6$ and $k = 5$, then $5 + n = 6$.
    d. If $3p = 9$, then $p = 3$.
    e. If $a = 3 + 4$, and $3 + 4 = y$, then $a = y$.
    f. If $a + 4 = 9$, then $(a + 4) - 4 = 5$.
    g. If $a = 7$, then $5a = 35$.
22. Justify each statement below by a property of the order relation "less than." The variables in each case represent whole numbers.
    a. If $a + 6 < 8$, $a < 2$.
    b. If p and q are whole numbers, $p < q$, $p = q$ or $p > q$.
    c. If $n - 9 < 16$, then $a < 25$
    d. If $2p < 8$, then $p < 4$.
    e. If $5n + 9 < 20$, then $5n < 11$.
    f. If $4r - 6 < 3$, then $4r < 9$.

In each problem below (Ex. 23–26) the replacement set of the variable is the set of whole numbers. List the elements in the following sets.

23. $\{x \mid x < 8\}$
24. $\{x \mid x > 14\}$
25. $\{x \mid x > 9 \text{ and } x < 12\}$
26. $\{x \mid x > 40 \text{ and } x < 50\}$

Give all the whole number values of n that make the following true statements. (Ex. 27–30)

27. $n + 2 < 8$
28. $n + 7 < 15$
29. $n + 4 < 12$
30. $2n + 5 < 15$

3.6.   Addition of Whole Numbers

We defined the whole numbers as cardinal numbers of sets. We define the binary operation of addition in terms of the union of disjoint sets.

DEFINITION 3.3. *If a and b are two whole numbers, their* **sum,** *denoted by a + b, is defined as follows. Let A and B be two disjoint sets such that $n(A) = a$ and $n(B) = b$. Then $a + b = n(A \cup B)$. The operation of finding the sum is called* **addition.**

If

$$A = \{a, b\}$$
and
$$B = \{k, m, s\}$$
then
$$A \cup B = \{a, b, k, m, s\}$$

We see that $n(A) = 2$, $n(B) = 3$ and $n(A \cup B) = 5$. By Definition 3.3 we have $2 + 3 = 5$. We read this "two plus three equals five."

When we start with two disjoint sets and form their union we are operating on sets. When we start with two whole numbers and find their sum we are operating on numbers. Addition of whole numbers is a binary operation that associates with each ordered pair (a, b) of whole numbers the unique whole number $a + b$. The whole numbers a and b are called **addends,** the whole number $a + b$ is called the **sum.** Thus in $6 + 7 = 13$, 6 and 7 are addends and 13 is the sum.

3.7. Properties of Addition

Addition is a binary operation performed on two whole numbers called addends which produces a unique whole number called the sum of the two whole numbers. The addends may be thought of as the cardinal numbers of each of two disjoint sets. The sum of these numbers is the cardinal number of the union of these two disjoint sets.

Addition is a binary operation because it is performed on just two numbers at a time.

Since the sum of two whole numbers is defined as the cardinal number of the union of two disjoint sets, and since cardinal numbers are whole numbers, the sum of two whole numbers is always a whole number. We describe this situation by saying that the set of whole numbers is **closed** under the operation of addition. This property of whole numbers and addition is called the **closure property of addition.**

> **W-1. Closure Property of Addition.** If a and b are any two whole numbers, then their sum, $a + b$, is a whole number.

Let us consider the two disjoint sets

$$A = \{a, b, c, d\}$$
$$B = \{p, s, t\}$$

Then

$$A \cup B = \{a, b, c, d, p, s, t\}$$
and
$$B \cup A = \{p, s, t, a, b, c, d\}$$

We see that $A \cup B = B \cup A$, and $n(A \cup B) = n(B \cup A)$.

Since addition of whole numbers is associated with the union of two disjoint sets, we see that

$$7 + 6 = 6 + 7$$
$$9 + 4 = 4 + 9$$
$$10 + 14 = 14 + 10,$$

and in general, that if a and b are any whole numbers, then $a + b = b + a$. This property of addition of whole numbers is called the **commutative property of addition.** It states that the order of the addends in addition may be changed without changing the sum.

> **W-2. Commutative Property of Addition.** If a and b are any two whole numbers, then $\mathbf{a + b = b + a.}$

Addition is a binary operation, that is it can be performed on just two numbers at a time. If we are asked to add three numbers, for example, $3 + 4 + 5$, without changing their order, there are two possible ways to do the addition. We may add 3 and 4, to produce the sum 7. To this sum we add 5 to get the sum 12. We write this

$$(3 + 4) + 5$$

On the other hand, we may add 4 and 5 first to produce the sum, 9. We then add 3 and this sum to get 12. We write this

$$3 + (4 + 5)$$

We see that whichever way we associate the addends, the sum is 12. Thus

$$(3 + 4) + 5 = 3 + (4 + 5)$$

This is true for any three whole numbers, as illustrated by the following examples:

$$(1 + 2) + 3 = 1 + (2 + 3)$$
$$(5 + 7) + 6 = 5 + (7 + 6)$$
$$(8 + 9) + 4 = 8 + (9 + 4)$$

In general, if a, b, and c are any whole numbers, then $(a + b) + c = a + (b + c)$. This property is called the **associative property of addition.**

W-3. Associative Property of Addition. If a, b, and c are any whole numbers, then $\mathbf{(a + b) + c = a + (b + c)}$.

Because $(a + b) + c = a + (b + c)$, it is not necessary to use parentheses in writing the sum of three addends. We merely write $a + b + c$. The associative property of addition may be generalized to hold for four or more addends. For example, $(2 + 3) + (4 + 7) = 2 + [3 + (4 + 7)]$. Again we usually omit the parentheses and write $2 + 3 + 4 + 7$.

The number zero plays a special role with respect to addition. Notice that

$$0 + 16 = 16 + 0 = 16$$
$$0 + 9 = 9 + 0 = 9$$
$$0 + 3 = 3 + 0 = 3.$$

In fact, if a is any whole number, than $a + 0 = 0 + a = a$; that is, *the sum of zero and any whole number is that whole number.* Zero is called the **identity element for addition** or the **additive identity.**

W-4. Identity Property of Addition. If a is any whole number, then $a + 0 = 0 + a = a$. The number 0 is called the **identity element for addition** or the **additive identity**.

Exercise 3.2

1. What properties of addition of whole numbers are illustrated by each of the following statements?
   a. $3 + 4 = 4 + 3$
   b. $8 + 0 = 8$
   c. $3 + (9 + 6) = (3 + 9) + 6$
   d. $0 + 17 = 17$
   e. $8 + (7 + 4) = 8 + (4 + 7)$
   f. $m + 3 = 3 + m$, m a whole number

2. What sum is associated with each ordered pair of whole numbers below?
   a. $(178, 193)$ d. $(342, 76)$
   b. $(786, 97)$ e. $(193, 178)$
   c. $(76, 342)$ f. $(99, 107)$

3. Which ordered pairs in Problem 2 give the same sum? Why?

4. Which of the following activities are commutative?
   a. To put on a hat and coat
   b. To mix equal parts of blue paint and red paint
   c. To put on a sweater and a shirt
   d. To cook dinner and eat it
   e. To put on a swimming suit and jump into a pool

5. Which of the following statements are true for all whole numbers a and b?
   a. $(a + b) + 0 = a + b$
   b. $(a + b) + 6 = a + (b + 6)$
   c. $a + b = a$
   d. $a + 12 = 6 + b$
   e. $a + 7 = 7 + a$
   f. $12 + a + b = (8 + a) + (4 + b)$

6. By inspection find all the whole number replacements for n that make the following true statements.
   a. $n + 7 = 9$ e. $(n + 3) + 4 = n + 7$
   b. $n + 6 = 6$ f. $n + 2 < 5$

---

c. $8 + n = 12$ g. $n + 7 < 15$

d. $n + 4 < 8$ h. $n + 3 > 5$

7. Which of the following sets are closed under the operation of addition?

   a. $\{0\}$

   b. $\{1, 3, 5, 7, 9, \ldots\}$

   c. $\{0, 2, 4, 6, 8, \ldots\}$

   d. $\{0, 1, 2, 3, 4\}$

   e. $\{0, 1\}$

8. Give six pairs of activities from everyday life which are commutative under the operation of "followed by." For example, the activities of putting on a coat and putting on a hat are commutative under the operation of "followed by" because the result of putting on a coat followed by putting on a hat is the same as the result of putting on a hat followed by putting on a coat.

9. Give six pairs of activities from everyday life that are not commutative under the operation of "followed by."

10. What is the identity element of addition for the set of whole numbers?

3.8. Multiplication of Whole Numbers

Multiplication of whole numbers is a binary operation that associates with an ordered pair of whole numbers, each of which is called a **factor,** a unique whole number called the **product.** When the number 12 is associated with the ordered pair (6, 2) by multiplication, 6 and 2 are factors and 12 is the product.

We define the binary operation of multiplication of whole numbers in terms of the Cartesian product of two sets.

DEFINITION 3.4. *If a and b are whole numbers, their* **product,** *denoted by* $a \times b$, $a \cdot b$, *or* ab *is defined as follows. Let A and B be two sets such that* $n(A) = a$ *and* $n(B) = b$, *then* $ab = n(A \times B)$. *The operation of finding the product is called* **multiplication.**

Let us consider the two sets A and B:

$$A = \{a, b, c\}$$
$$B = \{r, s, t, v\}$$

We find the Cartesian product, $A \times B$, by forming all the possible ordered pairs whose first components are elements of A and whose second components are elements of B. A schematic way of finding all of the ordered pairs of $A \times B$ is shown in Figure 3.2a. This schematic representation is called an **array.**

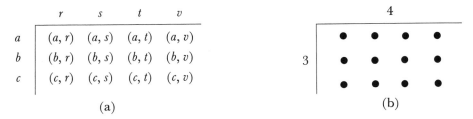

	r	s	t	v
a	(a, r)	(a, s)	(a, t)	(a, v)
b	(b, r)	(b, s)	(b, t)	(b, v)
c	(c, r)	(c, s)	(c, t)	(c, v)

(a)

(b)

Figure 3.2

We see that $n(A) = 3$, $n(B) = 4$ and $n(A \times B) = 12$. Then, by Definition 3.4 we have $3 \times 4 = 12$. We read this "three times four equals twelve."

We can represent products of whole numbers by arrays. We represent the product 3×4 by the array shown in Figure 3.2b. We see that the product 3×4 is represented by a rectangular array of dots; there are three rows and four columns in the array. When a product is represented by an array, the first factor gives the number of rows in the array and the second factor gives the number of columns.

Note that the array representing the product 3×4 may be interpreted as the union of three disjoint sets each having four elements (Figure 3.3a). In other words, 3×4 may be thought of as the sum of three fours; hence

$$3 \times 4 = 4 + 4 + 4.$$

The product 3×4 may also be interpreted as the union of four disjoint sets each having three elements (Figure 3.3b). Thus 3×4 may be thought of as the sum of four threes:

$$3 \times 4 = 3 + 3 + 3 + 3.$$

We see then that 3×4 may be interpreted as three fours or as four threes.

Every product may be represented by an array. Examples are shown in Figure 3.4. Again we note that when a product is represented by an array, the first factor tells the number of rows in the array and the second

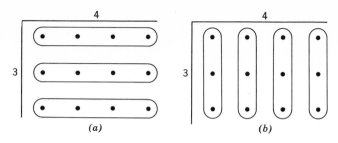

Figure 3.3

factor tells the number of columns. Thus 5×4 is represented by an array of five rows and four columns. In general, ab is represented by an array of a rows and b columns; ba is represented by an array of b rows and a columns.

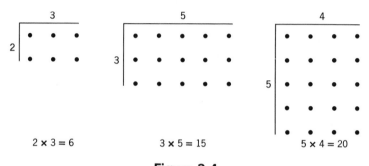

$2 \times 3 = 6$ $3 \times 5 = 15$ $5 \times 4 = 20$

Figure 3.4

3.9. Properties of Multiplication

Since the product of two whole numbers is defined as the cardinal number of the Cartesian product of two sets and since cardinal numbers are whole numbers, the product of two whole numbers is a whole number. We describe this situation by saying that the set of whole numbers is **closed** under the operation of multiplication. This property of whole numbers and multiplication is called the **closure property of multiplication.**

W-5. **Closure Property of Multiplication.** If a and b are any two whole numbers, then their product ab is a whole number.

Since $n(A \times B) = n(B \times A)$ for any two finite sets A and B and since the product of two whole numbers is defined as the cardinal number of the

Cartesian product of two sets, we see that $ab = ba$ for all whole numbers a and b. This is the **commutative property of multiplication.**

W-6. Commutative Property of Multiplication. If a and b are any two whole numbers, then $ab = ba$.

Figure 3.5 demonstrates the truth of the commutative property of multiplication for two particular sets of factors. An array representing 3×4 can be changed to an array representing 4×3 by simply turning the page 90° in a clockwise direction.

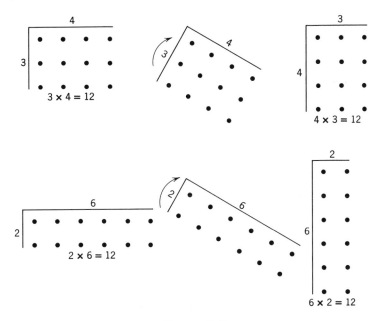

Figure 3.5

Multiplication is a binary operation, that is, it is an operation on two numbers. To find the product of three numbers, for example $4 \times 5 \times 6$, without changing the order of the factors, we may multiply 4 and 5 to get the product 20. We then multiply this product and 6 to get the product 120. We write this

$$(4 \times 5) \times 6.$$

We could also multiply 5 and 6 and obtain the product 30, and then multiply this product and 4 to get 120. We write this

$$4 \times (5 \times 6).$$

Whichever way we group the factors, the product is the same. That is

$$(4 \times 5) \times 6 = 4 \times (5 \times 6).$$

Observation of several examples, such as those below, helps convince us that the order in which we associate the factors in multiplication does not affect the product.

$$(3 \times 4) \times 2 = 3 \times (4 \times 2)$$
$$12 \times 2 = 3 \times 8$$
$$24 = 24$$
$$(7 \times 5) \times 9 = 7 \times (5 \times 9)$$
$$35 \times 9 = 7 \times 45$$
$$315 = 315$$
$$(2 \times 8) \times 5 = 2 \times (8 \times 5)$$
$$16 \times 5 = 2 \times 40$$
$$80 = 80$$

In general, if a, b, and c are any whole numbers, then $(ab)c = a(bc)$. This is called the **associative property of multiplication.**

W-7. Associative Property of Multiplication. If a, b, and c are any whole numbers, than **(ab)c = a(bc).**

We can illustrate the associative property of multiplication by considering a rectangular solid made up of cubical blocks. Let us use a solid with dimensions 2 by 3 by 4. The number of blocks used to construct the solids in Figure 3.6a and Figure 3.6b is 24 in each case. In Figure 3.6a, the shading shows 2×3 blocks in each vertical slice and four such vertical slices. Hence Figure 3.6a shows $(2 \times 3) \times 4$. Figure 3.6b shows, by shading, 3×4 blocks in each horizontal slice and two such horizontal slices. Since the number of blocks in each solid is 24, we see that $(2 \times 3) \times 4 = 2 \times (3 \times 4)$.

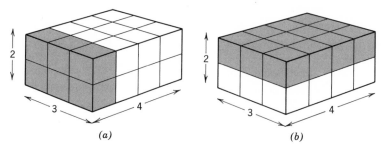

(a)   (b)

Figure 3.6

This same idea may be generalized with a solid of dimensions a by b by c.

Since $(2 \times 3) \times 4 = 2 \times (3 \times 4)$ we generally omit the parentheses and write $2 \times 3 \times 4$ when multiplying three whole numbers.

The associative property of multiplication may be generalized for four or more factors. For example, $(2 \times 3) \times (6 \times 8) = 2 \times [3 \times (6 \times 8)]$. Again, because of this generalized associative property we omit the parentheses and write $2 \times 3 \times 6 \times 8$.

The number 1 plays the same role for the operation of multiplication as the number 0 plays with respect to addition. Observe that

$$1 \times 6 = 6 \times 1 = 6$$
$$1 \times 9 = 9 \times 1 = 9$$
$$1 \times 26 = 26 \times 1 = 26$$

In general, if a is any whole number, $1 \times a = a \times 1 = a$.

> **W-8. Identity Property of Multiplication.** If a is any whole number, then $a \cdot 1 = 1 \cdot a = a$. The number 1 is called the **identity element for multiplication** or the **multiplicative identity**.

3.10. The Distributive Property

There is a very important property of whole numbers that relates the operations of addition and multiplication. This property is called the **distributive property of multiplication with respect to addition** or, simply, the **distributive property.** The easiest way to discover this property is to study some examples:

$$(2 \times 3) + (2 \times 5) = 6 + 10$$
$$= 16$$
$$= 2 \times 8$$
$$= 2 \times (3 + 5)$$
$$(5 \times 4) + (5 \times 9) = 20 + 45$$
$$= 65$$
$$= 5 \times 13$$
$$= 5 \times (4 + 9)$$
$$(8 \times 7) + (8 \times 6) = 56 + 48$$
$$= 104$$

$$= 8 \times 13$$
$$= 8 \times (7 + 6)$$
$$(5 \times 4) + (9 \times 4) = 20 + 36$$
$$= 56$$
$$= 14 \times 4$$
$$= (5 + 9) \times 4$$
$$(7 \times 2) + (6 \times 2) = 14 + 12$$
$$= 26$$
$$= 13 \times 2$$
$$= (7 + 6) \times 2$$
$$(8 \times 5) + (9 \times 5) = 40 + 45$$
$$= 85$$
$$= 17 \times 5$$
$$= (8 + 9) \times 5$$

These examples make plausible the statements that if a, b, and c are whole numbers, then

(1) $\qquad\qquad\qquad$ $ab + ac = a(b + c)$

(2) $\qquad\qquad\qquad$ $ba + ca = (b + c)a$

Because of the symmetric property of equality, statements (1) and (2) may be written

(3) $\qquad\qquad\qquad$ $a(b + c) = ab + ac$

(4) $\qquad\qquad\qquad$ $(b + c)a = ba + ca.$

These statements are called the **distributive property of multiplication over addition** or, simply, the **distributive property.**

Statements (1) and (2) are equivalent because of the commutative property of multiplication. Similarly, statements (3) and (4) are equivalent because of the same reason. It is important that we recognize the distributive property whether it is written in the form of statements (1), (2), (3), or (4).

> **W-9. Distributive Property of Multiplication over Addition.** If a, b, and c are whole numbers, then $a(b + c) = ab + ac$ and $(b + c)a = ba + ca.$

We can convince ourselves of the truth of the distributive property by using arrays. For example, let us demonstrate that $2 \times (3 + 4) = (2 \times 3) + (2 \times 4)$. The array representing 2×7 in Figure 3.7a can be

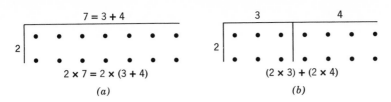

$2 \times 7 = 2 \times (3 + 4)$ $(2 \times 3) + (2 \times 4)$

(a)   (b)

Figure 3.7

separated by a vertical line segment into a 2×3 and a 2×4 array as shown in Figure 3.7b. This does not change the number of dots in the array, hence $2 \times (3 + 4) = (2 \times 3) + (2 \times 4)$.

EXAMPLE i.   Use the distributive property to write the product $6 \times (4 + 2)$ as the sum of two addends.
Solution: $6 \times (4 + 2) = (6 \times 4) + (6 \times 2)$

EXAMPLE ii.   Use the distributive property to write the product $(5 + 8) \times 3$ as the sum of two addends.
Solution: $(5 + 8) \times 3 = (5 \times 3) + (8 \times 3)$.

Exercise 3.3

1. What product is associated with each of the following ordered pairs?
    a. $(8, 4)$ e. $(4, 3)$ i. $(304, 61)$
    b. $(7, 9)$ f. $(12, 16)$ j. $(72, 83)$
    c. $(3, 4)$ g. $(24, 23)$ k. $(61, 304)$
    d. $(4, 8)$ h. $(9, 7)$ l. $(23, 24)$
2. Which ordered pairs in problem 1 give the same products? Why?
3. Draw arrays illustrating the following products.
    a. 2×3 d. 5×2
    b. 3×4 e. 8×2
    c. 1×6 f. 7×3
4. What properties of multiplication are illustrated by the following?
    a. $8 \times 4 = 4 \times 8$
    b. $3 \times (6 \times 7) = (3 \times 6) \times 7$
    c. $18 \times 1 = 18$
    d. $4 \times (7 \times 9) = 4 \times (9 \times 7)$
    e. $(3 \times 4) \times (6 \times 8) = 3 \times [4 \times (6 \times 8)]$
    f. $(18 \times 3) \times (4 \times 6) = (4 \times 6) \times (18 \times 3)$

g. $1 \times 27 = 27$

h. $(5 \times 6) \times (7 \times 3) = [(5 \times 6) \times 7] \times 3$

5. Use the distributive property to write the following products as the sum of two addends.

a. $8 \times (12 + 26)$

b. $15 \times (98 + 36)$

c. $(24 + 45) \times 38$

d. $(67 + 93) \times 107$

6. Which of the following sets are closed under the operation of multiplication?

a. $\{0, 1\}$

b. $\{1, 3, 5, 7, \ldots\}$

c. $\{0, 1, 2, 3, 4\}$

d. $\{0, 3, 6, 9, 12, \ldots\}$

e. $\{0, 1, 2, 4, 6\}$

7. By inspection find all the whole number replacements for n that make the following true statements.

a. $3n = 6$ e. $5n < 35$

b. $4n = 24$ f. $2n < 8$

c. $5n = 0$ g. $6n < 48$

d. $7n = 7$ h. $5n > 200$

8. Draw an array to verify that $7 \times (4 + 5) = (7 \times 4) + (7 \times 5)$

9. Why is $2 + 2$ equal to $(2 \times 1) + (2 \times 1)$?

10. Why is $(1 \times 5) + (1 \times 5)$ equal to $(1 + 1) \times 5$?

11. A **multiple** of a whole number is the product of that whole number and another whole number. Thus 6 is a multiple of 2 since $6 = 2 \times 3$. Write 24 as a multiple of

a. 2   c. 6   e. 12

b. 4   d. 8   f. 24

12. Write zero as a multiple of each of the following whole numbers. (See Problem 11).

a. 8   c. 6   e. 3

b. 12   d. 2   f. 5

13. A girl has 4 skirts and 5 blouses. How many different outfits does she have assuming that every skirt can be worn with every blouse?

14. A automobile company manufactures 4 different models. Each model comes in any one of 6 colors and with or without white wall

tires. How many different automobiles can be pictured in the company brochure?

15. Use the distributive property to find the following products. Write the larger factor as the sum of two addends one of which is a multiple of ten (the multiples of ten are 10, 20, 30, 40, 50, . . .). For example

$$2 \times 24 = 2 \times (20 + 4)$$
$$= (2 \times 20) + (2 \times 4)$$
$$= 40 + 8$$
$$= 48$$

a. 2×13 e. 9×58
b. 4×15 f. 6×94
c. 5×26 g. 8×39
d. 7×81 h. 3×76

3.11. Proving Theorems about Whole Numbers

Up to this point we have merely stated propositions that are true for whole numbers and their operations. We called these statements properties of the set of whole numbers and the operations of addition and multiplication. We did not prove these statements to be true; we accepted them as true. We call these properties the **axioms** or **postulates** of the whole number system. We now use these axioms together with the axioms of the equality relation to prove theorems about whole numbers.

THEOREM 3.1. **If a is any whole number, then a $\times$ 0 = 0.**

Proof: $a \times 0$ is a whole number   closure property of multiplication

$(a \times 0) + 0 = a \times 0$ identity element for addition

$a \times 0 = (a \times 0) + 0$ symmetric property of equality

$0 + 0 = 0$ identity property for addition

$a \times (0 + 0) = (a \times 0) + 0$ substitution property of equality

$a \times (0 + 0) = (a \times 0) + (a \times 0)$ distributive property

$(a \times 0) + 0 = (a \times 0) + (a \times 0)$ symmetric and transitive properties of equality

$$0 = a \times 0 \qquad \text{subtraction property of equality}$$
$$a \times 0 = 0 \qquad \text{symmetric property of equality}$$

Theorem 3.1 states that the product of any whole number and zero is zero. This is sometimes called **the multiplication property of zero.** This theorem is usually accepted without proof at the elementary levels of instruction.

THEOREM 3.2. **For all whole numbers a and b, a $\neq$ 0, if ab = 0, then b = 0.**

Proof:
$$ab = 0 \qquad \text{hypothesis (given)}$$
$$a \times 0 = 0 \qquad \text{Theorem 3.1}$$
$$ab = a \times 0 \qquad \text{symmetric and transitive properties of equality}$$
$$b = 0 \qquad \text{division property of equality}$$

Theorem 3.2 states that if the product of two whole numbers is zero and one of them is not zero, then the other must be zero. In other words, if the product of two whole numbers is zero, one or the other or both of the factors is zero.

THEOREM 3.3. **The additive identity is unique.**

Proof: We are trying to show that there is only one additive identity. Let us suppose that there are two additive identities, 0 and 0*. Then

$$0 + 0 = 0 \qquad \text{identity element for addition}$$
$$0 + 0^* = 0 \qquad \text{identity element for addition}$$
$$0 + 0^* = 0 + 0 \qquad \text{substitution property of equality}$$
$$0^* = 0 \qquad \text{subtraction property of equality}$$

We see that 0 and 0* are names for the same number. That is, the additive identity is unique.

THEOREM 3.4. **The multiplicative identity is unique.**

Proof: Let us assume that there are two multiplicative identities, 1 and 1'. Then

$$1 \times 1 = 1 \qquad \text{identity element for multiplication}$$
$$1 \times 1' = 1 \qquad \text{identity element for multiplication}$$
$$1 \times 1' = 1 \times 1 \qquad \text{substitution property of equality}$$
$$1' = 1 \qquad \text{division property of equality}$$

Since $1' = 1$, the multiplicative identity is unique.

When we perform addition computation we usually check our work by adding the numbers in the opposite order. Thus if we find the sum $(a + b) + c$, we check our work by adding $(c + b) + a$. We now show why this method of checking addition works.

THEOREM 3.5. For all whole numbers a, b, and c, $(a + b) + c = (c + b) + a$.

Proof:
$$\begin{aligned}(a + b) + c &= a + (b + c) && \text{associative property of addition} \\ &= a + (c + b) && \text{commutative property of addition} \\ &= (c + b) + a && \text{commutative property of addition}\end{aligned}$$

3.12. Inverse Operations

We often do something and then undo what has been done. For example, we sit down and then stand up; we put on a pair of gloves and then take them off; we walk three blocks north and then walk three blocks south. These "undoing" activities are called the **inverses** of the doing activities.

Activity	Inverse Activity
wrapping	unwrapping
raising	lowering
putting on	taking off
opening	closing

Not every activity has an inverse. For example, the activity of scrambling eggs has no inverse. There is no possible way to unscramble eggs.

Mathematical operations also have inverses. The inverse of adding 7 to a number is subtracting 7; the inverse of subtracting 6 from a number is adding 6. *The operation of subtraction is the inverse of the operation of addition.*

Addition is the operation on two known addends to produce a unique number called the sum. Subtraction is the operation of finding an unknown addend when the sum and one of the addends is known.

The operation of division is the inverse of the operation of multiplication. The inverse of multiplying a number by 5 is dividing by 5; the inverse of dividing a number by 4 is multiplying by 4. Multiplication is the operation on two known factors to produce a unique number called the product. Division is the operation of finding an unknown factor when the product and one factor are known.

Exercise 3.4

1. What is the inverse of each of the following activities?
   a. Walking five blocks south
   b. Opening a window
   c. Putting on a sweater
   d. Dropping a pencil on the floor
   e. Raising the left hand
   f. Writing "4" on the chalkboard
   g. Corking a bottle of wine
2. What is the inverse operation of addition?
3. What is the inverse operation of multiplication?
4. Replace the symbol * with numerals to make true statements.

   a. $(4 + 3) - * = 4$
   b. $(8 \times 6) \div * = 8$
   c. $(7 - 4) + * = 7$
   d. $(* + 8) - 8 = 12$
   e. $(8 \div 2) \times * = 8$
   f. $(* \times 4) \div 4 = 9$
   g. $(16 \times *) \div 5 = 16$
   h. $(23 + *) - 7 = 23$

Prove the following theorems. In each theorem a, b, c, d, and e are whole numbers.

5. $a(bc) = (cb)a$
6. $ac + cb = c(b + a)$
7. $(a + b)(c + d) = ac + bc + ad + bd$
8. If $a = b$ and $c = d$, then $a + c = b + d$
9. If $a = b$ and $c = d$, then $ac = bd$
10. $a(b + c + d + e) = ab + ac + ad + ae$

3.13. Subtraction

We define subtraction of whole numbers in terms of the complement of a set.

DEFINITION 3.5. *Let a and b be two whole numbers, then their* **difference,** *denoted by a − b, is defined as follows. Let U and B be two sets, $B \subseteq U$, such that $n(U) = a$ and $n(B) = b$, then $a - b = n(B')$. The operation of finding the difference is called* **subtraction.**

Let us consider the sets U and B, $B \subseteq U$:

$$U = \{a, b, c, d, e\}$$
$$B = \{b, e\}$$

The complement, B', of B is

$$B' = \{a, c, d\}.$$

We know that $n(U) = 5$, $n(B) = 2$ and $n(B') = 3$. By Definition 3.5 $5 - 2 = 3$. We read this "five minus two equals three."

Generally in mathematics we define the difference of two whole numbers in terms of their sum.

DEFINITION 3.6. *If a and b are whole numbers, $a \geq b$,* then their **difference** $a - b$ is defined to be that whole number c such that $b + c = a$.*

We see that

$$6 - 4 = 2 \text{ because } 4 + 2 = 6$$
$$8 - 3 = 5 \text{ because } 3 + 5 = 8$$
$$7 - 1 = 6 \text{ because } 1 + 6 = 7.$$

We note that in Definition 3.6 we stated that a must be greater than or equal to b. If a is less than b, the difference $a - b$ is not a whole number. We can verify this. For example, $6 - 9$ does not name a whole number because there is no whole number c such that $9 + c = 6$. Similarly, $1 - 6$ does not name a whole number, because there is no whole number c such that $6 + c = 1$. Note that we said that $6 - 9$ and $1 - 6$ did not name whole numbers. We did not say that these symbols did not name numbers. These symbols are names for numbers as we shall see in Chapter 11, but these numbers are not whole numbers.

For emphasis, we again note that if a and b are whole numbers $a - b$ names a whole number if and only if $a \geq b$. If $a < b$, $a - b$ does not name a whole number.

Subtraction is the inverse operation of addition. The subtraction $a - b$ asks the question: What number added to b gives a? Subtraction is described as finding an unknown addend, called the difference, in a sum when one of the addends and the sum are known. Thus

$$6 - 2 = 4 \text{ because } 2 + 4 = 6$$
$$5 - 3 = 2 \text{ because } 3 + 2 = 5.$$

---

* The symbol $\geq$ means "is greater than or equal to"; it is a combination of the symbols $>$ and $=$. Similarly, the symbol $\leq$ means "is less than or equal to."

A statement such as $6 + 8 = 14$ is called an **addition sentence.** A statement such as $6 - 2 = 4$ is called a **subtraction sentence.** With every addition sentence there are associated two subtraction sentences. Thus with the addition sentence

$$6 + 2 = 8$$

are associated the subtraction sentences

$$8 - 2 = 6 \quad \text{and} \quad 8 - 6 = 2.$$

With every subtraction sentence there is associated an addition sentence. Thus with the subtraction sentence

$$9 - 2 = 7$$

is associated the addition sentence

$$2 + 7 = 9.$$

3.14. Division

The operation of division of whole numbers is defined in terms of multiplication.

DEFINITION 3.7. *If a and b are whole numbers, $b \neq 0$, their* **quotient,** *denoted by $a \div b$, is the whole number c such that $bc = a$. The operation of finding the quotient is called* **division.**

The operation of division applied to a pair of whole numbers, for example 12 and 2, in that order, means that we must determine an unknown factor, called the **quotient,** in this case 6, such that the product of 2 and this factor is 12. If we go back to arrays, we see that in the case of the division $12 \div 2$, we are asking the question: If a set of twelve elements is arranged in two rows (or columns) with the same number of elements in each row (or column), how many elements will there be in each row (or column)? These arrays are shown in Figure 3.8.

We note that $a \div b$ does not always name a whole number. For example, $3 \div 5$ does not name a whole number since there is no whole number c such that $5c = 3$. Similarly, $1 \div 2$ does not name a whole number since there is no whole number c such that $2c = 1$. Although $3 \div 5$ and $1 \div 2$ do not name whole numbers, these symbols are names for numbers as we shall see in Chapter 10.

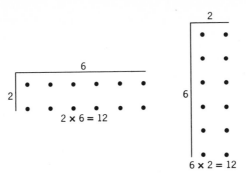

Figure 3.8

Division is the inverse operation of multiplication. The division $15 \div 3$ asks the question: What number multiplied by 3 gives the product 15? Division may be described as finding an unknown factor, called the quotient, in a product when the product and one factor are known. Thus

$$6 \div 3 = 2 \text{ because } 3 \times 2 = 6$$
$$15 \div 5 = 3 \text{ because } 5 \times 3 = 15.$$

In division we are given a product and a factor and asked to find the unknown factor, the quotient.

Notice that in Definition 3.7 we specified that b cannot be zero. Zero presents a problem in the operation of division. Observe that

$$0 \div 2 = n \text{ means } 2n = 0$$
$$0 \div 3 = n \text{ means } 3n = 0$$
$$0 \div 9 = n \text{ means } 9n = 0.$$

These statements are true if and only if $n = 0$. In general for all whole numbers $a \neq 0$, $0 \div a = 0$. We see that *zero divided by a natural number is zero*.

Now let us look at the following examples:

$$2 \div 0 = n \text{ means } 0 \times n = 2$$
$$3 \div 0 = n \text{ means } 0 \times n = 3$$
$$5 \div 0 = n \text{ means } 0 \times n = 5.$$

Since zero times any number is zero, there are no whole number replacements for n that make the above statements true. We conclude that division by zero is impossible. When we define $a \div b = c$ as $bc = a$, we must state that $b \neq 0$.

There is another special case that we must consider; that is $0 \div 0$. In this case we are looking for an unknown factor, n, such that $0 \times n = 0$. Since the product of any whole number and zero is zero, the unknown factor n may be any whole number and the statement $0 \times n = 0$ is true. Thus $0 \div 0$ does not name a unique whole number.

We conclude that *division by zero is impossible.* That is, $a \div 0$ is a meaningless symbol and does not name a whole number.

Statements such as $2 \times 3 = 6$ are called **multiplication sentences.** Sentences such as $6 \div 3 = 2$ are called **division sentences.** Every multiplication sentence is associated with two division sentences. Thus with

$$4 \times 2 = 8$$

is associated

$$8 \div 2 = 4 \quad \text{and} \quad 8 \div 4 = 2$$

Every division sentence is associatied with one multiplication sentence. With

$$8 \div 4 = 2$$

is associated the multiplication sentence

$$4 \times 2 = 8.$$

3.15.   Properties of Subtraction and Division

We now ask ourselves if the operations of subtraction and division possess any of the properties of addition and multiplication. Since $a - b$ names a whole number c if and only if $a \geq b$, we conclude that the set of whole numbers is not closed under subtraction.

That subtraction is neither commutative nor associative is exhibited in the examples below:

$$6 - 4 \neq 4 - 6$$
$$8 - 7 \neq 7 - 8$$
$$9 - 3 \neq 3 - 9$$
$$(9 - 7) - 3 \neq 9 - (7 - 3)$$
$$(12 - 8) - 2 \neq 12 - (8 - 2)$$

If $a - b = a$, it is clear that b must be equal to 0. But although $a - 0 = a$, $0 - a$ is not the name of a whole number, and we conclude that zero is not an identity element for the operation of subtraction.

Is the set of whole numbers closed under division? Since $7 \div 3$ is not the name of a whole number, we can conclude that the operation of division does not possess the closure property.

Division is neither commutative nor associative, as is exhibited in the examples below:

$$8 \div 2 \neq 2 \div 8$$
$$12 \div 3 \neq 3 \div 12$$
$$(24 \div 6) \div 2 \neq 24 \div (6 \div 2)$$
$$(18 \div 6) \div 3 \neq 18 \div (6 \div 3)$$

Since $6 \div 1 = 6$ but $1 \div 6$ is not a whole number, 1 is not an identity element for division.

Division and subtraction do have some important properties. Observe the following examples:

(a) $6 \times (4 - 2) = 6 \times 2 = 12$ $\left.\right\}$ hence $6 \times (4 - 2) =$
    $(6 \times 4) - (6 \times 2) = 24 - 12 = 12$ $(6 \times 4) - (6 \times 2)$

(b) $4 \times (8 - 3) = 4 \times 5 = 20$ $\left.\right\}$ hence $4 \times (8 - 3) =$
    $(4 \times 8) - (4 \times 3) = 32 - 12 = 20$ $(4 \times 8) - (4 \times 3)$

(c) $5 \times (9 - 6) = 5 \times 3 = 15$ $\left.\right\}$ hence $5 \times (9 - 3) =$
    $(5 \times 9) - (5 \times 6) = 45 - 30 = 15$ $(5 \times 9) - (5 \times 6)$

The examples above illustrate a relationship of multiplication to subtraction. This relationship is called the **distributive property of multiplication over subtraction.**

W-10. Distributive Property of Multiplication over Subtraction. If a, b, and c are whole numbers, $b \geq c$, then $\mathbf{a(b - c) = ab - ac}$.

Now observe the following:

(a) $(8 + 4) \div 2 = 12 \div 2 = 6$ $\left.\right\}$ hence $(8 + 4) \div 2 =$
    $(8 \div 2) + (4 \div 2) = 4 + 2 = 6$ $(8 \div 2) + (4 \div 2)$

(b) $(24 + 16) \div 4 = 40 \div 4 = 10$ $\left.\right\}$ hence $(24 + 16) \div 4 =$
    $(24 \div 4) + (16 \div 4) = 6 + 4 = 10$ $(24 \div 4) + (16 \div 4)$

(c) $(30 + 15) \div 5 = 45 \div 5 = 9$ } hence $(30 + 15) \div 5 =$
$(30 \div 5) + (15 \div 5) = 6 + 3 = 9$ } $(30 \div 5) + (15 \div 5)$

These examples are illustrations of **the distributive property of division over addition.**

W-11. Distributive Property of Division over Addition. If a, b, and c are whole numbers, $c \neq 0$, then $(\mathbf{a} + \mathbf{b}) \div \mathbf{c} = (\mathbf{a} \div \mathbf{c}) + (\mathbf{b} \div \mathbf{c})$ provided $(a + b) \div c$, $a \div c$ and $b \div c$ are whole numbers.

The examples below are illustrations of a property called the **distributive property of division over subtraction.**

(a) $(12 - 4) \div 2 = 8 \div 2 = 4$ } hence $(12 - 4) \div 2 =$
$(12 \div 2) - (4 \div 2) = 6 - 2 = 4$ } $(12 \div 2) - (4 \div 2)$

(b) $(27 - 9) \div 3 = 18 \div 3 = 6$ } hence $(27 - 9) \div 3 =$
$(27 \div 3) - (9 \div 3) = 9 - 3 = 6$ } $(27 \div 3) - (9 \div 3)$

(c) $(35 - 10) \div 5 = 25 \div 5 = 5$ } hence $(35 - 10) \div 5 =$
$(35 \div 5) - (10 \div 5) = 7 - 2 = 5$ } $(35 \div 5) - (10 \div 5)$

W-12. Distributive Property of Divison over Subtraction. If a, b, and c are whole numbers, $a \geq b$ and $c \neq 0$, then $(\mathbf{a} - \mathbf{b}) \div \mathbf{c} = (\mathbf{a} \div \mathbf{c}) - (\mathbf{b} \div \mathbf{c})$ provided $(a - b) \div c$, $a \div c$, and $b \div c$ are whole numbers.

Exercise 3.5

1. Which of the following symbols are names for whole numbers?
   a. $3 + 18$ g. $67 - 127$
   b. $18 \div 9$ h. 14×27
   c. $7 - 9$ i. $69 \div 38$
   d. $46 + 9$ j. $136 \div 7$
   e. 3×17 k. $84 - 17$
   f. $112 \div 2$ l. $267 - 488$

2. Verify each statement below by a distributive property.
   a. $(2 \times 5) + (2 \times 8) = 2 \times (5 + 8)$
   b. $(8 - 6) \div 2 = (8 \div 2) - (6 \div 2)$
   c. $(12 + 8) \times 3 = (12 \times 3) + (8 \times 3)$

d. $(24 + 16) \div 4 = (24 \div 4) + (16 \div 4)$

e. $(4 \times 40) + (4 \times 60) = 4 \times 100$

f. $[(2 + 3) \times 6] + [(2 + 3) \times 4] = (2 + 3) \times (6 + 4)$

3. Which of the following are meaningless symbols in the whole number system?

a. $6 \div 0$ e. $15 \div 0$

b. $7 \div 0$ f. $0 \div 26$

c. $0 \div 8$ g. $0 \div 9$

d. $0 \div 12$ h. $26 \div 0$

4. Write two subtraction sentences associated with each addition sentence.

a. $8 + 16 = 24$ e. $93 + 77 = 170$

b. $14 + 12 = 26$ f. $46 + 98 = 144$

c. $46 + 87 = 133$ g. $190 + 320 = 510$

d. $82 + 56 = 138$ h. $179 + 266 = 445$

5. Write an addition sentence for each subtraction sentence.

a. $9 - 3 = 6$ f. $87 - 29 = 58$

b. $12 - 7 = 5$ g. $96 - 38 = 58$

c. $14 - 6 = 8$ h. $117 - 29 = 88$

d. $19 - 9 = 10$ i. $437 - 116 = 321$

e. $26 - 12 = 14$ j. $189 - 89 = 100$

6. Write two division sentences for each multiplication sentence.

a. $3 \times 4 = 12$ e. $36 \times 5 = 180$

b. $9 \times 6 = 54$ f. $16 \times 4 = 64$

c. $7 \times 9 = 63$ g. $24 \times 7 = 168$

d. $12 \times 7 = 84$ h. $25 \times 15 = 375$

7. Write a multiplication sentence for each division sentence.

a. $54 \div 9 = 6$ e. $3348 \div 54 = 62$

b. $27 \div 3 = 9$ f. $420 \div 28 = 15$

c. $36 \div 9 = 4$ g. $1134 \div 14 = 81$

d. $72 \div 8 = 9$ h. $828 \div 69 = 12$

8. By inspection tell what whole number replacement for n makes a true statement.

a. $12 - 8 = n$ e. $n - 18 = 26$

b. $15 - n = 7$ f. $n - 72 = 18$

c. $36 - n = 24$ g. $75 - n = 50$

d. $n - 42 = 64$ h. $49 - n = 36$

9. By inspection tell what whole number replacement for n makes a true statement.

a. $6 \div 3 = n$ e. $56 \div n = 8$

b. $n \div 2 = 8$ f. $60 \div n = 10$

c. $n \div 5 = 30$ g. $n \div 8 = 14$

d. $27 \div n = 9$ h. $n \div 16 = 2$

10. By inspection tell all whole number replacements for n that make true statements.

a. $12 - n < 8$ d. $n - 5 < 7$

b. $n - 6 < 4$ e. $16 - n < 9$

c. $9 - n < 3$ f. $n - 5 < 7$

11. Use the distributive property of division over addition to find the following quotients. Write each product as the sum of two addends the larger of which is a multiple of 10. For example

$$36 \div 2 = (30 + 6) \div 2$$
$$= (30 \div 2) + (6 \div 2)$$
$$= 15 + 3$$
$$= 18$$

a. $12 \div 3$ e. $96 \div 8$

b. $24 \div 4$ f. $64 \div 4$

c. $75 \div 5$ g. $88 \div 4$

d. $78 \div 3$ h. $96 \div 4$

4.1. Ancient Systems of Numeration

A **numeration system** is a means of naming numbers. It involves a set of symbols representing numbers and some rules or **principles** for combining these symbols to form names for numbers. The earliest numeration system used by prehistoric man probably involved but a single symbol, the tally, I, marked on a stick or the wall of a cave.

Rather than dwelling on the history of man's development of a system of numeration, we shall study briefly two ancient systems of numeration, the Egyptian and the Roman, because they offer an opportunity to learn by contrast. The study of these systems, appraising their weaknesses as well as their strong points, and noticing their basic principles, furthers our ability to understand our own system of numeration.

In presenting these ancient systems of numeration, we shall consider their respective principles, for these contain the essence of the system. From these principles a system derives its structure and organi-

Systems of

Numeration

IV

zation. The symbols used to name the numbers are only of passing interest. The symbols and the choice of compounding point, called the **base,** may be referred to as the accidentals of the system. **Accidentals** are those things about the system that are not fundamentally necessary to the system for it to function. If the accidentals are changed, the basic principles of the system remain unaffected. On the other hand, to change the principles is to change the system itself. We shall see this more clearly later when we introduce systems of numeration in bases other than ten.

The ancient Egyptians attained a high degree of civilization many thousands of years ago. Among their various accomplishments was a form of picture writing known as hieroglyphics. Included in this, to satisfy their number requirements, was a set of numerical symbols. The Egyptians evolved a system of repetition of symbols which permitted coverage of a wide range of values with limited symbolism. As we study the Egyptian system, we note the systematic way in which new symbols were introduced.

In its operation the Egyptian system followed a simple pattern. The symbol for one was a stroke, I; the symbol for two was the symbol for one written twice, II; the symbol for three was the symbol for one written three times, III; and so on to nine, IIIIIIIII. Up to this point the system was hardly more than a tally method of one-to-one correspondence. At ten a compounding took place and the heel bone symbol, ∩, replaced what would have been ten strokes. As the Egyptian continued to count, his notation grew by the principle of **addition.** Thus the symbol for twelve was ∩II*; the symbol for eighteen was ∩IIII IIII; the symbol for twenty was ∩∩; and so on.

At one hundred, or ten tens, a new symbol, the scroll, **9**, was introduced for what otherwise would have been ten heel bones. Essentially, the value of any Egyptian numeral was precisely the sum of its parts. The Egyptian system of notation was governed by the principles of **addition** and **repetition.**

Repeating the basic symbols usually took care of cases of multiplicity for the Egyptians, except in those instances where a new symbol was substituted for ten symbols of a lower order. Table 4.1 shows the symbols for numbers in both the Hindu-Arabic and the Egyptian systems of numeration.

Using the principles of addition and repetition, the number named by any particular set of symbols was found by adding the value of each sym-

* The Egyptians commonly wrote from right to left, but they also wrote from left to right. All the Egyptian numerals in this chapter are written from left to right.

Table 4.1

Hindu-Anabic Symbol	Egyptian Symbol	Number Name
1	|	One
10	∩	Ten
100	9	One hundred
1,000	⚡	One thousand
10,000	⌢	Ten thousand
100,000	⌣	One hundred thousand
1,000,000	𓀀	One million

bol represented. Thus

$$∩∩||| = 10 + 10 + 1 + 1 + 1 = 23$$
$$99∩|| = 100 + 100 + 10 + 1 + 1 = 212$$
$$⌣ \; \lceil|||| = 100,000 + 10,000 + 1 + 1 + 1 + 1 = 110,004$$

If the symbol for any value was written more than four times, the Egyptians saved lateral space by writing these symbols in two or more rows; thus:

$$\begin{matrix}∩∩∩ \\ ∩∩∩\end{matrix} = 60, \qquad \begin{matrix}\lceil\lceil\lceil \quad 9999 \\ \lceil\lceil\lceil \quad 9999\end{matrix} \quad ∩∩∩ \; \begin{matrix}|||| \\ ||||\end{matrix} = 60,939$$

The position of the symbols in a numeral could be written from right to left or from left to right without changing the value of the number named, although it was customary for the symbols of greater value to precede the symbols for a lesser value. We see then, that the numeration system of the Egyptians was not essentially a positional system as is our own decimal system. The Egyptian numeration system was basically decimal (base ten) in nature, but without the concept of place-value.

The second ancient system of notation that we shall study is the Roman system. Like the Egyptians, the Roman system of notation employed the principle of **addition.** That is, the value of any numeral is equal to the sum of its parts.

There are seven basic symbols in the Roman system of numeration as we know it today. They are:

I	V	X	L	C	D	M
1	5	10	50	100	500	1000

The Roman system may be characterized as a decimal system. Special symbols represent powers of ten:

$$I = 1$$
$$X = 10$$
$$C = 10 \times 10 = 100$$
$$M = 10 \times 10 \times 10 = 1000$$

An economy of symbolism was in effect in the Roman system with the introduction of mid-values for each of these powers of ten. Thus:

$$V = 5 \text{ (mid-ten)}$$
$$L = 50 \text{ (mid-hundred)}$$
$$D = 500 \text{ (mid-thousand)}.$$

The Roman system uses the principles of **addition** and **repetition** for writing all numbers except those involving fours and nines. Thus

$$XXII = 10 + 10 + 1 + 1 = 22$$
$$MDCCVIII = 1000 + 500 + 100 + 100 + 5 + 1 + 1 + 1 = 1,708$$

To handle large numbers, a third principle, **multiplication,** was put to use. A bar drawn over portions of the numeral indicated that the number represented by those symbols covered was to be multiplied by one thousand, then added to the remaining symbols of the expression to effect its total value. Thus:

$$\overline{XX}XII = (1000 \times 20) + 12 = 20,012$$
$$\overline{XXXV} = 1000 \times 35 = 35,000$$
$$\overline{MMMD} = 1000 \times 3500 = 3,500,000$$

Still a fourth principle, that of **subtraction,** came into being at a later date. It is claimed by some historians that this was an invention of the early clockmakers, who were hard pressed for space on the faces of their clocks! Whatever the case, by substituting IV for IIII and IX for VIIII, a more economical notation was certainly achieved. Had this contraction been restricted only to specific values of four and nine we would not have a principle but merely a special case. In today's version of the Roman system of numeration we see in effect the principle of subtraction. It operates specifically with respect to the fours and nine of each order. Thus:

$IV = 5 - 1 = 4$	$XL = 50 - 10 = 40$	$CD = 500 - 100 = 400$
$IX = 10 - 1 = 9$	$XC = 100 - 10 = 90$	$CM = 1000 - 100 = 900$

This economy of symbolism produces a new problem. In the Egyptian system, the value of the number expressed was not materially affected by the order in which the symbols in the numeral were written. For example, the Egyptians could write twenty-three either as ∩∩||| or as |||∩∩. In the Roman system coupling I and V could mean either four or six, depending upon whether the addition or subtraction principle was to operate. Thus we see in the Roman system a rule of order. It is this: *Symbols are to be written from left to right in order of decreasing value, and the principle of addition applies. The only exceptions to this order are the pairs I before V or X, X before L or C, and C before D or M, in which case the subtraction of the number represented by the left numeral from the number represented by the right numeral is indicated.*

EXAMPLE i. Write a decimal numeral for the Egyptian numeral

$$\text{𓂀} \; \text{𓆼} \text{𓆼} \; ∩∩∩ \; \overset{|||}{||||}$$

Solution: $\text{𓂀} \; \text{𓆼} \text{𓆼} \; ∩∩∩ \; \overset{|||}{||||} = 1{,}000{,}000 + 200{,}000 + 30 + 7$
$$= 1{,}200{,}037$$

EXAMPLE ii. Write a decimal numeral for the Roman numeral MCMXLVI.

Solution: Since C is on the left of M, the CM combination represents $1000 - 100 = 900$. Since X is on the left of L the XL combination represents $50 - 10 = 40$. Then MCMXLVI $= 1000 + 900 + 40 + 6 = 1946$

EXAMPLE iii. Write a Roman numeral for the decimal numeral 3,969.

Solution: $3969 = 3000 + 900 + 60 + 9$
$$= \text{MMMCMLXIX}$$

Exercise 4.1

1. Write a decimal numeral for each of the following Egyptian numerals.
   a. ∩∩|||||
   b. 99∩ ‖‖
   c. ⌐ 𓆼 99|||
   d. 𓂀 𓂀 𓆼 ⌐⌐⌐ ∩∩∩ ‖‖‖
2. Write an Egyptian numeral for each of the following decimal numerals.
   a. 36
   b. 372
   c. 5,583
   d. 14,341
   e. 128,339
   f. 1,000,750

3. Write a decimal numeral for each of the following Roman numerals.
   a. MDX
   b. MCMLVI
   c. MCMLXIX
   d. MCMXIX
   e. MDCCLXXVI
   f. MCDLXXXVIII
   g. MCMIX
   h. MMDCCLXIV

4. Write a decimal numeral for each of the following Roman numerals.
   a. $\overline{\text{VIII}}$
   b. $\overline{\text{XXVII}}$
   c. $\overline{\overline{\text{XXVII}}}$
   d. $\overline{\text{CLV}}$
   e. $\overline{\text{MDCC}}$
   f. $\text{CCL}\overline{\text{XIX}}$

5. Write a Roman numeral for each of the following decimal numerals.
   a. 1776
   b. 1989
   c. 1952
   d. 1919
   e. 1999
   f. 2001
   g. 2509
   h. 1066

6. Write a Roman numeral for each of the following decimal numerals.
   a. 42,000
   b. 26,000
   c. 19,000
   d. 29,000
   e. 46,000
   f. 592,000
   g. 1,000,000
   h. 3,000,000

7. Why do you think the Egyptian and Roman numeration systems have no symbol for zero?

8. How many basic symbols are needed to write numerals in the Roman system of numeration as we know it today? What are they?

9. What are the four principles of the Roman system of numeration?

10. Write the Roman numeral for the number that is one less than each of the following.
    a. C
    b. LXV
    c. XCIV
    d. MCMLIX

11. Write a Roman numeral for the year in each date below.
    a. May 7, 1429 (Joan of Arc's Victory at Orleans)
    b. August 4, 1492 (First Voyage of Columbus)
    c. April 19, 1775 (Battle of Lexington)
    d. April 30, 1803 (Louisiana Purchase)
    e. Dec. 17, 1903 (First Airplane Flight)
    f. Aug. 1, 1914 (Beginning of World War I)

4.2. The Hindu-Arabic System of Numeration

The Hindu-Arabic system of numeration is a place-value system with symbols for the numbers zero, one, two, three, four, five, six, seven, eight, and nine. The next number, ten, is the compounding point in the system and is called the **base** of the system. Because the base of the system is ten, it is called the **decimal** system from the Latin word "decem" meaning ten.

The symbols for the numbers zero, one, two, three, four, five, six, seven, eight, and nine, that is, the symbols in the set

$$\{0, 1, 2, 3, 4, 5, 6, 7, 8, 9\}$$

are called **digits.** To write numerals for all numbers greater than nine, we use a combination of these digits. These combinations are formed according to a pattern determined by our system of **place-value.**

Let us observe a number named in the decimal system and analyze what is meant by a place-value system. The number four hundred eighty-six is written

② ① ⓪
4 8 6 = (four × one hundred) + (eight × ten) + (six × one).

The numerals written in circles above the digits in the numeral indicate that there are three positions involved in writing a numeral with three digits. Similarly, there are four positions involved in writing a numeral with four digits, and so on. In the numeral 486, the digit 6 occupies the 0 position; the digit 8, the 1 position; and the digit 4, the 2 position. To each position we assign a number which is the place-value of that position. The first seven place-values (reading from right to left in a numeral) of the decimal system are shown in Table 4.2.

Table 4.2

Position	6	5	4	3	2	1	0
Place-value	one million 1,000,000 10 × 100,000	hundred thousand 100,000 10 × 10,000	ten thousand 10,000 10 × 1000	one thousand 1000 10 × 100	one hundred 100 10 × 10	ten 10 10 × 1	one 1 1

Place-value is a number assigned to a position and in no way depends on the digit in that position. In any decimal numeral the number represented by a digit, called the **value** of the digit (such as "4" in 486), is a product. This product is the number represented by the digit in the position and the place-value assigned to the position. For the numeral 486:

the value of 4 is 4 × 100
the value of 8 is 8 × 10
the value of 6 is 6 × 1.

The number named by the numeral is the sum of the products. Thus

$$486 = (4 \times 100) + (8 \times 10) + (6 \times 1) = 400 + 80 + 6.$$

The basic principles of the Hindu-Arabic system of numeration are the **place-value** principle and the **additive** principle. The principle of place-value involves two basic ideas: (1) there is a number, called the place-value, assigned to each position in a numeral; and (2) each digit in a numeral represents the product of the number it names and the place-value of its position.

EXAMPLE i. What is the place-value of the digit "4" in 14,639,178?
Solution: The place value of "4" is 1,000,000.

EXAMPLE ii. Write 3,694 as the sum of products.
Solution: 3,694 = (3 × 1000) + (6 × 100) + (9 × 10) + (4 × 1).

EXAMPLE iii. Write a decimal numeral for the sum:
(4 × 10,000) + (6 × 1000) + (4 × 100) + (0 × 10) + (9 × 1).
Solution: 46,409

EXAMPLE iv. What is the value of the digit "5" in 25,869?
Solution: The place-value of the "5" is 1000. The value of the "5" is 5 × 1000 = 5000.

4.3. Exponents

There is a way of expressing a product of equal factors, for example the product 10 × 10, in a simpler form. We write 10 × 10 as

$$10^2$$

Here the superscript, 2, is called an **exponent.** It indicates that 10 is to be used as a factor two times. The number 10 in 10^2 is called the **base** and the number 10^2 is called the **power.** We read the symbol 10^2 as "ten to the second power" or "ten squared." Similarly

$$10^3 = 10 \times 10 \times 10 = 1000$$
$$10^4 = 10 \times 10 \times 10 \times 10 = 10,000$$
$$10^5 = 10 \times 10 \times 10 \times 10 \times 10 = 100,000$$
$$10^6 = 10 \times 10 \times 10 \times 10 \times 10 \times 10 = 1,000,000.$$

We read 10^3 as "ten to the third power" or "ten cubed"; we read 10^4 as "ten to the fourth power"; we read 10^5 as "ten to the fifth power"; we read 10^6 as "ten to the sixth power."

We call 10^2, 10^3, 10^4, . . . , **powers of ten.**

The base can be any whole number, thus:

$$2^3 = 2 \times 2 \times 2 = 8$$
$$5^4 = 5 \times 5 \times 5 \times 5 = 625$$
$$3^2 = 3 \times 3 = 9.$$

If the exponent is one, the power in defined as the base:

$$3^1 = 3$$
$$5^1 = 5$$
$$10^1 = 10.$$

If the exponent is zero, the power is defined to be one:

$$2^0 = 1$$
$$5^0 = 1$$
$$10^0 = 1.$$

DEFINITION 4.1. *If a is any whole number, and n is any whole number then we define the* **nth power of a,** *denoted by a^n, as follows:*

$$a^0 = 1$$
$$a^1 = a$$
$$a^n = \underbrace{a\,a\,a\ldots a}_{n \text{ factors}} \quad n > 1.$$

In a^n, a is called the **base,** n is called the **exponent** and a^n is called the **nth power of a.**

The powers of ten present a very interesting pattern. Observe the following:

$$10^0 = 1$$
$$10^1 = 10$$
$$10^2 = 10 \times 10 = 100$$
$$10^3 = 10 \times 10 \times 10 = 1000$$
$$10^4 = 10 \times 10 \times 10 \times 10 = 10{,}000$$
$$10^5 = 10 \times 10 \times 10 \times 10 \times 10 = 100{,}000, \text{ etc.}$$

Notice that 10^5 is another name for 100,000. The numeral 100,000 consists of "1" followed by five "0s." Similarly

$$10^6 = 10 \times 10 \times 10 \times 10 \times 10 \times 10 = 1{,}000{,}000.$$

The numeral naming 10^6 consists of "1" followed by six "0s."

This is true in general: The numeral naming 10^n consists of "1" followed by n "0s."

EXAMPLE i. Write a standard numeral for 8^2.
Solution: $8^2 = 8 \times 8 = 64$

EXAMPLE ii. What whole number value of n makes $2^n = 64$ a true statement?
Solution: Since $2 \times 2 \times 2 \times 2 \times 2 \times 2 = 64 = 2^6$, the value of n is 6.

EXAMPLE iii. What whole number value of n makes $n^3 = 125$?
Solution: Since $125 = 5 \times 5 \times 5 = 5^3$, the value of n is 5.

4.4. Expanded Notation

The decimal system of numeration has the number ten as its base. Starting at the position 0, every place to the left of position 0 has a place-value ten times as great as the position to its right. Thus:

Position	Number Name	Place-Value
0	one	$1 = 1 = 10^0$
1	ten	$10 = 10 \times 1 = 10^1$
2	one hundred	$100 = 10 \times 10 = 10^2$
3	one thousand	$1{,}000 = 10 \times 10 \times 10 = 10^3$
4	ten thousand	$10{,}000 = 10 \times 10 \times 10 \times 10 = 10^4$
⋮	⋮	⋮

Every decimal numeral may be written in a form involving exponents. For example

$$673 = (6 \times 100) + (7 \times 10) + (3 \times 1)$$
$$= (6 \times 10^2) + (7 \times 10^1) + (3 \times 10^0).$$
$$4387 = (4 \times 1000) + (3 \times 100) + (8 \times 10) + (7 \times 1)$$
$$= (4 \times 10^3) + (3 \times 10^2) + (8 \times 10^1) + (7 \times 10^0).$$

This form is called **expanded notation** of the numeral.

Notice that when a numeral has, for example, four digits, the place-value of the first digit (on the left) is 10^3, that is, the exponent of the base, 10, is one less than the number of digits in the numeral. This is true in general. If a numeral has n digits, the place-value of the first digit (on the left) is 10^{n-1}.

EXAMPLE i.   Write 39,642 in expanded notation.

Solution: There are five digits in the numeral; so the place-value of the "3" is 10^4. Then

$$39,642 = (3 \times 10^4) + (9 \times 10^3) + (6 \times 10^2) + (4 \times 10^1) + (2 \times 10^0).$$

EXAMPLE ii.   Write 8,000,000 in expanded notation.

Solution: Since the numeral has seven digits, the place-value of the "8" is 10^6. Then

$$8,000,000 = (8 \times 10^6) + (0 \times 10^5) + (0 \times 10^4) + (0 \times 10^3)$$
$$+ (0 \times 10^2) + (0 \times 10^1) + (0 \times 10^0).$$

Because zero times any number is zero (see Section 3.10) and the sum of zero and any whole number is that whole number (additive identity property), we may write

$$8,000,000 = 8 \times 10^6$$

EXAMPLE iii.   Write a standard numeral for $(3 \times 10^5) + (2 \times 10^3) + (1 \times 10^0)$

Solution: $(3 \times 10^5) + (2 \times 10^3) + (1 \times 10^0) = (3 \times 10^5) + (0 \times 10^4) + (2 \times 10^3) + (0 \times 10^2) + (0 \times 10^1) + (1 \times 10^0) = 302,001.$

Exercise 4.2

1. Write each of the following using exponents.
    a. $2 \times 2 \times 2 \times 2$
    b. $3 \times 3 \times 3 \times 3 \times 3$

c. $5 \times 5 \times 5$

d. $7 \times 7 \times 7 \times 7 \times 7 \times 7 \times 7$

e. $4 \times 4 \times 4 \times 4 \times 4 \times 4 \times 4 \times 4$

f. $9 \times 9 \times 9 \times 9 \times 9 \times 9 \times 9 \times 9 \times 9 \times 9$

2. Write each of the following as a power of ten.

    a. one hundred   d. ten thousand

    b. one million   e. one billion (1000 million)

    c. ten million   f. one

3. Write each of the following as a power of ten.

    a. 1000   c. 1,000,000

    b. 100,000   d. 100,000,000

4. Give a standard numeral for each of the following.

    a. 10^3 c. 10^4 e. 10^8

    b. 10^2 d. 10^7 f. 10^{10}

5. Give a standard numeral for each of the following.

    a. 8^0 c. 9^0 e. 10^0

    b. 12^0 d. 2^0 f. 14^0

6. Write the decimal numeral for each of the following.

    a. nine   d. four thousand sixty-four

    b. seventeen   e. ten thousand sixteen

    c. three hundred six   f. one million thirty-four thousand six

7. What is the value of the "4" in each of the following?

    a. 36,904   d. 50,341

    b. 234,162   e. 4,073,800

    c. 127,416   f. 10,487,000

8. What is the place-value of the "5" in each of the following?

    a. 50   e. 57,000,306

    b. 105   f. 5,809,766

    c. 5,822   g. 5,000,809,799

    d. 52,876   h. 3,054,019,111

9. Write each of the following in expanded notation using exponents.

    a. 486   e. 4,000,000

    b. 1,392   f. 4,369,214

    c. 84,000   g. 8,261,004,311

    d. 12,006   h. 9,101,016,777

10. Write in decimal notation

    a. $(3 \times 10^5) + (2 \times 10^4) + (1 \times 10^3) + (7 \times 10^2) + (6 \times 10^1) + (0 \times 10^0)$.

b. $(4 \times 10^4) + (2 \times 10^2) + (1 \times 10^0)$.

c. $(5 \times 10^5) + (3 \times 10^4)$.

d. $(7 \times 10^3) + (1 \times 10^1)$.

e. (8×10^5).

11. Write a standard numeral for each of the following.

    a. 2^3 d. 12^2

    b. 3^4 e. 2^5

    c. 5^3 f. 15^2

12. What whole number values of n make the following true statements?

    a. $2^n = 4$ d. $13^2 = n$

    b. $n^2 = 9$ e. $3^n = 27$

    c. $10^n = 1$ f. $5^n = 625$

4.5 Place-Value Systems with Bases other than Ten

The selection of ten as the base or compounding point of our system of numeration was more a physiological accident of nature than a rational choice. Actually, the choice has little to commend it. Twelve would have made a better selection because of its greater divisibility. The choice of ten as the base is undoubtedly due to the fact that man was born with ten fingers.

There are advantages to be gained from speculation on what might have been had man actually been born with, for example, eight, twelve, five or two fingers instead of ten. Such conjectures lead to new scales of notation and offer an excellent opportunity to learn more about our own decimal system. In such instances, things which are essential to our system, the principles, are not changed, although those things which are only accidental to the system differ from base to base. Moreover, consideration of these new systems with bases other than ten carries additional value for teachers in that it points out many difficulties which a young child encounters as he attempts to gain control over what is, for him, an equally strange system, the decimal system.

Let us construct a few systems of notation involving bases other than ten. These systems will be designed in accordance with the principles of addition and place-value characteristic of the Hindu-Arabic system of notation. Instead of ten distinct symbols, however, we shall sometimes need more, sometimes less, but the symbol for zero is always necessary. The compounding point will vary with the base selected. We could introduce a new set of symbols for each new base, but that would only complicate matters unnecessarily. After all, the digits are merely symbols which represent certain concepts and are applicable in any base. We shall continue to use the familiar Arabic numerals, 0, 1, 2, 3, 4, 5, 6, 7, 8, 9, augmenting this set or eliminating from it with the choice of base.

In building place-value systems of numeration with bases other than ten we shall encounter semantic difficulties. It is important here that we know the difference between a number word and its symbolic concept. Familiarity with the decimal system leads us to pair the number word and the symbol for a number without distinction. Actually, they are independent to each other. We tend to think "ten" as always being expressed by the symbol 10. We are so familiar with the decimal system that we are not conscious of the fact that the "1" stands for one ten and the "0" stands for an absence of ones. In a system of numeration with base twelve, for example, the symbol "10" means one dozen and no ones, that is, twelve.

In working with systems of numeration with bases other than ten, let us agree to read the symbol "16" as "one-six." This, of course, means one of the base and six more.

In the Hindu-Arabic system, it will be recalled, each digit contributes to the total value of the number in relation to the number it names and its place-value. Since we shall continue to use the Arabic symbols, the number named by a digit will be immediately understood, except in cases where the base exceeds ten. Here additional symbols must be defined, which we shall do at the time they are needed. In the matter of place-value, we recall that there exists a relationship between position and power of the base. In other bases this relationship between position and power of the base will remain constant, an essential "principle" of the system; on the other hand, the choice of the base, an accidental, will vary. That is, whereas in base ten the place-values from right to left are successive powers of ten— ones (10^0), tens (10^1), hundreds (10^2), thousands (10^3), and so forth—, in base five, for instance, they are successive powers of five—ones (5^0), fives (5^1), twenty-fives (5^2), one hundred twenty-fives (5^3), and so forth; similarly, in base eight, they are successive powers of eight—ones (8^0), eights (8^1), sixty-fours (8^2), five hundred twelves (8^3), and so forth.

Now let us use the principles of our decimal system and build a system of numeration with base eight, called the **octal system.** Since the base of this system is eight, we need eight digits representing zero, one, two, three, four, five, six, and seven. Let us use the familiar digits of our base ten system, that is the numerals 0, 1, 2, 3, 4, 5, 6, and 7.

In the decimal system (base ten) each succeeding position, reading from right to left, has a place-value ten times the preceding position. In the base eight system, each succeeding position reading from right to left will have a place-value eight times the preceding position.

The place-values in base eight have positions 0, 1, 2, and so forth, as in base ten, but the place-value assigned to each position is a power of eight. The first four place-values (reading from right to left) in base eight are given in Table 4.3.

Table 4.3

Position	3	2	1	0
Place-value	five hundred twelve eight × eight × eight $eight^3$	sixty-four eight × eight $eight^2$	eight eight $eight^1$	one one $eight^0$

In the octal system, the numeral 43 means $(4 \times eight^1) + (3 \times eight^0)$. That is, 43 in base eight names the same number as

$$(4 \times eight^1) + (3 \times eight^0) = (4 \times 8) + (3 \times 1)$$
$$= 32 + 3$$
$$= 35$$

in base ten.

Similarly,

$$524 \text{ (base eight)} = (5 \times eight^2) + (2 \times eight^1) + (4 \times eight^0)$$
$$= (5 \times 64) + (2 \times 8) + (4 \times 1)$$
$$= 340 \text{ (base ten)}.$$
$$1037 \text{ (base eight)} = (1 \times eight^3) + (0 \times eight^2) + (3 \times eight^1)$$
$$+ (7 \times eight^0)$$
$$= (1 \times 512) + (0 \times 64) + (3 \times 8) + (7 \times 1)$$
$$= 543 \text{ (base ten)}.$$

From now on we shall use a subscript to denote the base to which a numeral is written. For example $543_{(eight)}$ means that the numeral is written in the octal (base eight) system. If no subscript is used, the numeral is understood to be in the decimal system.

It is quite simple to convert a numeral in the octal system to a numeral naming the same number in the decimal system. We need only apply the principle of place-value in our system. The number represented by each digit in an octal numeral is a product as it is in a decimal numeral. The value of the digit 7 in a numeral is the product of seven and the place-value assigned to its position; the value of the digit 6 is the product of six and the place-value assigned to its position; and so on. Finally, the number is the sum of the particular products. For example:

$$234_{(eight)} = (2 \times eight^2) + (3 \times eight^1) + (4 \times eight^0)$$
$$= (2 \times 8^2) + (3 \times 8^1) + (4 \times 8^0) \quad \text{(converting to decimal notation)}$$
$$= (2 \times 64) + (3 \times 8) + (4 \times 1)$$
$$= 128 + 24 + 4$$
$$= 156$$

We see from the above that $234_{(eight)}$ and 156 name the same number.

EXAMPLE i. Convert $177_{(eight)}$ to a decimal numeral.
Solution: $177_{(eight)} = (1 \times eight^2) + (7 \times eight^1) + (7 \times eight^0)$
$$= (1 \times 8^2) + (7 \times 8^1) + (7 \times 8^0)$$
$$= 64 + 56 + 7$$
$$= 127$$

EXAMPLE ii. Convert $1023_{(eight)}$ to a base ten numeral.
Solution: $1023_{(eight)} = (1 \times eight^3) + (0 \times eight^2) + (2 \times eight^1) +$
$$(3 \times eight^0)$$
$$= (1 \times 8^3) + (0 \times 8^2) + (2 \times 8^1) + (3 \times 8^0)$$
$$= 512 + 0 + 16 + 3$$
$$= 531$$

We use a similar method to convert any place-value numeral to a decimal numeral.

EXAMPLE iii. Convert $123_{(five)}$ to a decimal numeral.
Solution: The numeral $123_{(five)}$ is written in the base five system.
The place-values in the base five system (from right to left) are $five^0$, $five^1$, $five^2$, $five^3$, and so on.

$$123_{(five)} = (1 \times five^2) + (2 \times five^1) + (3 \times five^0)$$
$$= (1 \times 5^2) + (2 \times 5^1) + (3 \times 5^0)$$
$$= 25 + 10 + 3 = 38$$

To change a decimal numeral to an equivalent octal numeral, we think of the number of objects in the set whose cardinal number is named by this numeral as grouped in sets of eight^0, eight^1, eight^2, and so forth. Suppose we wish to find the octal numeral equivalent to the decimal numeral 465.

We ask ourselves: What is the largest power of eight less than 465? Since $8^3 = 512$, we cannot take a set of 8^3 from 465; $8^2 = 64$ is less than 465 so to find the digit in position 2 we determine the greatest multiple of 64 less than 465.

$$
\begin{array}{r}
7 \\
64)\overline{465} \\
448 \\
\hline
17
\end{array}
$$

We see that we can subtract $7 \times 8^2 = 448$ from 465 hence we write a 7 in position 2.

2	1	0
7		

To find the digit in position 1 we determine the largest multiple of $8^1 = 8$ less than $465 - 448 = 17$.

$$
\begin{array}{r}
2 \\
8)\overline{17} \\
16 \\
\hline
1
\end{array}
$$

We see that we can subtract $2 \times 8^1 = 16$ from 17, hence we write a 2 in position 1.

2	1	0
7	2	

Since there is a remainder of 1 when we subtract $2 \times 8^1 = 16$ from 17, we put a 1 in position 0.

2	1	0
7	2	1

Hence

$$465 = 721_{\text{(eight)}}$$

The above discussion is shown below in shortened form.

$$465 = 448 + 16 + 1$$
$$= (7 \times 64) + (2 \times 8) + (1 \times 1)$$
$$= (7 \times \text{eight}^2) + (2 \times \text{eight}^1) + (1 \times \text{eight}^0)$$
$$= 721_{\text{(eight)}}$$

Similarly,

$$284 = 256 + 24 + 4$$
$$= (4 \times 64) + (3 \times 8) + (4 \times 1)$$
$$= (4 \times \text{eight}^2) + (3 \times \text{eight}^1) + (4 \times \text{eight}^0)$$
$$= 434_{\text{(eight)}}$$

and

$$1364 = 1024 + 320 + 16 + 4$$
$$= (2 \times 512) + (5 \times 64) + (2 \times 8) + (4 \times 1)$$
$$= (2 \times \text{eight}^3) + (5 \times \text{eight}^2) + (2 \times \text{eight}^1) + (4 \times \text{eight}^0)$$
$$= 2524_{\text{(eight)}}$$

Another method of converting decimal numerals to equivalent octal numerals involves repeated division. Suppose we wish to change 284 to an equivalent octal numeral. We divide 284 by 8:

$$
\begin{array}{r}
35 \\
8)\overline{284} \\
24 \\
\hline
44 \\
40 \\
\hline
4
\end{array}
$$

We see that 284 divided by 8 gives a quotient of 35 and a remainder of 4. This means

1) $284 = (8 \times 35) + 4.$

Now let us divide 35 (the quotient in the above division) by 8:

$$\begin{array}{r} 4 \\ 8\overline{)35} \\ 32 \\ \hline 3 \end{array}$$

Then

2) $\qquad\qquad 35 = (8 \times 4) + 3$

If we replace 35 in equality 1) by $(8 \times 4) + 3$ we have

3) $\quad 284 = 8 \times [(8 \times 4) + 3] + 4 \qquad$ substitution

$\qquad = (8 \times 8 \times 4) + (8 \times 3) + 4 \qquad$ distributive property

$\qquad = (8^2 \times 4) + (8 \times 3) + (4 \times 1) \qquad 8 \times 8 = 8^2$ and identity property of multiplication

$\qquad = (4 \times 8^2) + (3 \times 8^1) + (4 \times 8^0) \qquad$ commutative property of multiplication; $8 = 8^1$, $1 = 8^0$

$\qquad = 434_{\text{(eight)}} \qquad\qquad\qquad\qquad$ place-value notation

We usually shorten this process by the abbreviated form:

$$\begin{array}{ll} 8\overline{)284} & \\ 8\overline{)35} & \text{R} \uparrow 4 \times 8^0 \\ 8\overline{)4} & \text{R} \mid 3 \times 8^1 \\ 0 & \text{R} \; 4 \times 8^2 \end{array}$$

If we carry out the division until we have a zero quotient, the remainders, reading up, in the process above give the digits from left to right in the octal numeral.

EXAMPLE i. Change 1364 to an equivalent octal numeral using the repeated division method.

Solution:

$$\begin{array}{ll} 8\overline{)1364} & \\ 8\overline{)170} & \text{R} \uparrow 4 \times 8^0 \\ 8\overline{)21} & \text{R} \mid 2 \times 8^1 \\ 8\overline{)2} & \text{R} \mid 5 \times 8^2 \\ 0 & \text{R} \; 2 \times 8^3 \end{array}$$

$$1364 = 2524_{\text{(eight)}}$$

In horizontal form we have

$$
\begin{aligned}
1364 &= (8 \times 170) + 4 \\
&= 8 \times [(8 \times 21) + 2] + 4 \\
&= (8 \times 8 \times 21) + (8 \times 2) + 4 \\
&= (8^2 \times 21) + (8 \times 2) + 4 \\
&= \{8^2 \times [(8 \times 2) + 5]\} + (8 \times 2) + 4 \\
&= (8^2 \times 8 \times 2) + (8^2 \times 5) + (8 \times 2) + 4 \\
&= (2 \times 8^3) + (5 \times 8^2) + (2 \times 8^1) + (4 \times 8^0) \\
&= 2524_{\text{(eight)}}
\end{aligned}
$$

EXAMPLE ii.   Change 2123 to an equivalent octal numeral.

Solution:

$$
\begin{array}{lll}
8)\,2123 & & \\
\overline{8)\,265} & \text{R} & 3 \times 8^0 \\
\overline{8)\,33} & \text{R} & 1 \times 8^1 \\
\overline{8)\,4} & \text{R} & 1 \times 8^2 \\
\overline{0} & \text{R} & 4 \times 8^3 \\
\end{array}
$$

$$2123 = 4113_{\text{(eight)}}$$

4.6.   The Duodecimal System

We now construct a system of numeration with a base greater than ten. The **duodecimal system** with base twelve is just such a system. We recall that in a place value system of numeration we need symbols to represent each of the whole numbers less than the base, hence in the duodecimal system we need twelve symbols. If we use, as previously, the symbols 0, 1, 2, 3, 4, 5, 6, 7, 8, 9 for the numbers zero, one, two, three, four, five, six, seven, eight, and nine, we need two more symbols to represent ten and eleven. Let us use T for ten and E for eleven.* Our set of digits for the duodecimal system is then

$$\{0, 1, 2, 3, 4, 5, 6, 7, 8, 9, T, E\}$$

The place-values in the duodecimal system are the ones (twelve^0), the twelves (twelve^1), the one hundred forty-fours (twelve^2) and so on.

* The symbols χ and ϵ are often used for ten and eleven respectively. We use T and E since these are the symbols most often used for ten and eleven in the duodecimal system in elementary and junior high school mathematics texts.

The duodecimal numeral for the number of x's shown below is $15_{(\text{twelve})}$

xxxxxxxxxxxxxxxxx

(read: "one-five, base twelve" or "one dozen and five"). Table 4.4 shows some numerals in the decimal and duodecimal systems.

Table 4.4

Decimal Notation	Duodecimal Notation
0	0
1	1
2	2
3	3
4	4
5	5
6	6
7	7
8	8
9	9
10	T
11	E
12	10
13	11
14	12
. . .	. . .
24	20
25	21
26	22
. . .	. . .
34	2T
35	2E
36	30
. . .	. . .

We use the same method to convert duodecimal numerals to decimal numerals as we did to convert octal numerals to decimal numerals (Section 4.5).

EXAMPLE i. Convert $34_{(\text{twelve})}$ to a decimal numeral.

Solution: $34_{(\text{twelve})} = (3 \times \text{twelve}^1) + (4 \times \text{twelve}^0)$
$= (3 \times 12^1) + (4 \times 12^0)$
$= 36 + 4 = 40.$

EXAMPLE ii. Convert $12E_{(twelve)}$ to a decimal numeral.

Solution: $12E_{(twelve)} = (1 \times twelve^2) + (2 \times twelve^1) + (E \times twelve^0)$
$$= (1 \times 12^2) + (2 \times 12^1) + (11 \times 12^0)$$
$$= 144 + 24 + 11$$
$$= 179.$$

We use the methods previously explained for converting decimal numerals to octal numerals to change decimal numerals to duodecimal numerals.

EXAMPLE iii. Convert 164 to a duodecimal numeral.

Solution: $164 = 144 + 12 + 8$
$$= (1 \times 12^2) + (1 \times 12^1) + (8 \times 12^0)$$
$$= (1 \times twelve^2) + (1 \times twelve^1) + (8 \times twelve^0)$$
$$= 118_{(twelve)}.$$

Using the repeated division method we have:

$$12)\overline{164}$$
$$12)\overline{13} \quad R \uparrow 8 \times 12^0$$
$$12)\overline{1} \quad R \mid 1 \times 12^1 \qquad 164 = 118_{(twelve)}$$
$$0 \quad R \mid 1 \times 12^2$$

EXAMPLE iv. Convert 35 to a base twelve numeral.

Solution: $35 = 24 + 11$
$$= (2 \times 12^1) + (11 \times 12^0)$$
$$= (2 \times twelve^1) + (E \times twelve^0)$$
$$= 2E_{(twelve)}.$$

Using the repeated division method we have

$$12)\overline{35}$$
$$12)\overline{2} \quad R \uparrow E \times 12^0 \qquad 35 = 2E_{(twelve)}$$
$$0 \quad R \mid 2 \times 12^1$$

In using the repeated division method to convert decimal numerals to duodecimal numerals, we must always remember to record the remainders in the duodecimal system of notation since these are the digits in the duodecimal numeral.

EXAMPLE v. Convert 1582 to an equivalent duodecimal numeral.

Solution: 12) 1582
 $\overline{\quad 12)131\quad}$ R ↑ T × 12⁰ (the remainder ten is written T in
the duodecimal system)

$\qquad$ 12) 10 R │ E × 12¹ (the remainder eleven is written E
in the duodecimal system)

$\qquad$ 0 R │ T × 12² (the remainder ten is written T in
the duodecimal system)

$$1582 = TET_{(twelve)}$$

Exercise 4.3

1. Write the octal numerals naming the numbers from 1 to 25 inclusive.
2. Write the octal numerals naming the numbers from 26 through 40 inclusive.
3. Write the octal numeral naming the number of *'s shown below.

* * * * * * * * * * * * *

4. Change the following octal numerals to decimal numerals.
   a. $22_{(eight)}$ d. $707_{(eight)}$ g. $1111_{(eight)}$
   b. $71_{(eight)}$ e. $1464_{(eight)}$ h. $3007_{(eight)}$
   c. $134_{(eight)}$ f. $2031_{(eight)}$ i. $6061_{(eight)}$
5. Change the following decimal numerals to octal numerals.
   a. 25   d. 89   g. 1034
   b. 76   e. 126   h. 2367
   c. 69   f. 667   i. 4017
6. Replace the * by < or > to make true statements.
   a. $164_{(eight)}$ * $14_{(eight)}$ c. $4317_{(eight)}$ * $4371_{(eight)}$
   b. $707_{(eight)}$ * $716_{(eight)}$ d. $1011_{(eight)}$ * $1000_{(eight)}$
7. Write the following duodecimal numerals as decimal numerals.
   a. $91_{(twelve)}$ c. $TOE_{(twelve)}$ e. $107_{(twelve)}$
   b. $TE_{(twelve)}$ d. $316_{(twelve)}$ f. $TEE_{(twelve)}$
8. Write the following decimal numerals as duodecimal numerals.
   a. 34   c. 211   e. 275
   b. 179   d. 285   f. 369
9. How many digits are used to write numerals in the following numeration systems?

a. decimal   c. duodecimal   e. base seven

b. octal   d. base five   f. base nine

10. Write the base five numerals naming the numbers from 1 to 25 inclusive.

11. Write the base five numerals naming the numbers from 25 to 40 inclusive.

12. Write the following base five numerals as decimal numerals.

    a. $32_{(five)}$ c. $214_{(five)}$ e. $1011_{(five)}$

    b. $100_{(five)}$ d. $1344_{(five)}$ f. $1014_{(five)}$

13. Write the following decimal numerals as base five numerals.

    a. 7   d. 23   g. 136

    b. 9   e. 25   h. 260

    c. 16   f. 74   i. 707

14. Write the numeral naming the number of toe's you have in the following systems of numeration.

    a. decimal system   d. base five system

    b. octal system   e. base four system

    c. duodecimal system   f. base nine system

15. Write the four digit numeral that names the greatest possible number in the following systems.

    a. base eight system

    b. base seven system

    c. base twelve system

16. Change the following numerals to decimal notation.

    a. $34_{(five)}$ d. $42_{(twenty)}$ g. $26_{(nine)}$

    b. $301_{(seven)}$ e. $202_{(four)}$ h. $1111_{(five)}$

    c. $111_{(three)}$ f. $130_{(four)}$ i. $364_{(seven)}$

17. Replace the * with $=$, $<$, or $>$ to make true statements.

    a. $26_{(eight)}$ * $1011_{(eight)}$ d. $100_{(five)}$ * $120_{(four)}$

    b. $5_{(seven)}$ * $11_{(four)}$ e. $TE_{(twelve)}$ * $1234_{(five)}$

    c. $100_{(eight)}$ * $120_{(four)}$ f. $1000_{(seven)}$ * $1000_{(twelve)}$

18. What is the value of the "4" in each of the following?

    a. 432   d. $304_{(seven)}$

    b. $142_{(eight)}$ e. $403_{(six)}$

    c. $124_{(five)}$ f. $4102_{(twelve)}$

19. What is the place-value of the "3" in each of the following?

    a. 532   d. $1132_{(twelve)}$

    b. $352_{(seven)}$ e. $3000_{(five)}$

    c. $3011_{(eight)}$ f. $30,167_{(nine)}$

20. What is the numeral for the next number after each of the following?

    a. $66_{(seven)}$ c. $444_{(five)}$ e. $777_{(eight)}$

    b. $22_{(three)}$ d. $EEE_{(twelve)}$ f. $8888_{(nine)}$

21. Write 234 as a

    a. base four numeral   d. base three numeral

    b. base seven numeral   e. base nine numeral

    c. base six numeral   f. base twelve numeral

4.7. The Binary System

Since the binary system of numeration is important in the computer field, special attention will be given to it. The **binary**, or **base two system**, requires only two symbols, 0 and 1. The place-values of this system are powers of two some of which are given in Table 4.5.

Table 4.5

Position	5	4	3	2	1	0
Place-value	thirty-two 2^5	sixteen 2^4	eight 2^3	four 2^2	two 2^1	one 2^0

The binary numeral for two is

$$10_{(two)} = (1 \times two^1) + (0 \times two^0)$$

The binary system is based on sets of two, just as the decimal system is based on sets of ten, the octal system on sets of eight and the duodecimal system on sets of twelve. Table 4.6 shows some binary numerals and their decimal equivalents.

Let us examine some binary numerals and their decimal equivalents.

$$1,001_{(two)} = (1 \times two^3) + (0 \times two^2) + (0 \times two^1) + (1 \times two^0)$$
$$= (1 \times 2^3) + (0 \times 2^2) + (0 \times 2^1) + (1 \times 2^0)$$
$$= 8 + 0 + 0 + 1 = 9$$
$$11,011_{(two)} = (1 \times two^4) + (1 \times two^3) + (0 \times two^2) + (1 \times two^1)$$
$$+ (1 \times two^0)$$
$$= (1 \times 2^4) + (1 \times 2^3) + (0 \times 2^2) + (1 \times 2^1) + (1 \times 2^0)$$
$$= 16 + 8 + 0 + 2 + 1$$
$$= 27$$

Table 4.6

Decimal Numerals	Binary Numerals
0	0
1	1
2	10
3	11
4	100
5	101
6	110
7	111
8	1000
9	1001
10	1010

We convert decimal numerals to equivalent binary numerals using the same method as previously discussed for the octal and duodecimal systems. Study the following:

$$18 = 16 + 2$$
$$= (1 \times 16) + (0 \times 8) + (0 \times 4) + (1 \times 2) + (0 \times 1)$$
$$= (1 \times two^4) + (0 \times two^3) + (0 \times two^2) + (1 \times two^1) + (0 \times two^0)$$
$$= 10{,}010_{(two)}$$

Using the repeated division method we have:

$$
\begin{array}{lll}
2)\,\overline{18} & & \\
2)\,\overline{9} & R & 0 \times 2^0 \\
2)\,\overline{4} & R & 1 \times 2^1 \\
2)\,\overline{2} & R & 0 \times 2^2 \qquad 18 = 10{,}010_{(two)} \\
2)\,\overline{1} & R & 0 \times 2^3 \\
\quad 0 & R & 1 \times 2^4 \\
\end{array}
$$

4.8.   Binary-Octal Relation

Binary numerals are cumbersome because they require many digits. For example:

$$64 = 1{,}000{,}000_{(two)}$$
$$46 = 101{,}110_{(two)}$$

Binary numerals convert easily to octal numerals which require fewer digits than the corresponding binary numerals. For example,

$$1,000_{(\text{two})} = 10_{(\text{eight})}$$
$$101,110_{(\text{two})} = 56_{(\text{eight})}$$

Let us examine a method for converting a binary numeral to an octal numeral. First we insert commas in the binary numeral, starting from the right in the customary fashion in groups of three, thus: $1,101,111_{(\text{two})}$. We then convert the numeral to the equivalent octal numeral as follows:

$$1,101,111_{(\text{two})} = [1 \times \text{two}^6] + [(1 \times \text{two}^5) + (0 \times \text{two}^4) + (1 \times \text{two}^3)]$$
$$+ [(1 \times \text{two}^2) + (1 \times \text{two}^1) + (1 \times \text{two}^0)]$$

Notice the first bracket [ ] encloses the number represented by "1"; the second bracket, the number represented by "101"; and the third bracket the number represented by "111."

Let us observe separately the three expressions enclosed in brackets:

$$[1 \times \text{two}^6] = \text{sixty-four}$$
$$= 1 \times \text{eight}^2$$
$$[(1 \times \text{two}^5) + (0 \times \text{two}^4) + (1 \times \text{two}^3)] = \text{thirty-two} + \text{zero} + \text{eight}$$
$$= \text{forty}$$
$$= 5 \times \text{eight}^1$$
$$[(1 \times \text{two}^2) + (1 \times \text{two}^1) + (1 \times \text{two}^0)] = \text{four} + \text{two} + \text{one}$$
$$= \text{seven}$$
$$= 7 \times \text{eight}^0$$

Putting these together we have

$$1,101,111_{(\text{two})} = 157_{(\text{eight})}$$

This example suggests a method for converting the binary numeral $1,101,111_{(\text{two})}$ to the octal numeral $157_{(\text{eight})}$. Notice the three groups of digits in the binary numeral: 1; 101; and 111. Let us consider each group separately:

1 names one in base two
101 names five in base two
111 names seven in base two.

Notice that we found that $1,101,111_{(\text{two})}$ names the same number as $157_{(\text{eight})}$. This demonstrates a pattern for converting binary numerals to octal numerals. Observe the following examples:

$$11, 101, 111, 001_{\text{(two)}} = 3{,}571_{\text{(eight)}}$$

names 3   names 5   names 7   names 1

$$111, 000, 110, 111_{\text{(two)}} = 7{,}067_{\text{(eight)}}$$

names 7   names 0   names 6   names 7

We see then, to convert a binary numeral to an equivalent octal numeral:

(1) We separate the binary numeral into groups of three digits by commas, starting at the right.
(2) We find the decimal numerals named by each group of binary digits.
(3) The digits, in order, of the numerals found in (2) are the digits in the octal numeral.

We convert an octal numeral to an equivalent binary numeral by reversing this process. In this case we write a binary numeral of three digits for each digit in the octal numeral. Thus:

octal numeral	4	6	7	3
binary numeral	100	110	111	011

Then

$$4673_{\text{(eight)}} = 100{,}110{,}111{,}011_{\text{(two)}}.$$

EXAMPLE i. Convert $110{,}001{,}111_{\text{(two)}}$ to an equivalent octal numeral.
Solution: We find the number named by each group of digits in the binary numeral:

110,001,111

6   1   7

Then $$110,001,111_{(two)} = 617_{(eight)}$$

EXAMPLE ii. Convert $412_{(eight)}$ to a binary numeral.

Solution: We name each digit in the octal numeral by a binary numeral with three digits:

$$\underbrace{4}_{100} \quad \underbrace{1}_{001} \quad \underbrace{2}_{010}$$

Then

$$412_{(eight)} = 100,001,010_{(two)}$$

4.9. Summary

We have studied place-value systems of numeration using base ten, base eight, base twelve, and base two. Place-value systems may be constructed using as a base any whole number greater than one. All place-value systems have some features in common:

(1) Every place-value system of numeration has a base which may be any whole number greater than one.
(2) The number of symbols necessary in any place-value system is the same as the base.
(3) Each position in a numeral is assigned a place-value which is a power of the base.

Table 4.7 shows the numeral sequence for several place-value systems of numeration.

In examining Table 4.7 we make the following observations:

(1) The numeral naming the base of the system is always 10.
(2) In writing a numeral for any number greater than the base, more than one digit must be used.
(3) Each digit in a numeral represents a number. This number is the product of a number less than the base and the place-value assigned to the position in which the digit is written.
(4) The number named by any numeral is the sum of products defined in (3) above.

Table 4.7

Base Numerals	Twelve	Ten	Eight	Five	Four	Three	Two
	0	0	0	0	0	0	0
	1	1	1	1	1	1	1
	2	2	2	2	2	2	10
	3	3	3	3	3	10	11
	4	4	4	4	10	11	100
	5	5	5	10	11	12	101
	6	6	6	11	12	20	110
	7	7	7	12	13	21	111
	8	8	10	13	20	22	1000
	9	9	11	14	21	30	1001
	T	10	12	20	22	31	1010
	E	11	13	21	23	32	1011
	10	12	14	22	30	40	1100
	11	13	15	23	31	41	1101
	12	14	16	24	32	42	1110
	13	15	17	30	33	50	1111
	14	16	20	31	40	51	10000
	15	17	21	32	41	52	10001

Exercise 4.4

1. Write the binary numerals naming the numbers from 1 through 20 inclusive.
2. Write the binary numerals naming the numbers from 32 to 40 inclusive.
3. Write the following binary numerals as decimal numerals.

    a. $111_{(two)}$ d. $101,110,111_{(two)}$

    b. $1,111_{(two)}$ e. $1,101,101,110_{(two)}$

    c. $101,111_{(two)}$ f. $101,111,111,101_{(two)}$

4. Write the following decimal numerals as binary numerals.

    a. 6   c. 27   e. 136

    b. 16   d. 86   f. 272

5. Write the following binary numerals as octal numerals.

    a. $101,100_{(two)}$ e. $10,101,111,000_{(two)}$

    b. $100,001,001_{(two)}$ f. $11,001,101,011_{(two)}$

    c. $100,001,110_{(two)}$ g. $10,100,110,001_{(two)}$

    d. $111,111,111_{(two)}$ h. $1,000,000,111,111_{(two)}$

6. Write the following octal numerals as binary numerals.

    a. $73_{(eight)}$ e. $346_{(eight)}$

    b. $64_{(eight)}$ f. $107_{(eight)}$

    c. $30_{(eight)}$ g. $432_{(eight)}$

    d. $100_{(eight)}$ h. $2234_{(eight)}$

7. Replace the symbol * by $=$, $<$, or $>$ to make true statements.

    a. $101,111_{(two)}$ * $57_{(eight)}$

    b. $111,000_{(two)}$ * $77_{(eight)}$

    c. $101,001,110_{(two)}$ * $500_{(eight)}$

    d. $111,111,000_{(two)}$ * $1346_{(eight)}$

    e. $567_{(eight)}$ * $101,110,111_{(two)}$

8. What disadvantages are there in the use of a system of numeration with a small number for the base?

9. What is the numeral for the base in the binary system? in the octal system?

10. How can you tell the difference between numerals naming odd numbers and numerals naming even numbers in the binary system?

11. What numeral names the number after $101_{(two)}$ in the binary system?

12. A group of four people want to share equally a hotel bill of $120_{(four)}$. How much will each person pay?

13. Let us make up a base four place-value system of numeration using symbols other than the decimal numerals. The symbols in this system are as follows:

$$n(\phi) = \ominus$$
$$n(\{-\}) = -$$
$$n(\{-, \wedge\}) = \wedge$$
$$n(\{-, \wedge, \triangle\}) = \triangle$$

    a. Write the numerals naming the numbers one through sixteen in this system.

b. What is the decimal numeral that names the same number as △∧∧?

c. What number is one more than △△△?

d. What numeral names the same number in this system as the decimal numeral 36?

14. A simple device for illustrating binary numerals may be made by using a cardboard box and a string of Christmas tree lights wired in parallel (that is, when one light goes out the others stay on). When a light is on, indicated in the figure below by ○, it corresponds to the numeral 1 in the binary system. What a light is off, indicated by ● in the figure, it corresponds to the numeral 0. The binary numeral indicated in the figure is $100,101,011_{(two)}$.

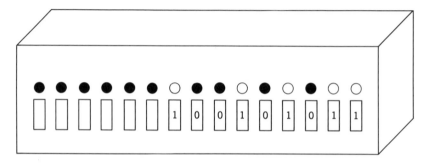

a. Write a binary numeral for the following arrangement of lights on the device described above.

(1) ○○○●●●

(2) ○●●●○○○●

(3) ○○○●●●○○

(4) ○●●○○○●●○○●○

b. Write the decimal numeral for the binary numerals in (a).

15. Write a (1) duodecimal numeral; (2) an octal numeral; (3) a base five numeral; for the year in each date given below.

a. Oct. 14, 1066 (Conquest of England by William)

b. Sept. 24, 1493 (Second voyage of Columbus)

c. June 18, 1815 (Battle of Waterloo)

d. Nov. 19, 1863 (Gettysburg Address)

e. April 6, 1917 (U.S.A.'s entry into World War I)

f. Oct. 29, 1929 (Stock Market Crash)

5.1.   Introduction

In about 825 A.D. a book on the use of Hindu-Arabic numerals was written by Al Khowarizmi. This book was translated into Latin and the Latin version was translated into English. The English translation reads: "Spoken has Algorithmi." In translation the name Al Khowarizmi became Algorithmi from which we get the word algorithm. The word **algorithm** means a special process for finding the result of an operation on two numbers when the result is not obvious.

The algorithms we know depend upon two things: (1) the properties of the set of whole numbers and their operations, and (2) the properties of our place-value system of numeration.

The Algorithms for Operations with Whole Numbers

Although the properties of the whole numbers and their operations are true regardless of how the numbers are recorded, the familiar algorithms work only when these numbers are recorded in a place-value system of notation.

In discussing the algorithms we shall assume that we have at our disposal the following:

(1) The set of whole numbers.
(2) The sum of any pair of whole numbers named by the digits in our system of numeration. We shall call these sums the **addition facts** or **basic sums.**
(3) The product of any pair of whole numbers named by the digits in our system of numeration. We shall call these products the **multiplication facts** or **basic products.**
(4) The properties of the whole numbers and their operations.
(5) A place-value system of numeration.

5.2. The Addition Algorithm

Although the history of the evolution of the addition algorithm as we know it today is very interesting, it is not our purpose to discuss this evolution. We shall, instead, discuss the reasons why these patterns with which we are all familiar work, that is, provide a logical basis for the addition algorithm.

We usually record the addition of 18 and 27 as

$$
\begin{array}{r}
18 \\
27 \\
\hline
45
\end{array}
$$

or, at the elementary level, as

$$
\begin{array}{r}
18 = 10 + 8 \\
27 = 20 + \ 7 \\
\hline
30 + 15 = (30 + 10) + 5 = 45
\end{array}
$$

Let us consider this rather elementary addition problem and state the logical reasons, that is, the properties and principles that justify adding ones and ones, and tens and tens, and the process of "carrying." The reason for each step in the addition example above is stated at the right of each step in the explanation below.

$18 + 27$

$= [(1 \times 10^1) + (8 \times 10^0)] + [(2 \times 10^1) + (7 \times 10^0)],$ Expanded notation

$= (1 \times 10^1) + \{(8 \times 10^0) + [(2 \times 10^1) + (7 \times 10^0)]\},$ Associative property of addition

$= (1 \times 10^1) + \{[(2 \times 10^1) + (7 \times 10^0)] + (8 \times 10^0)\},$ Commutative property of addition

$= (1 \times 10^1) + \{(2 \times 10^1) + [(7 \times 10^0) + (8 \times 10^0)]\},$ Associative property of addition

$= [(1 \times 10^1) + (2 \times 10^1)] + [(7 \times 10^0) + (8 \times 10^0)],$ Associative property of addition

$= [(1 + 2) \times 10^1] + [(7 + 8) \times 10^0],$ Distributive property

$= (3 \times 10^1) + (15 \times 10^0),$ Basic sums

$= (3 \times 10^1) + \{[(1 \times 10^1) + (5 \times 10^0)] \times 10^0\},$ Expanded notation

$= (3 \times 10^1) + \{[(1 \times 10^1) + (5 \times 10^0)] \times 1\},$ $10^0 = 1$

$= (3 \times 10^1) + [(1 \times 10^1) + (5 \times 10^0)],$ Multiplicative identity

$= [(3 \times 10^1) + (1 \times 10^1)] + (5 \times 10^0),$ Associative property of addition

$= [(3 + 1) \times 10^1] + (5 \times 10^0),$ Distributive property

$= (4 \times 10^1) + (5 \times 10^0),$ Basic sums

$= 45.$ Place-value notation

Notice that in this long explanation the distributive property justifies the addition of tens and tens and ones and ones. The associative and commutative properties of addition are used over and over to group and order the terms so that the distributive property may be used. When we use column addition, that is when we use the familiar addition algorithm, the adding of the numbers named by the digits in the various columns is justified by the reasons stated in the long form above. The illustrated example shows the "carrying" of one ten. The same method is used to "carry" hundreds, thousands, etc.

The explanation above used numerals in the familiar decimal system. The same algorithm may be used with numerals written in any place-value systems regardless of the base.

The place-values of the octal system are $eight^0$ (ones), $eight^1$ (eights), $eight^2$ (sixty-fours), and so forth. The basic sums recorded in octal numerals are given in Table 5.1.

Table 5.1

+	0	1	2	3	4	5	6	7
0	0	1	2	3	4	5	6	7
1	1	2	3	4	5	6	7	10
2	2	3	4	5	6	7	10	11
3	3	4	5	6	7	10	11	12
4	4	5	6	7	10	11	12	13
5	5	6	7	10	11	12	13	14
6	6	7	10	11	12	13	14	15
7	7	10	11	12	13	14	15	16

Let us add $34_{(eight)}$ and $42_{(eight)}$. Writing $34_{(eight)}$ and $42_{(eight)}$ in expanded notation we have

$$34_{(eight)} + 42_{(eight)} = [(3 \times eight^1) + (4 \times eight^0)] + [(4 \times eight^1) + (2 \times eight^0)].$$

If the commutative and associative properties of addition are applied in the same manner as before when we were adding two numbers named by decimal numerals, we have:

$$[(3 \times eight^1) + (4 \times eight^1)] + [(4 \times eight^0) + (2 \times eight^0)].$$

Using the distributive property we obtain:

$$[(3 + 4) \times eight^1] + [(4 + 2) \times eight^0] = (7 \times eight^1) + (6 \times eight^0)$$
$$= 76_{(eight)}$$

since $3 + 4 = 7$, $4 + 2 = 6$ (Table 5.1) and $(7 \times eight^1) + (6 \times eight^0)$ may be written $76_{(eight)}$ because of place-value notation.

Using the addition algorithm the above addition is written:

$$34_{(eight)}$$
$$42_{(eight)}$$
$$\overline{76_{(eight)}}$$

We can use the familiar addition algorithm to add numbers named in any place-value system of notation. In each case it is necessary to know the basic sums. Hence, to use the addition algorithm when the addends are recorded in, for example, the base five system of numeration, we construct an addition table for this system as we did for the octal system (Table 5.1).

Study the examples below. When all numerals in a problem are in the same base of numeration, we usually omit the subscripts in that particular problem and merely state at the beginning the base used.

EXAMPLE i. Add $23_{(eight)} + 46_{(eight)} + 72_{(eight)}$.
Solution: All numerals are in the octal system of notation.

$$23$$
$$46$$
$$72$$
$$\overline{163}$$

Explanation: $3 + 6 + 2 = 13$ (ones)
13 ones equal 1 eight and 3 ones
Write the "3" in the ones column and "carry" 1 eight to the eights column

$$1 + 2 + 4 + 7 = 16 \text{ (eights)}$$

16 eights equal 1 eight^2 and 6 eights
Write the "6" in the eights column and the "1" in the eight^2 column.
The sum is $163_{(eight)}$

EXAMPLE ii. Add $31_{(five)} + 14_{(five)} + 32_{(five)}$.
Solution: All numerals are in the base five system.

$$31$$
$$14$$
$$32$$
$$\overline{132}$$

Explanation: $1 + 4 + 2 = 12$ (ones)
12 ones equal 1 five and 2 ones

Write the "2" in the ones column and "carry" 1 five to the fives column

$$1 + 3 + 1 + 3 = 13 \text{ (fives)}$$

13 fives equal 1 five^2 and 3 fives
Write the "3" in the fives column and the "1" in the five^2 column.
The sum is $132_{(five)}$

EXAMPLE iii. Add $ET_{(twelve)} + 123_{(twelve)} + 6T3_{(twelve)}$
Solution: All numerals are in the duodecimal system.

$$
\begin{array}{r}
ET \\
123 \\
6T3 \\
\hline
904
\end{array}
$$

Explanation: $T + 3 + 3 = 14$ (ones)
14 ones equal 1 twelve and 4 ones
Write the "4" in the ones column and "carry" the "1" to the twelves column

$$1 + E + 2 + T = 20 \text{ (twelves)}$$

20 twelves equal 2 twelve^2 and 0 twelves
Write the "0" in the twelves column and "carry" 2 twelve^2 to the twelve^2 column

$$2 + 1 + 6 = 9 \text{ (twelve}^2\text{)}$$

Write the "9" in the twelve^2 column
The sum is $904_{(twelve)}$

5.3. The Subtraction Algorithm

The operation of subtraction is the inverse operation of addition. That is, if a and b are whole numbers and $a \geq b$, then $a - b = c$ means $b + c = a$.

An important subtraction property is illustrated in the statements below:

$$(9 + 8) - (4 + 3) = (9 - 4) + (8 - 3)$$
$$17 - 7 = 5 + 5$$
$$10 = 10$$

$$(7 + 3) - (5 + 1) = (7 - 5) + (3 - 1)$$
$$10 - 6 = 2 + 2$$
$$4 = 4$$

In general, if a, b, c, and d are whole numbers, $a \geq c$ and $b \geq d$, then $(a + b) - (c + d) = (a - c) + (b - d)$.

Subtraction Property. For all whole numbers a, b, c, and d, $a \geq c$ and $b \geq d$, $(a + b) - (c + d) = (a - c) + (b - d)$.

The subtraction property is vital in the explanation of the subtraction algorithm.

Let us subtract $83 - 47$. When we use the subtraction algorithm we write this problem

$$
\begin{array}{r}
83 \\
47 \\
\hline
\end{array}
$$

and subtract ones from ones and tens from tens. Since we cannot subtract 7 ones from 3 ones (that is there is no whole number n such that $7 + n = 3$), we rename 83 so that it is possible to subtract ones from ones.

Let us justify this **renaming** or **regrouping** commonly called "borrowing."

$83 = (8 \times 10^1) + (3 \times 10^0),$	Expanded notation
$= (8 \times 10^1) + (3 \times 1),$	$10^0 = 1$
$= [(7 + 1) \times 10^1] + (3 \times 1),$	Basic sums
$= [(7 \times 10^1) + (1 \times 10^1)] + (3 \times 1),$	Distributive property
$= (7 \times 10^1) + [(1 \times 10^1) + (3 \times 1)],$	Associative property of addition
$= (7 \times 10^1) + [(10 \times 1) + (3 \times 1)],$	Commutative property of multiplication and $10^1 = 10$
$= (7 \times 10^1) + [(10 + 3) \times 1],$	Distributive property
$= (7 \times 10^1) + (13 \times 1),$	Place-value notation
$= 70 + 13.$	Place-value notation and multiplicative identity

Now that we have renamed 83 as $70 + 13$, we justify each step in the subtraction $83 - 47$:

$$83 - 47 = (70 + 13) - [(4 \times 10^1) + (7 \times 10^0)]$$

Renaming and expanded notation

$$= (70 + 13) - (40 + 7)$$ Place-value notation

$$= (70 - 40) + (13 - 7)$$ Subtraction property

$$= [(7 \times 10) - (4 \times 10)] + [13 - 7]$$ Place-value notation

$$= [(7 - 4) \times 10] + 6$$ Distributive property of multiplication over subtraction and basic sums

$$= (3 \times 10) + (6 \times 1)$$ Basic sums and multiplicative identity

$$= 36$$ Place-value notation

Using the familiar subtraction algorithm we may write this in the shortened form:

$$
\begin{array}{rcccl}
83 & = & 80 + 3 & = & 70 + 13 \\
47 & = & 40 + 7 & = & 40 + 7 \\
\hline
& & & & 30 + 6 = 36.
\end{array}
$$

As we become proficient using the subtraction algoritm we usually shorten this even more and write

$$
\begin{array}{l}
\overset{7\ 13}{\rlap{/}8\rlap{/}3} \\
47 \\
\hline
36
\end{array}
$$
     regrouping line

We can use the subtraction algorithm to find the difference of two whole numbers when the sum and addend are named in any place-value system of numeration. Some examples are shown below.

EXAMPLE i. Subtract $476_{(eight)} - 237_{(eight)}$

Solution: All numerals are base eight.

$$
\begin{array}{rcl}
476 & = & 400 + 60 + 16 \\
237 & = & 200 + 30 + 7 \\
\hline
& & 200 + 30 + 7 = 237
\end{array}
$$

Explanation: We cannot subtract 7 from 6; so we rename the 7 eights in 476 as 6 eights and 10 ones which we add to the 6 ones making 16 ones.

$$16 - 7 = 7 \text{ (ones)}$$

Write the "7" in the ones column

$$6 - 3 = 3 \text{ (eights)}$$

Write the "3" in the eights column

$$4 - 2 = 2 \text{ (eight}^2\text{)}$$

Write the "2" in the eight^2 column.
The difference or missing addend is $237_{\text{(eight)}}$
Using the shortened form we have:

$$
\begin{array}{l}
\overset{6\ 16}{4\cancel{7}\cancel{6}} \qquad \underline{\hspace{3cm}} \text{ regrouping line} \\
237 \\
\hline
237
\end{array}
$$

EXAMPLE ii.   Subtract $413_{\text{(five)}} - 124_{\text{(five)}}$

Solution: All numerals are in base five

$$
\begin{array}{l}
\overset{3\ 10}{\underset{}{\cancel{0}\ 13}} \\
\cancel{4}\cancel{1}\cancel{3} \qquad \underline{\hspace{3cm}} \text{ regrouping line} \\
124 \\
\hline
234
\end{array}
$$

Explanation: We cannot subtract 4 from 3 so we rename 1 five in 413 as 10 ones which we add to the 3 ones making 13 ones

$$13 - 4 = 4 \text{ (ones)}$$

Write the "4" in the ones column
We cannot subtract 2 from 0 so we rename the 4 five^2 in 413 as 3 five^2 and 10 fives

$$10 - 2 = 3 \text{ (fives)}$$

Write the "3" in the fives column

$$3 - 1 = 2 \text{ (five}^2\text{)}$$

Write the "2" in the five^2 column
The difference or missing addend is $234_{\text{(five)}}$

EXAMPLE iii. Subtract $924_{(twelve)} - 4TE_{(twelve)}$

Solution: All the numerals are duodecimal numerals.

$$
\begin{array}{l}
8\;11 \\
\cancel{1}14 \\
\hline
\cancel{9}\,\cancel{2}\,\cancel{4} \\
4\,T\,E \\
\hline
4\,3\,5
\end{array}
\qquad \text{regrouping line}
$$

Explanation: We cannot subtract E from 4 so we rename the 2 twelves in 924 as 1 twelve and 10 ones which we add to the 4 ones making 14 ones

$$14 - E = 5 \text{ (ones)}$$

Write the "5" in the ones column
We cannot subtract T from 1 so we rename the 9 twelve^2 in 924 as 8 twelve^2 and 10 twelves which we add to the 1 twelve making 11 twelves

$$11 - T = 3 \text{ (twelves)}$$

Write the "3" in the twelves column

$$8 - 4 = 4 \text{ (twelve}^2\text{)}$$

Write the "4" in the twelve^2 column
The difference or missing addend is $435_{(twelve)}$

EXAMPLE iv. Subtract $900_{(twelve)} - 138_{(twelve)}$

Solution: All numerals are duodecimal numerals.

$$
\begin{array}{l}
E\;10 \\
8\;\cancel{10} \\
\hline
\cancel{9}\,\cancel{0}\,\cancel{0} \\
1\,3\,8 \\
\hline
7\,8\,4
\end{array}
\qquad \text{regrouping line}
$$

Explanation: We rename 9 twelve^2 as 8 twelve^2 and 10 twelves. We now rename 10 twelves as E twelves and 10 ones
Now that the renaming is completed, the subtraction is easy

$$10 - 8 = 4 \text{ (ones)}$$

Write the "4" in the ones column

$$E - 3 = 8 \text{ (twelves)}$$

Write the "8" in the twelves column

$$8 - 1 = 7 \text{ (twelve}^2\text{)}$$

Write the "7" in the twelve² column
The difference or missing addend is $784_{(twelve)}$

Exercise 5.1

1. Fill in the blanks with numerals to make true statements.
    a. $300 + 20 + 16 = 300 + \underline{\hspace{1cm}} + 6$
    b. $500 + 140 + 8 = \underline{\hspace{1cm}} + 40 + 8$
    c. $200_{(eight)} + 30_{(eight)} + 17_{(eight)} = 200_{(eight)} + \underline{\hspace{1cm}}_{(eight)} + 7_{(eight)}$
    d. $500_{(twelve)} + 1TO_{(twelve)} + E_{(twelve)} = \underline{\hspace{1cm}}_{(twelve)} + TO_{(twelve)} + E_{(twelve)}$
    e. $200_{(five)} + 130_{(five)} + 2_{(five)} = \underline{\hspace{1cm}}_{(five)} + 30_{(five)} + 2_{(five)}$
2. Fill in the blanks with numerals to make true statements.
    a. $500 + 80 + 6 = 500 + 70 + \underline{\hspace{1cm}}$
    b. $200 + 20 + 7 = \underline{\hspace{1cm}} + 120 + 7$
    c. $600_{(eight)} + 30_{(eight)} + 6_{(eight)} = 600_{(eight)} + \underline{\hspace{1cm}}_{(eight)} + 16_{(eight)}$
    d. $E00_{(twelve)} + TO_{(twelve)} + 9_{(twelve)} = T00_{(twelve)} + \underline{\hspace{1cm}}_{(twelve)} + 9_{(twelve)}$
    e. $400_{(five)} = 300_{(five)} + \underline{\hspace{1cm}}_{(five)} + 10_{(five)}$
3. Add. All numerals are in base eight.
    a. 46   b. 27   c. 134
        72   37   675
4. Add. All numerals are in base eight.
    a. 73   b. 436   c. 715
        67   707   674
5. Make an addition table. Record all sums in base twelve numerals.
6. Made an addition table. Record all sums in base five numerals.
7. Add. All numerals are in base twelve.
    a. 4 9T   b. 1 3 4   c. TE6
        7E 4   TEE   9E6
8. Add. All numerals are in base five.
    a. 321   b. 104   c. 143
        144   234   131
        442   441   443

---

9. Add. All numerals are in base twelve.

a. 7 3 6
   4 1 2
   1ET
   ―――

b. 7EE
   3TE
   T 9 9
   ―――

c. T 0E
   9 8T
   5T 7
   ―――

10. Add. All numerals are in base two.

a. 101
   110
   ―――

b. 1011
   1111
   ―――

c. 1011
   1001
   1001
   ―――

11. Add. All numerals are in base twelve.

a. 1 0 9
   5TT
   8EE
   ―――

b. 10 7
   89E
   T8 2
   ―――

c. 94 4
   T1 1
   61T
   ―――

12. Add. All numerals are in base five.

a. 213
   403
   442
   134
   ―――

b. 324
   121
   444
   234
   ―――

c. 333
   414
   112
   222
   ―――

13. Add. All numerals are in base two.

a. 101
   111
   100
   ―――

b. 1011
   101
   1000
   ―――

c. 1101
   111
   1010
   ―――

14. a. Check your results for Problem 9 by converting all numerals to the decimal system of notation.

    b. Check your results for Problem 11 by converting all numerals to the decimal system.

15. Subtract. All numerals are in base eight.

a. 732
   611
   ―――

b. 762
   473
   ―――

c. 712
   444
   ―――

16. Subtract. All numerals are in base twelve.

a. 69 3
   12T
   ―――

b. T3E
   48 7
   ―――

c. 9 16
   5T9
   ―――

17. Subtract. All numerals are in base two.

a. 101
   100
   ―――

b. 1000
   101
   ―――

c. 1010
   111
   ―――

18. Subtract. All numerals are in base twelve.
    a. E92 6   b. T7 2 0   c. 934 2
        438T   36TE   389T

19. Subtract. All numerals are in base eight.
    a. 7000   b. 5000   c. 6000
        1234   3654   2376

20. Subtract. All numerals are in base twelve.
    a. 60 00   b. T00 0   c. 800 0
        34T2   365E   428T

21. Subtract. All numerals are in base two.
    a. 1000   b. 10,000   c. 10,011
         111   111   1,000

22. Subtract. All numerals are in base five.
    a. 3014   b. 4301   c. 4012
        1221   2444   1421

23. Check your results for Problem 16 by converting all numerals to the decimal system.
24. Check your results for Problem 21 by converting all numerals to the decimal system.
25. Check your results for Problem 22 by converting all numerals to the decimal system.

5.4. The Multiplication Algorithm

Explanation of the multiplication algorithm is greatly simplified if we use the familiar rule of annexing a "0" to a numeral to name the product of a whole number and ten; by annexing two "0's" to the numeral to name the product of a whole number and one hundred; and so forth. Everyone is familiar with this rule of multiplying a number by a power of ten.

Let us discover why this rule works. The rule states that to multiply 10×58 we annex a zero to the numeral 58 and the product is 580. Here is the justification:

10×58
$= 58 \times 10$ Commutative property
                                              of multiplication

$= [(5 \times 10^1) + (8 \times 10^0)] \times 10^1$ Expanded notation and
                                                   $10 = 10^1$

$= [(5 \times 10^1) \times 10^1] + [(8 \times 10^0) \times 10^1]$ Distributive property

$= [5 \times (10^1 \times 10^1)] + (8 \times 10^1)$ Associative property of
      multiplication; $10^0 = 1$;
      multiplicative identity

$= (5 \times 10^2) + (8 \times 10^1)$ $10^1 \times 10^1 = 10 \times 10 = 10^2$

$= (5 \times 10^2) + (8 \times 10^1) + (0 \times 10^0)$ Additive identity;
      Theorem 3.1

$= 580$ Place-value notation

The same technique is used to prove the rules for multiplication by 100, 1000, and so forth.

Just as the addition algorithm depends primarily upon the commutative and associative properties of addition and the distributive property, the multiplication algorithm depends on the commutative and associative properties of multiplication and the distributive property.

Let us consider the problem of finding the product 63×34.

63×34

$= 63 \times [(3 \times 10^1) + (4 \times 10^0)]$ Expanded notation

$= [63 \times (3 \times 10^1)] + [63 \times (4 \times 10^0)]$ Distributive property

$= [(63 \times 3) \times 10^1] + [(63 \times 4) \times 10^0]$ Associative property of multiplication

$= [(3 \times 63) \times 10^1] + [(4 \times 63) \times 10^0]$ Commutative property of multiplication

$= (\{3 \times [(6 \times 10^1) + (3 \times 10^0)]\} \times 10^1) +$
$(\{4 \times [(6 \times 10^1) + (3 \times 10^0)]\} \times 10^0)$ Expanded notation

$= (\{[3 \times (6 \times 10^1)] + [3 \times (3 \times 10^0)]\} \times 10^1) +$
$(\{[4 \times (6 \times 10^1)] + [4 \times (3 \times 10^0)]\} \times 10^0)$ Distributive property

$= (\{[(3 \times 6) \times 10^1] + [(3 \times 3) \times 10^0]\} \times 10^1) +$
$(\{[(4 \times 6) \times 10^1] + [(4 \times 3) \times 10^0]\} \times 10^0$ Associative property of multiplication

$= \{[(18 \times 10^1) + (9 \times 10^0)] \times 10^1\} +$
$\{[(24 \times 10^1) + (12 \times 10^0)] \times 10^0\}$ Basic products

$= ([(18 \times 10^1) \times 10^1] + [(9 \times 10^1]) +$ $[(24 \times 10^1) + (12 \times 10^0)]$	Distributive property; $10^0 = 1$; multiplicative identity
$= \{[18 \times (10^1 \times 10^1)] + (9 \times 10^1)\}$ $+ \{(24 \times 10^1) + (12 \times 10^0)\}$	Associative property of multiplication
$= [18 \times (10^1 \times 10^1)] + [(9 \times 10^1) +$ $(24 \times 10^1)] + (12 \times 10^0)$	Associative property of multiplication
$= (18 \times 100) + [(9 + 24) \times 10^1] + (12 \times 10^0)$	Distributive property; $10^1 \times 10^1 = 100$
$= 1800 + (33 \times 10^1) + (12 \times 10^0)$	Multiplication by 100; addition algorithm
$= 1800 + 330 + 12$	Multiplication by 10; $10^0 = 1$; multiplication identity
$= 2142$	Addition algorithm

Now let us look at the multiplication algorithm in a more familiar form.

$$
\begin{array}{r}
34 \\
\underline{63} \\
102 = \ 3 \times 34 \\
\underline{2040} = 60 \times 34 \\
2142
\end{array}
$$

There are really four products involved, as shown in the longer form below.

$$
\begin{array}{r}
34 \\
\underline{63} \\
12 = \ 3 \times 4 \\
90 = \ 3 \times 30 \\
240 = 60 \times 4 \\
\underline{1800} = 60 \times 30 \\
2142
\end{array}
$$

$\left.\begin{array}{l}12 \\ 90\end{array}\right\} 12 + 90 = 102$

$\left.\begin{array}{l}240 \\ 1800\end{array}\right\} 240 + 1800 = 2040$

Notice that

$$12 = 12 \times 10^0$$
$$90 = 9 \times 10^1$$
$$240 = 24 \times 10^1$$
$$1800 = 18 \times 10^2 = 18 \times 10^1 \times 10^1$$

These products are easily found in the lengthy explanation already given. We usually shorten the procedure still further as shown below.

$$
\begin{array}{r}
34 \\
63 \\
\hline
102 \\
204 \\
\hline
2142
\end{array}
$$

Note that we indent the 204 one place to the left because we are actually multiplying by 60, not by 6.

In the multiplication algorithm above 102 and 2040 are called **partial products.**

Just as the addition and subtraction algorithms may be used when the numbers are named by numerals in any place-value system of numeration, so may the multiplication algorithm.

The basic products when the products are recorded in the octal system are given in Table 5.2.

With the basic products given in Table 5.2 we can now use the multiplication algorithm when all numbers are named in the octal system. Study the examples below.

Table 5.2

×	0	1	2	3	4	5	6	7
0	0	0	0	0	0	0	0	0
1	0	1	2	3	4	5	6	7
2	0	2	4	6	10	12	14	16
3	0	3	6	11	14	17	22	25
4	0	4	10	14	20	24	30	34
5	0	5	12	17	24	31	36	43
6	0	6	14	22	30	36	44	52
7	0	7	16	25	34	43	52	61

EXAMPLE i. Multiply $32_{(eight)} \times 5_{(eight)}$

Solution: All numerals are in the octal system.

$$
\begin{array}{r}
32 \\
5 \\
\hline
12 \\
170 \\
\hline
202
\end{array}
$$

Explanation:

$$5 \times 2 = 12$$
$$5 \times 30 = 170$$
$$12 + 170 = 202$$

We usually shorten the above as follows.

$$
\begin{array}{r}
\underline{1} \qquad \text{carry line} \\
32 \\
5 \\
\hline
202
\end{array}
$$

Explanation: $5 \times 2 = 12$ (ones)
12 ones equal 1 eight and 2 ones
Write the "2" in the ones place and "carry" 1 eight

$$5 \times 3 = 17 \text{ (eights)}$$
$$17 + 1 = 20 \text{ (eights)}$$

20 eights equal 2 eight^2 and 0 eights
Write the 0 in the eights column and the 2 in the eight^2 column
The product is $202_{(eight)}$

EXAMPLE ii. Multiply $432_{(eight)} \times 25_{(eight)}$

Solution: All numerals are in the octal system

$$
\begin{array}{r}
\underline{21} \qquad \text{carry line} \\
432 \\
25 \\
\hline
2602 \\
1064 \\
\hline
13442
\end{array}
$$

Explanation: Multiply by 5:

$$5 \times 2 = 12 \text{ (ones)}$$

12 ones equal 1 eight and 2 ones
Write the "2" in the ones column and "carry" 1 eight

$$5 \times 3 = 17 \text{ (eights)}$$
$$17 + 1 = 20 \text{ (eights)}$$

20 eights equal 1 eight^2 and 0 eights
Write the "0" in the eights column and "carry" the "2" to the eight^2 column

$$5 \times 4 = 24 \text{ (eight}^2\text{)}$$
$$24 + 2 = 26 \text{ (eight}^2\text{)}$$

Write the "6" in the eight^2 column and the "2" in the eight^3 column
Multiply by 2: this time we indent the partial product one place to the left because we are multiplying by 20 (2 eights) not by 2.

$$2 \times 2 = 4 \text{ (eights)} \quad \text{[because we are multiplying by 20]}$$

Write the "4" in the eights column

$$2 \times 3 = 6 \text{ (eight}^2\text{)} \quad \text{[because we are multiplying by 20]}$$

Write the "6" in the eight^2 column

$$2 \times 4 = 10 \text{ (eight}^3\text{)} \quad \text{[because we are multiplying by 20]}$$

10 eight^3 equals 1 eight^4 and 0 eight^3
Write the "0" in the eight^3 column and the "1" in the eight^4 column.
We now add the two partial products.
The product is $13,442_{\text{(eight)}}$

We can use the multiplication algorithm to multiply numbers named in any place-value system of notation. In each case it is necessary to know the basic products. Hence, to use the multiplication algorithm when the factors are recorded in any place-value system of numeration, we construct a multiplication table for the system needed.

EXAMPLE i.   Multiply $T_{\text{(twelve)}} \times 16_{\text{(twelve)}}$
Solution: All numerals are in the duodecimal system.

$$\begin{array}{r} 5 \\ \hline 1\,6 \\ \mathrm{T} \\ \hline 130 \end{array}$$ carry line

Explanation: $\mathrm{T} \times 6 = 50$ (ones)

$$50 \text{ ones} = 5 \text{ twelves} + 0 \text{ ones}$$

Write the "0" in the ones column and "carry" the "5" to the twelves column

$$\mathrm{T} \times 1 = \mathrm{T} \text{ (twelves)}$$
$$\mathrm{T} + 5 = 13 \text{ (twelves)}$$
$$13 \text{ twelves} = 1 \text{ twelve}^2 + 3 \text{ twelves}$$

Write the "3" in the twelves column and the "1" in the twelve^2 column. The product is $130_{\text{(twelve)}}$

EXAMPLE ii. Multiply $23_{\text{(five)}} \times 43_{\text{(five)}}$
Solution: All numerals are in the base-five system of numeration.

$$\begin{array}{r} 1 \\ 1 \\ \hline 43 \\ 23 \\ \hline 234 \\ 141 \\ \hline 2144 \end{array}$$ carry line

Explanation: $3 \times 3 = 14$ (ones)

$$14 \text{ ones} = 1 \text{ five} + 4 \text{ ones}$$

Write the "4" in the ones column and "carry" 1 five

$$3 \times 4 = 22 \text{ (fives)}$$
$$22 + 1 = 23 \text{ (fives)}$$
$$23 \text{ fives} = 2 \text{ five}^2 + 3 \text{ fives}$$

Write the "3" in the fives column and the "2" in the five^2 column
We now multiply by 2. We indent one place to the left because we are multiplying by 20 (2 fives).

$$2 \times 3 = 11 \text{ (fives)}$$
$$11 \text{ fives} = 1 \text{ five}^2 + 1 \text{ five}$$

Write the "1" in the fives column and "carry" 1 five^2

$$2 \times 4 = 13 \text{ (five}^2)$$
$$13 + 1 = 14 \text{ (five}^2)$$
$$14 \text{ five}^2 = 1 \text{ five}^3 + 4 \text{ five}^2$$

Write the "4" in the five^2 column and the "1" in the five^3 column. We now add the two partial products:

$$234 + 1410 = 2144$$

The product is $2144_{\text{(five)}}$.

5.5. The Division Algorithm

The operation of division is related to both the operation of multiplication and the operation of subtraction. We recall that division is the inverse operation of multiplication. Thus

$$12 \div 6 = 2 \text{ because } 6 \times 2 = 12.$$

Division may also be interpreted as repeated subtraction. Thus $12 \div 6$ may be interpreted as asking "How many times can 6 be subtracted from 12?" We see from the subtractions below, that 6 can be subtracted from 12 two times.

$$
\begin{array}{rl}
12 & \\
\underline{6} & 1 \\
6 & \\
\underline{6} & 1 \\
0 &
\end{array}
\Big\} \, 2 \text{ subtractions}
$$

The set of whole numbers is not closed under the operation of division. That is, the division of one whole number by another (not zero) does not always give a whole number. For example $12 \div 2 = 6$ since $2 \times 6 = 12$, but $15 \div 4$ is not a whole number since there is no whole numbers n such that $4n = 15$.

Although division of one whole number by another is not always possible, that is, it does not produce a whole number, there is a division property for any ordered pair (a, b), $b \neq 0$, of whole numbers.

Let $b \neq 0$ be any whole number. Every other whole number a, will either be a **multiple** of b, that is of the form bk where k is a whole number, or will fall between two consecutive multiples of b. That is, there is a unique whole number q such that

$$a = bq$$

or

$$bq < a < b(q + 1).$$

Geometrically, this may be visualized as follows. Starting at the point labeled 0, the number line may be partitioned into intervals b units in length (Figure 5.1). Any point with the whole number a for coordinate either lies within one of these intervals or is an endpoint of one of these intervals.

Figure 5.1

As an example, let $a = 15$ and $b = 6$. Since 15 is not a multiple of 6, it must fall between two consecutive multiples of 6. In fact,

$$(6 \cdot 2) < 15 < 6(2 + 1)$$
$$12 < 15 < 18.$$

Thus for any ordered pair (a, b), $b \neq 0$, of whole numbers we may write

$$a = bq + r$$

where q is a whole number and r is one of the whole numbers $0, 1, 2, \ldots, b - 1$. That is, $r \geq 0$ and $r < b$. We usually write $r \geq 0$ and $r < b$ as $0 \leq r < b$, which is read "r is greater than or equal to zero and less than b." The whole numbers a and b are called the **dividend** and **divisor**, respectively, the whole number q is called the **quotient** and the whole number r is called the **remainder**. This division property is called the **division algorithm**. We shall accept the division algorithm without proof.

Division Algorithm. For every ordered pair (a, b) of whole numbers, b $\neq$ 0, there exist unique whole numbers q and r such that a = bq + r, $0 \leq$ r $<$ b.

When we find q and r given whole numbers a and b, we say that we are **dividing** a by b. If $r = 0$ we say that a is **divisible** by b or that a is a **multiple** of b, and that b is a **factor** or a **divisor** of a.

If a and b are small numbers, it is easy to find q and r. If a and b are large, we use the familiar pattern shown below for $a = 4826$ and $b = 36$.

$$
\begin{array}{r}
134 \\
\hline
36)\,4826 \\
3600 \\
\hline
1226 \\
1080 \\
\hline
146 \\
144 \\
\hline
2
\end{array}
\qquad 4826 = (36 \times 134) + 2
$$

Let us analyze this familiar pattern. We want to find the greatest multiple of 36 that can be subtracted from 4826. We may subtract any multiple of 36 that is not greater than 4826. We usually consider first the multiples of 36 that have 10, 100, 1000, and so forth (that is, the powers of 10) as factors, because it is easy to find these products. First we see that 36×1000 cannot be subtracted from 4826 because $36{,}000 > 4826$. We see that $36 \times 100 = 3600$ can be subtracted from 4826, the result being 1226. Can 36×100 be subtracted again? Obviously it cannot because $3600 > 1226$. Next we use the product 36×10. We subtract 36×10 as many times as possible and then use the product 36×1.

$$
\begin{array}{r}
\hline
36)\,4826 \\
3600 \qquad 36 \times 100 \\
\hline
1226 \\
360 \qquad 36 \times 10 \\
\hline
866 \\
360 \qquad 36 \times 10 \\
\hline
506 \\
360 \qquad 36 \times 10 \\
\hline
146 \\
36 \qquad 36 \times 1 \\
\hline
110
\end{array}
$$

$$\frac{36}{74} \quad 36 \times 1$$

$$\frac{36}{38} \quad 36 \times 1$$

$$\frac{36}{2} \quad 36 \times 1$$

We see that

$$4826 = [(36 \times 100) + (36 \times 10) + (36 \times 10) + (36 \times 10) + (36 \times 1)$$
$$+ (36 \times 1) + (36 \times 1) + (36 \times 1)] + 2$$
$$= [36 \times (100 + 10 + 10 + 10 + 1 + 1 + 1 + 1)] + 2$$
$$= (36 \times 134) + 2$$

Now let us do the same division using the more familiar shorter form. We see that $36 \times 1000 = 36,000$ is too great a multiple and that $36 \times 100 = 3600$ is the best basic multiple to use. What is the greatest multiple of 3600 that can be subtracted from 4826? Notice that

$$1 \times 3600 = 3600$$
$$2 \times 3600 = 7200.$$

Since 7200 is greater than 4826, the greatest multiple of 3600 that we can subtract from 4826 is (1×3600) with a remainder of 1226.

$$
\begin{array}{r}
1 \\
36)\overline{4826} \\
3600 \quad 36 \times 100 \\
\hline
1226
\end{array}
$$

We next inspect the multiples of $36 \times 10 = 360$:

$$1 \times 360 = 360 \qquad (36 \times 10)$$
$$2 \times 360 = 720 \qquad (36 \times 20)$$
$$3 \times 360 = 1080 \qquad (36 \times 30)$$
$$4 \times 360 = 1440. \qquad (36 \times 40)$$

The greatest multiple of 360 we can subtract from 1226 is $3 \times 360 = 1080$ (that is 30×36) with a remainder of 146:

$$\begin{array}{r} 13 \\ 36\overline{)4826} \\ 3600 \\ \hline 1226 \\ 1080 \\ \hline 146 \end{array}$$ $\begin{array}{l} \\ \\ 36 \times 100 \\ \\ 36 \times 30 \\ \end{array}$

Now we inspect the multiples of 36×1:

$$1 \times 36 = 36$$
$$2 \times 36 = 72$$
$$3 \times 36 = 108$$
$$4 \times 36 = 144$$
$$5 \times 36 = 180.$$

The greatest multiple of 36 we can subtract from 146 is $4 \times 36 = 144$ with a remainder of 2.

$$\begin{array}{r} 134 \\ 36\overline{)4826} \\ 3600 \\ \hline 1226 \\ 1080 \\ \hline 146 \\ 144 \\ \hline 2 \end{array}$$ $\begin{array}{l} \\ \\ 36 \times 100 \\ \\ 36 \times 30 \\ \\ 36 \times 4 \\ \end{array}$

Since $2 < 36$ the division process is complete. The quotient is 134 and the remainder is 2.

This process may be written in horizontal form as shown:

$$\begin{aligned} 4826 &= (36 \times 100) + 1226 \\ &= (36 \times 100) + [(36 \times 30) + 146] \\ &= [(36 \times 100) + (36 \times 30)] + 146 \\ &= [36 \times (100 + 30)] + 146 \\ &= (36 \times 130) + 146 \\ &= (36 \times 130) + [(36 \times 4) + 2] \\ &= [(36 \times 130) + (36 \times 4)] + 2 \\ &= [36 \times (130 + 4)] + 2 \\ &= (36 \times 134) + 2. \end{aligned}$$

Examination of the preceding shows that the division algorithm depends on (1) the associative property of addition and (2) the distributive property. The only other requirements are the renaming of numbers and the fact that the subtractions are possible.

Now let us use this same process to find the quotient $467_{(eight)} \div 36_{(eight)}$. We want to find the largest multiple of $36_{(eight)}$ that we can subtract from $467_{(eight)}$. We first try multiples of $36_{(eight)}$ that have the powers of $10_{(eight)}$ as a factor:

$$36_{(eight)} \times 100_{(eight)} = 3600_{(eight)} > 467_{(eight)}$$
$$36_{(eight)} \times 10_{(eight)} = 360_{(eight)} < 467_{(eight)}$$

We see that we can subtract $36_{(eight)} \times 10_{(eight)}$ from $467_{(eight)}$. Can we subtract $2_{(eight)} \times [36_{(eight)} \times 10_{(eight)}]$? Since

$$2_{(eight)} \times 36_{(eight)} \times 10_{(eight)} = 2_{(eight)} \times 360_{(eight)}$$
$$= 740_{(eight)} > 467_{(eight)}$$

we see that we cannot subtract $20_{(eight)} \times 36_{(eight)}$ from $467_{(eight)}$.

We now use the familiar form and write,

$$
\begin{array}{r}
1 \quad\quad \\
36_{(eight)} \overline{)\, 467_{(eight)}} \\
360_{(eight)} \\
\hline
107_{(eight)}
\end{array}
$$

We see from the above that when we subtract $36_{(eight)} \times 10_{(eight)}$ there are $107_{(eight)}$ left. We next try multiples of $36_{(eight)} \times 1_{(eight)}$:

$$36_{(eight)} \times 1_{(eight)} = 36_{(eight)} < 107_{(eight)}$$
$$36_{(eight)} \times 2_{(eight)} = 74_{(eight)} < 107_{(eight)}$$
$$36_{(eight)} \times 3_{(eight)} = 132_{(eight)} > 107_{(eight)}.$$

Now we subtract $36_{(eight)} \times 2_{(eight)} = 74_{(eight)}$ from $107_{(eight)}$.

$$
\begin{array}{r}
12_{(eight)} \quad\quad\quad\quad\quad\quad\quad \\
36_{(eight)} \overline{)\, 467_{(eight)}} \quad\quad\quad\quad\quad \\
360_{(eight)} = 36_{(eight)} \times 10_{(eight)} \\
\hline
107_{(eight)} \quad\quad\quad\quad\quad \\
74_{(eight)} = 36_{(eight)} \times 2_{(eight)} \\
\hline
13_{(eight)} \quad\quad\quad\quad\quad\quad
\end{array}
$$

Since $13_{(eight)} < 36_{(eight)}$, the division is completed. The quotient is $12_{(eight)}$ and the remainder is $13_{(eight)}$. Study the examples below.

EXAMPLE i. Divide $326_{(eight)}$ by $5_{(eight)}$

Solution: All numerals are in the octal system.

$$
\begin{array}{r}
52 \\
5)\overline{326} \\
\underline{310} = 5 \times 50 \\
16 \\
\underline{12} = 5 \times 2 \\
4
\end{array}
$$

The quotient is $52_{(eight)}$ and the remainder is $4_{(eight)}$

EXAMPLE ii. Divide $4366_{(eight)}$ by $37_{(eight)}$

Solution: All numerals are in the octal system.

$$
\begin{array}{r}
112 \\
37)\overline{4366} \\
\underline{3700} = 37 \times 100 \\
466 \\
\underline{370} = 37 \times 10 \\
76 \\
\underline{76} = 37 \times 2 \\
0
\end{array}
$$

The quotient is $112_{(eight)}$

EXAMPLE iii. Divide $9ET_{(twelve)}$ by $6T_{(twelve)}$

Solution: All numerals are in the duodecimal system.

$$
\begin{array}{r}
1\,5 \\
6T)\overline{9ET} \\
\underline{6T0} = 6T \times 10 \\
31T \\
\underline{2T2} = 6T \times 5 \\
38
\end{array}
$$

The quotient is $15_{(twelve)}$ and the remainder is $38_{(twelve)}$.

1. Give a reason for each step in the proof below.
    Prove: For every whole number a, $a + a = 2 \times a$.
    Proof: (a) $a + a = (a \times 1) + (a \times 1)$
         (b) $= a \times (1 + 1)$
         (c) $= a \times 2$
         (d) $= 2 \times a$

2. Give a reason for each step in the proof below.
    Prove: For all whole numbers a, b, and c, $(a \times b) \times c = b \times (c \times a)$
    Proof: (a) $(a \times b) \times c = c \times (a \times b)$
         (b) $= (c \times a) \times b$
         (c) $= b \times (c \times a)$

3. Find q and r of the division algorithm given the ordered pairs (a, b)
    below. That is, find q and r in the statement $a = bq + r$, $0 \leq r < b$.
    a. $(26, 5)$ c. $(266, 19)$ e. $(367, 37)$
    b. $(138, 7)$ d. $(844, 28)$ f. $(599, 13)$

4. Multiply. All numerals are in base eight.

    a.   234
           7

    b.   436
           5

    c.   567
           4

5. Multiply. All numerals are in base eight.

    a.   736
          34

    b.   527
          63

    c.   714
          72

6. Divide. All numerals are in base eight.

    a. $6\overline{)463}$ b. $7\overline{)534}$ c. $4\overline{)561}$

7. Divide. All numerals are in base eight.

    a. $73\overline{)246}$ b. $35\overline{)736}$ c. $57\overline{)774}$

8. Construct a multiplication table for the base five numeration system.
9. Multiply. All numerals are in base five.

    a.   23
          4

    b.   43
          2

    c.   44
          3

10. Multiply. All numerals are in base five.

    a.   342
          23

    b.   443
          14

    c.   214
          34

11. Divide. All numerals are in base five.

   a. $4\overline{)342}$ b. $3\overline{)433}$ c. $2\overline{)413}$

12. Divide. All numerals are in base five.

   a. $23\overline{)443}$ b. $34\overline{)2342}$ c. $42\overline{)4130}$

13. Construct a multiplication table for the duodecimal system of numeration.

14. Multiply. All numerals are in base twelve.

   a. 3T4   b. 976   c. T2E
      5   8   E

15. Multiply. All numerals are in base twelve.

   a. 936   b. 4E9   c. TE1
      45   TE   96

16. Divide. All numerals are in base twelve.

   a. $9\overline{)964}$ b. $8\overline{)4TE}$ c. $6\overline{)TTE}$

17. Divide. All numerals are in base twelve.

   a. $34\overline{)96T}$ b. $9E\overline{)TE94}$

18. Construct a multiplication table for the binary system of numeration.

19. Multiply. All numerals are in base two.

   a. 101   b. 11011
     10   101

20. Divide. All numerals are in base two.

   a. $11\overline{)1011}$ b. $101\overline{)1111}$

21. What is the largest multiple of 64 that can be subtracted from 3864?

22. What is the largest multiple of 94 that can be subtracted from 10,766?

23. What is the largest multiple of $24_{(eight)}$ that can be subtracted from $453_{(eight)}$?

24. What is the largest multiple of $32_{(five)}$ that can be subtracted from $143_{(five)}$?

25. What is the largest multiple of $26_{(twelve)}$ that can be subtracted from $T64_{(twelve)}$?

6.1. Mathematical Sentences

Mathematical sentences may be statements. For example, the following mathematical sentences are statements because they can be labeled True or False.

$$6 + 3 = 9 \qquad \text{True}$$
$$6 + 9 < 4 \qquad \text{False}$$
$$(7 \cdot 3) + 5 > 2 \qquad \text{True}$$
$$18 \div 6 = 15 + 3 \qquad \text{False}$$

Other mathematical sentences may have the appearance of statements, but nevertheless are not statements since they do not have a truth value. The mathematical sentences

$$n + 6 = 9$$
$$5n > 4$$
$$8 \div 2 < n + 2$$

have statements form but do not have truth value as they stand. Such sentences are called **open sentences,** or simply, **sentences.**

Mathematical sentences have verbs which express a relationship between numbers. We use symbols to represent these verbs or verb

Number Sentences

VI

phrases. The symbols that we shall most often use in mathematical sentences are:

Verb or verb phrase	Symbol
equals or is equal to	$=$
is less than	$<$
is greater than	$>$
is not equal to	$\neq$
is less than or equal to	$\leq$
is greater than or equal to	$\geq$

A mathematical sentence which contains the verb "equals" is called an **equation.** Mathematical sentences that contain the verb phrases "is less than," "is greater than," "is greater than or equal to," and "is less than or equal to" are called **inequalities.**

6.2. Open Sentences

An example of an open sentence is

$$n + 9 = 14.$$

The letter n, in this case, is a symbol that stands for an unspecified number. We call such a symbol a **variable.** This open sentence is true if we replace the symbol, n, by 5. It is false for all other replacements for n. A mathematical sentence may contain one or more variables. The symbol that is used for the variable in a sentence is immaterial. For example, in the open sentence $n + 9 = 14$, we might have used $x, y, ?, \triangle$, or $\square$ for the variable. Thus

$$x + 9 = 14$$
$$y + 9 = 14$$
$$? + 9 = 14$$
$$\triangle + 9 = 14$$
$$\square + 9 = 14$$

are all forms of the same open sentence; only the symbol for the variable is different. The symbols $\triangle$ and $\square$ are called **frames.** They are usually used as symbols for variables in the primary grades.

6.3 Solution Sets

A set of numbers either implied or explicitly stated as permissible replacements of the variable in an open sentence is called the **replacement set,** the **domain of the variable,** or, simply, the **domain.** The subset of the replacement set whose elements make the open sentence a true statement is called the **solution set** or **truth set** of the sentence. Every element in the solution set is called a **solution.**

Let us consider the open sentence

$$5n + 2 = 17.$$

Let the domain of the variable be

$$D = \{1, 2, 3, 4, 5\}.$$

We can find the solution set of the sentence by trying each replacement:

Replace n by 1: $(5 \times 1) + 2 = 17$
                   $5 + 2 = 17$
                   $7 = 17$ False

Replace n by 2: $(5 \times 2) + 2 = 17$
                   $10 + 2 = 17$
                   $12 = 17$ False

Replace n by 3: $(5 \times 3) + 2 = 17$
                   $15 + 2 = 17$
                   $17 = 17$ True

Replace n by 4: $(5 \times 4) + 2 = 17$
                   $20 + 2 = 17$
                   $22 = 17$ False

Replace n by 5: $(5 \times 5) + 2 = 17$
                   $25 + 2 = 17$
                   $27 = 17$ False

We see that 3 is the only member of the domain for which the sentence $5n + 2 = 17$ is true. Hence the solution set is $\{3\}$ and 3 is the only solution.

Now let us find the solution set of the inequality

$$5n - 6 > 20.$$

Let the domain be

$$D = \{4, 5, 6, 7\}.$$

Again, replacing the variable by each element of the domain in turn we have:

Replace n by 4: $(5 \times 4) - 6 > 20$

$20 - 6 > 20$

$14 > 20$ False

Replace n by 5: $(5 \times 5) - 6 > 20$

$25 - 6 > 20$

$19 > 20$ False

Replace n by 6: $(5 \times 6) - 6 > 20$

$30 - 6 > 20$

$24 > 20$ True

Replace n by 7: $(5 \times 7) - 6 > 20$

$35 - 6 > 20$

$29 > 20$ True.

The solution set is $\{6, 7\}$ and the solutions of $5n - 6 > 20$ are 6 and 7.

Before we discuss methods for finding solution sets of equations and inequalities for which the domain is an infinite set, we must define equivalent sentences. Open sentences that have the same solution set are called **equivalent sentences.** Thus

$$2k + 6 = 18$$
$$2k = 12$$
and $$k = 6$$

are equivalent sentences since they have the same solution set, $\{6\}$.

The inequalities

$$2n + 5 < 21$$
$$2n < 16$$
$$n < 8$$

are also equivalent sentences since they have the same solution set $\{0, 1, \ldots, 7\}$.

When we apply the addition, subtraction, multiplication, and division properties of equality (Chapter 3, Section 3.2) to a given equation, the resulting equations are equivalent to the given equation. When we apply the addition, subtraction, multiplication, and division properties of the order relations (Chapter 3, Section 3.3) "is less than" and "is greater than" to a given inequality, the resulting inequalities are equivalent to the given inequality.

We now restate, for convenience, the properties of the equality relation and the order relations that we shall be using here. In each statement given below a, b, and c are whole numbers.

E-4. Addition Property. If $a = b$, then $a + c = b + c$ and $c + a = c + b$.

E-5. Subtraction Property. If $a = b$, $a \geq c$ and $b \geq c$, then $a - c = b - c$.

E-6. Multiplication Property. If $a = b$, then $ac = bc$ and $ca = cb$.

E-7. Division Property. If $a = b$ and $c \neq 0$ is a divisor of a and b, then $a \div c = b \div c$.

O-1. Addition Property. If $a < b$, then $a + c < b + c$ and $c + a < c + b$.

O-2. Subtraction Property. If $a < b$, $a \geq c$ and $b \geq c$, then $a - c < b - c$.

O-3. Multiplication Property. If $a < b$, $c \neq 0$, then $ac < bc$ and $ca < cb$.

O-4. Division Property. If $a < b$ and $c \neq 0$ is a divisor of both a and b, then $a \div c < b \div c$.

Now let us consider the equation

$$(1) \qquad\qquad 3n + 5 = 20$$

where the domain is the set of whole numbers.

Since the domain is an infinite set, it is impossible to replace n by every element in the domain as we did in the examples above. By the subtraction property of equality, we may subtract 5 from each member of equation (1) obtaining the equivalent equation

$$(2) \qquad\qquad 3n = 15$$

By the division property of equality we may divide each member of equation (2) by 3 obtaining the equivalent equation

$$n = 5.$$

The solution set of each equation above is $\{5\}$.

Now let us consider the inequality

(3) $$x - 7 > 5.$$

Let the domain be the set of whole numbers.

Again, since the domain is an infinite set, it is impossible to replace the variable x by every element in the domain. We recall that the properties stated for the relation "is less than" are also true for the relation "is greater than". If we replace the symbol $<$ in each of the statements O-1 through O-4, above, by $>$, the statements obtained by the replacement are true.

Let us add 7 to each member of inequality (3). We obtain the equivalent inequality

$$x > 12.$$

This tells us that x may be any whole number greater than 12. The solution set of the given inequality is

$$\{13, 14, 15, \ldots\}.$$

Now let us find the solution set of the inequality

$$x + 2 < 2.$$

The domain is the set of whole numbers.

By O-2 we may subtract 2 from each member of the inequality, obtaining the equivalent inequality.

$$x < 0.$$

This inequality says that x may be any whole number less than 0. But the least whole number is zero, hence there are no numbers in the given domain, the set of whole numbers, that satisfy this inequality. Therefore, the solution set is the empty set, ϕ.

In this chapter, unless otherwise stated the domain is the set of whole numbers.

EXAMPLE i. Find the solution set of $2k + 6 = 18$.
Solution: Since the domain is not specified, it is the set of whole numbers. By the subtraction property of equality (E-5) we may subtract 6 from both members of the equation to obtain the equivalent equation

$$2k = 12$$

By the division property of equality (E-7) we may divide both members of $2k = 12$ by 2, obtaining the equivalent equation

$$k = 6$$

The solution set is $\{6\}$.

EXAMPLE ii.   Find the solution set of $3n + 7 < 18$.
Solution: Subtracting 7 from both members of the given inequality (O-2) we obtain the equivalent inequality

$$3n < 11.$$

Since $3 \times 4 = 12$, n must be a whole number less than 4. The solution set is $\{0, 1, 2, 3\}$.

EXAMPLE iii.   Jack and Ray went fishing. Ray owns the boat so it was agreed that he would receive 5 more of the catch than Jack. If the total catch was less than 21 fish, what was the greatest number of fish Jack received?
Solution: Let n represent the number of fish Jack received. Since Ray received 5 more fish than Jack, he received $(n + 5)$ fish. The sum of the number of Jack's fish and the number of Ray's fish is less than 21. The conditions of the problem are then given by the inequality

$$n + (n + 5) < 21.$$

Since $n + n = 2n$, we have

$$2n + 5 < 21$$

Subtracting 5 from each member of this inequality (O-2) we obtain the equivalent inequality

$$2n < 16$$

Dividing each member of this sentence by 2 (O-4) we have

$$n < 8$$

We see that the solution set of the sentence $n + (n + 5) < 21$ is $\{0, 1, 2, 3, 4, 5, 6, 7\}$. Jack could receive 0, 1, 2, 3, 4, 5, 6, or 7 fish. The greatest number of fish that he could receive is 7.
Check: If Jack received 7 fish, Ray received $7 + 5 = 12$ fish, and the catch was $7 + 12 = 19$ fish. If Jack had received 8 fish, Ray would have received

$5 + 8 = 13$ fish. The catch in this case would have been $8 + 13 = 21$, which is impossible since the catch was less than 21 fish.

Exercise 6.1

1. Which of the following statements are true?

    a. $4 + 7 > 9$
    b. $3 + 6 = 9 - 7$
    c. $(26 \times 18) - 42 > 72$
    d. $97 + 83 = 77 + 43$
    e. $46 + 19 \leq 65$
    f. $(53 \times 7) + 9 \geq 40$
    g. $26(3 + 18) < 26 \times 24$
    h. $7(3 + 97) = 7 \times 100$

2. Write the following as mathematical sentences. Use the letter n for the variable.

    a. Seven plus a number is equal to sixteen.
    b. The sum of nine and some number is equal to fourteen.
    c. Twenty less than some number is sixty.
    d. This number plus twelve is greater than forty-six.
    e. Five times this number added to seven is less than sixty-nine.

3. Find the solution sets of the sentences below. The domain is the set of whole numbers.

    a. $n - 14 = 77$
    b. $n + 6 > 9$
    c. $2 + n < 16$
    d. $4t = 16$
    e. $k + 4 < 15$
    f. $2p + 5 < 40$
    g. $m + 7 \leq 11$
    h. $4 + k > 16$

4. Which of the following are equivalent to $3n + 2 < 20$?

    a. $x + 1 < 6$
    b. $n + 3 = 4$
    c. $2k + 1 < 12$
    d. $p < 6$
    e. $5n + 9 > 4$
    f. $3n < 16$
    g. $p + 9 = 10$
    h. $4n + 1 \leq 8$

5. What is the solution set of $x + 3 < 1$?
6. What is the solution set of $x^2 + 1 = 0$?
7. Translate the following number sentences into English sentences.

    a. $12 + 6 = 18$
    b. $17 < 3 + n$
    c. $3n = 18$
    d. $5 + n = 26$
    e. $7 + k > 9$
    f. $k - 3 \leq 8$
    g. $3k + 6 \geq 11$
    h. $7p + 5 = 26$

8. Find the solution set of $3n + 9 > 18$

9. Find the solution set of $2n + 3 \leq 12$
10. Find the solution set of $2k + 6 = 40$
11. Find the solution set of $n - 6 > 4$
12. Find the solution set of $y + 4 = 144$
13. Find the solution set of $2p > 70$
14. Find the solution set of $2r + 3 < 20$
15. What is the solution set of $2n + 3 = 9$ when the domain is
    a. $\{1, 2, 3, 4, 5\}$?
    b. $\{4, 5, 6, 7\}$?
    c. $\{0, 1, 2, 3\}$?
    d. $\{0, 1, 2, 3, \ldots\}$?
16. What is the solution set of $2n + 3 > 9$ when the domain is
    a. $\{1, 2, 3, 4, 5\}$?
    b. $\{4, 5, 6, 7\}$?
    c. $\{0, 1, 2, 3\}$?
    d. $\{0, 1, 2, 3, \ldots\}$?
17. What is the solution set of $5y + 3 < 18$ when the domain is
    a. $\{0, 1, 2, 3\}$?
    b. $\{3, 4, 5, 6, 7\}$?
    c. $\{0, 1, 2, 3, 4, 5, 6\}$?
    d. $\{0, 1, 2, 3, \ldots\}$?
18. What is the solution set of $3k + 6 < 18$ when the domain is
    a. $\{0, 1, 2, 3, 4, 5, 6\}$?
    b. $\{3, 4, 5, 6, 7\}$?
    c. $\{7, 8, 9, 10, 11, 12\}$?
    d. $\{0, 1, 2, 3, \ldots\}$?
19. What is the solution set of $n + 2_{\text{(eight)}} = 14_{\text{(eight)}}$ when the domain is the set of whole numbers?
20. What is the solution set of $2_{\text{(five)}} \cdot n > 13_{\text{(five)}}$ when the domain is the set of whole numbers?
21. A man weighs 40 pounds more than his twelve-year-old son. Their combined weight is less than 280 pounds. How much does the boy weigh?
22. Mr. Wright and Mr. Carlson agree to furnish at least $10,000 capital to start a new business. If Mr. Carlson furnishes $500 more than Mr. Wright, what is the least Mr. Wright can contribute?
23. Bud's golf score was 3 less than Kirk's. Bud has a score of 89. What is Kirk's score?

24. Mrs. Fry weighs 50 pounds more than Mrs. Slender. Their combined weight is at least 270 pounds. What is the least Mrs. Slender can weigh?
25. A chocolate marshmallow has 5 more calories than a regular marshmallow. Together they contain at least 57 calories. Find the smallest number of calories in a regular marshmallow.

6.4. Compound Open Sentences

Sometimes we are asked to find solution sets of compound open sentences. For example, we may be asked to find the solution set of the conjunction

$$n > 2 \text{ and } n < 9.$$

We know that a conjunction is true if and only if both of its components are true. Hence, the sentence $n < 2$ and $n > 9$ is true only when $n < 2$ is true *and* when $n > 9$ is true. Since the domain is the set of whole numbers, the solution set of

$$n > 2$$

is

$$\{3, 4, 5, 6, \ldots\}$$

The solution set of

$$n < 9$$

is

$$\{0, 1, 2, 3, 4, 5, 6, 7, 8\}.$$

Notice that the solution set of $n > 2$ is an infinite set; the solution set of $n < 9$ is a finite set.

Since the compound sentence $n > 2$ and $n < 9$ is true if and only if both $n > 2$ and $n < 9$ are true, the solution set of the compound sentence is the intersection of the two solution sets. That is, the solution set is

$$\{3, 4, 5, \ldots\} \cap \{0, 1, 2, 3, 4, 5, 6, 7, 8\}.$$

We see that the solution set of the given sentence is

$$\{3, 4, 5, 6, 7, 8\}.$$

The conjunction

$$n > 2 \text{ and } n < 9$$

is generally written in the shortened form

$$2 < n < 9$$

which is read: "2 is less than n and n is less than 9" or "n is greater than 2 and less than 9."

Now let us solve the disjunction

$$x > 7 \text{ or } x = 7$$

when the domain is the set of whole numbers.

The solution set of $x > 7$ is

$$\{8, 9, 10, \ldots\}$$

The solution set of $x = 7$ is

$$\{7\}$$

Recalling that a disjunction is true if one or the other or both of its components are true, the solution set of the disjunction $x > 7$ or $x = 7$ is the union of the solution set of $x > 7$ and the solution set of $x = 7$. That is, the solution set is

$$\{8, 9, 10, \ldots\} \cup \{7\} = \{7, 8, 9, 10, \ldots\}$$

The disjunction

$$x > 7 \text{ or } x = 7$$

is usually written in the shortened form

$$x \geq 7$$

which is read "x is greater than 7 or x is equal to 7" or "x is greater than or equal to 7."

Often we encounter situations involving even more complex sentences. For example, consider the problem: "Candidates for the Civic Mathematics Fellowship must be between the ages of 20 and 35 inclusive." Suppose we let x represent the age of the eligible candidate. He must be at least 20 years old; that is

$$20 \leq x.$$

But he also must not be older than 35 years old; that is

$$x \leq 35.$$

These conditions are described in the conjunction

$$20 \leq x \text{ and } x \leq 35$$

or

$$20 \leq x \leq 35.$$

This shortened form of the sentence is read: "20 is less than or equal to x and x is less than or equal to 35" or "x is greater than or equal to 20 and less than or equal to 35." The conjunction $20 \leq x \leq 35$ is made up of two disjunctions, $20 \leq x$ and $x \leq 35$.

Consider the disjunction

$$20 \leq x.$$

This says $20 < x$ or $20 = x$. The solution set of $20 < x$ is $\{21, 22, 23, \ldots\}$. The solution set of $x = 20$ is $\{20\}$. Since $20 \leq x$ is a disjunction, the solution set is the union of $\{21, 22, 23, \ldots\}$ and $\{20\}$. That is, the solution set is

$$\{20, 21, 22, 23, \ldots\}.$$

The disjunction $x \leq 35$ says

$$x < 35 \text{ or } x = 35.$$

The solution set of $x < 35$ is $\{0, 1, 2, 3, \ldots, 34\}$. The solution set of $x = 35$ is $\{35\}$. The solution set of $x \leq 35$ is the union of these two sets:

$$\{0, 1, 2, 3, \ldots, 35\}$$

The solution set of the conjunction $20 \leq x \leq 35$ is the intersection of the solution sets of $20 \leq x$ and $x \leq 35$, that is

$$\{20, 21, \ldots\} \cap \{0, 1, 2, 3, \ldots, 35\} = \{20, 21, 22, \ldots, 35\}$$

EXAMPLE i.   Find the solution set of $9 < x < 15$.
Solution: The given sentence is a conjunction. The components of the conjunction are $9 < x$ and $x < 15$. The solution set of $9 < x$ is

$$\{10, 11, 12, \ldots\}$$

The solution set of $x < 15$ is

$$\{0, 1, 2, \ldots, 14\}$$

The solution set of the conjunction is

$$\{10, 11, 12, \ldots\} \cap \{0, 1, 2, \ldots, 14\} = \{10, 11, 12, 13, 14\}$$

Example ii.   Find the solution set of $x \geq 8$ or $x < 5$.

Solution: This sentence is a disjunction. The components of the disjunction are $x \geq 8$ and $x < 5$. The component $x \geq 8$ is also a disjunction whose components are $x > 8$ and $x = 8$. The solution set of $x \geq 8$ is

$$\{8, 9, 10, \ldots\}$$

The solution set of $x < 5$ is

$$\{0, 1, 2, 3, 4\}$$

The solution set of the given disjunction is the union of these two solution sets:

$$\{8, 9, 10, \ldots\} \cup \{0, 1, 2, 3, 4\} = \{0, 1, 2, 3, 4, 8, 9, \ldots\}$$

6.5.   Graphing Solution Sets

We can picture the solution set of a number sentence in one variable on the number line. This picture is called the **graph** of the solution set. The graph of the solution set is the collection of points on the number line that represent the elements of the solution set.

Let us consider the sentence

$$x \leq 6$$

The domain is the set of whole numbers.

The solution set of $x \leq 6$ is

$$\{0, 1, 2, 3, 4, 5, 6\}$$

The graph of this solution set is shown in Figure 6.1.

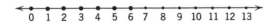

$$\begin{array}{ccccccccccccc} 0 & 1 & 2 & 3 & 4 & 5 & 6 & 7 & 8 & 9 & 10 & 11 & 12 & 13 \end{array}$$

Figure 6.1

Now let us graph the solution set of

$$2 < x < 9$$

The domain is the set of whole numbers.

The solution set of $2 < x < 9$ is

$$\{3, 4, 5, 6, 7, 8\}$$

The graph of this solution set is shown in Figure 6.2.

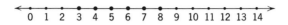

Figure 6.2

Let us graph the solution set of

$$x > 10$$

Again let the domain be the set of whole numbers.
The solution set of $x > 10$ is

$$\{11, 12, 13, 14, \ldots\}$$

The graph of this solution set is shown in Figure 6.3. Since this set is an infinite set it is impossible to show on the graph every point in the solution set. We graph a few points of the solution set and then draw an arrow as shown to indicate that all the points to the right which are graphs of whole numbers are also members of the solution set.

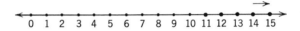

Figure 6.3

Exercise 6.2

Find the solution sets of the following compound sentences and graph each solution set. The domain is the set of whole numbers.

1. $x > 9$ and $x < 17$
2. $x \geq 8$ and $x \leq 14$
3. $x > 7$ and $x \leq 12$
4. $x > 9$ or $x = 9$
5. $x > 9$ or $x < 4$
6. $x \geq 8$ or $x \leq 15$
7. $x \geq 12$ or $x < 6$

8. $x \geq 5$
9. $x \leq 9$
10. $4 < x < 8$
11. $2 \leq x \leq 10$
12. $3 < n \leq 15$
13. $y + 11 \geq 27$
14. $n + 3 \leq 18$
15. $k + 5 \geq 7$
16. How do you read the following?
    a. $2 < x < 16$
    b. $3 \leq x \leq 15$
    c. $p \geq 17$
    d. $k \leq 24$

6.6. Sentences with Two Variables

Even at the primary level one encounters problems which are solved by sentences involving more than one variable. Suppose a child is told that he may have five cookies from a plate containing round chocolate cookies and square gingerbread cookies. Since there are two kinds of cookies, the number sentence used to solve this problem contains two variables. If x represents the number of chocolate cookies and y represents the number of gingerbread cookies, the number sentence representing this problem is

$$x + y = 5.$$

Since this sentence involves two variables, each solution in the solution set is an ordered pair of the form (x, y). Since the child is only permitted to have five cookies, the domain of each variable is $\{0, 1, 2, 3, 4, 5\}$. If the first choice is one chocolate cookie and four gingerbread cookies, we have the ordered pair $(x, y) = (1, 4)$ which is a solution of the sentence. All possible replacements for x and y that give solutions are shown in Table 6.1.

The solution set is the set of ordered pairs

$$\{(0, 5), (1, 4), (2, 3), (3, 2), (4, 1), (5, 0)\}.$$

Now let us solve the sentence

$$x + y = 9$$

Table 6.1

x	y
0	5
1	4
2	3
3	2
4	1
5	0

where the domain of each variable is the set of whole numbers. Each solution of this sentence is an ordered pair of the form (x, y).

The greatest value x can have is 9 since no whole number greater than 9 can be added to another whole number and give a sum of 9. The solution set of this sentence is

$$\{(0, 9),(1, 8), (2, 7), (3, 6), (4, 5), (5, 4), (6, 3), (7, 2), (8, 1), (9, 0)\}$$

Now let us find the solution set of

$$x + y < 8.$$

The domain of the variables is the set of whole numbers.

We see that x can never be greater than 7, hence the possible replacements for x are 0, 1, 2, 3, 4, 5, 6, and 7.

If x is replaced by 0, we have $0 + y < 8$. We see in this case that y can be any whole number less than 8. Thus when x is replaced by 0, y can be replaced by 0, 1, 2, 3, 4, 5, 6, and 7 and true statements result. These replacements of x and y give the ordered pairs.

$$(0, 0), (0, 1), (0, 2), (0, 3), (0, 4), (0, 5), (0, 6), (0, 7).$$

If x is replaced by 1, we have $1 + y < 8$. We see that in this case y can be replaced by any whole number less than 7. These replacements of x and y give the ordered pairs

$$(1, 0), (1, 2), (1, 3), (1, 4), (1, 5), (1, 6).$$

If x is replaced by 2, we have $2 + y < 8$. We see that with this replacement of x, y can be any whole number less than 6. These replacements of x and y give the ordered pairs

$$(2,0), (2, 1), (2, 2), (2, 3), (2, 4), (2, 5).$$

If x is replaced by 3, we have $3 + y < 8$. We see that with this replacement of x, y can be any whole number less than 5. These replacements of x and y give the ordered pairs

$$(3, 0), (3, 1), (3, 2), (3, 3), (3,4).$$

Continuing in this fashion we find the solution set:

$$\left\{\begin{array}{l} (0, 7), (0, 6), (0, 5), (0, 4), (0, 3), (0, 2), (0, 1), (0, 0), \\ (1, 6), (1, 5), (1, 4), (1, 3), (1, 2), (1, 1), (1, 0), \\ (2, 5), (2, 4), (2, 3), (2, 2), (2, 1), (2, 0) \\ (3, 4), (3, 3), (3, 2), (3, 1), (3, 0), \\ (4, 3), (4, 2), (4, 1), (4, 0), \\ (5, 2), (5, 1), (5, 0), \\ (6, 1), (6, 0), \\ (7, 0) \end{array}\right\}$$

6.7. Graphing Ordered Pairs

In graphing ordered pairs we need two number lines, one placed horizontally and the other vertically, which are perpendicular to each other and intersect at the point of each line that has coordinate 0. (Figure 6.4.) These number lines are called **axes**. The horizontal axis is called the **first axis** or the **x-axis**. The vertical axis is called the **second axis** or the **y-axis**. We now draw lines parallel to the two axes to form a **lattice** called a **lattice plane**, as shown in Figure 6.4. The intersections of these lines are called **lattice points**. Each lattice point represents an ordered pair (x, y) of whole numbers. Each lattice point is then the graph of an ordered pair of whole numbers.

To find the graph of an ordered pair, for example, $(2, 5)$, we find the graph of 2 on the x-axis and the graph of 5 on the y-axis. The vertical line through the graph of 2 on the x-axis intersects the horizontal line through the graph of 5 on the vertical axis in the lattice point that is the graph of the ordered pair $(2, 5)$.

The points of the axes themselves have two different coordinates: the

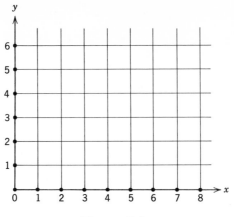

Figure 6.4

whole numbers which were their coordinates on the number lines and the ordered pairs of whole numbers which are their coordinates as lattice points of the lattice plane. All the points on the vertical axis have coordinates whose first components are 0. All the points on the horizontal axis have coordinates whose second components are 0.

Some of the ordered pairs represented on the lattice in Figure 6.5 are labeled. Using a lattice plane, we can graph the solution set of a sentence containing two variables.

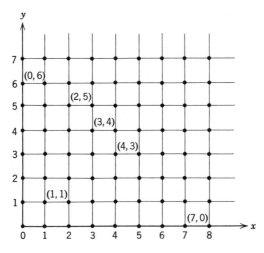

Figure 6.5

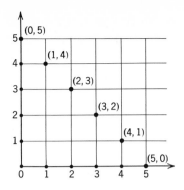

Figure 6.6

The solution set of

$$x + y = 5$$

was found to be

$$\{(5, 0), (4, 1), (3, 2), (2, 3), (1, 4), (0, 5)\}.$$

The graph of this solution set is shown in Figure 6.6.

Let us solve the sentence

$$x = y$$

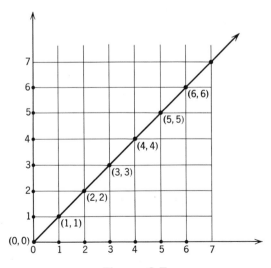

Figure 6.7

and graph the solution set. Since $x = y$, the ordered pairs which are solutions of this sentence must have the first components equal to the second components. Some solutions are (2, 2), (5, 5), and (8, 8).

The solution set consists of every ordered pair of the form (a, a), a a whole number. The solution set of this sentence is an infinite set.

Now let us graph this solution set. Since the solution set is an infinite set it is impossible to show every element of the solution set on the graph. We graph some of the solutions. We observe that they all lie on a ray whose endpoint is the graph of (0, 0). The graph of the solution set is the set of all lattice points on the ray drawn in the lattice in Figure 6.7.

EXAMPLE i.   Find the solution set of $3x = y$ and graph.

Solution: Since $3x = y$, all ordered pairs in the solution set are of the form $(a, 3a)$, a a whole number. Some solutions are (0, 0), (1, 3), (2, 6), (3, 9). The graph of the solution set (Figure 6.8) is the set of all lattice points on the ray marked on the lattice.

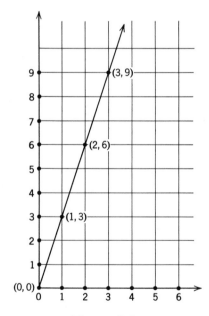

Figure 6.8

EXAMPLE ii.   Find the solution set $x > y$ and graph.

Solution: The solution set is the set of all ordered pairs of whole numbers whose x component is greater than the y component. This set of ordered pairs is an infinite set. Some of the solutions of the sentence are

$$(1, 0), (2, 0), (2, 1), (3, 0), (3, 1)$$

The graph consists of all the lattice points below the dotted line in Figure 6.9. That is all the lattice points in the shaded area.

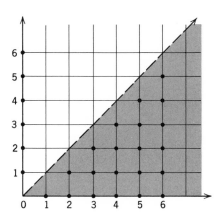

Figure 6.9

Exercise 6.3

1. Graph the following ordered pairs.
   a. $(7, 3)$ f. $(5, 9)$
   b. $(1, 1)$ g. $(9, 2)$
   c. $(2, 1)$ h. $(1, 7)$
   d. $(3, 1)$ i. $(0, 0)$
   e. $(4, 7)$ j. $(3, 4)$

   Find the solution sets and graph. The domain of the variables is $\{1, 2, 3, 4, 5\}$ (Ex. 2–6).
2. $x + y = 9$
3. $x + y = 3$
4. $x + y > 5$
5. $2x + y > 11$

6. $2x > y$

Find the solutions sets and graph. The domain of the variables is the set of whole numbers (Ex. 7–20).

7. $x + y = 8$
8. $2x + y < 12$
9. $x + y < 10$
10. $3x + y < 14$
11. $5x + 2y < 18$
12. $2x = y$
13. $2x + 4y < 15$

14. $xy = 4$
15. $3x + 2y < 10$
16. $y + x = 1$
17. $2x + y = 8$
18. $x = 2y$
19. $3x + 2y = 18$
20. $3x + 5y < 13$

7.1. Figurate Numbers

Number patterns intrigue both mathematicians and people who claim to know no mathematics at all. Even though they may not attach any special significance to them, it amuses most people to see a car with a license number ABC-123 or IAM-023, the telephone number 234-5678 or a dollar bill with the serial number G56565656W.

The ancient Greeks were similarly fascinated by special numbers. Since they were particularly interested in geometry, they singled out those numbers associated with geometric figures. Such numbers are called **figurate** or **polygonal numbers.**

Numbers are called **triangular** if they can be pictured as triangles. The first few triangular numbers, 1, 3, 6, 10 and 15, are shown in Figure 7.1.

Square numbers are those that can be represented as squares as shown in Figure 7.2. The first four square numbers are 1, 4, 9 and 16.

Topics from
Number Theory

VII

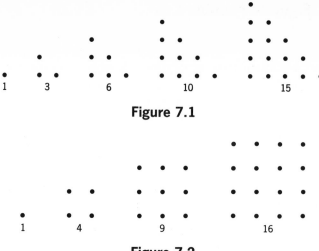

Figure 7.1

Figure 7.2

Pentagonal numbers are those that can be pictured in the form of a pentagon as shown in Figure 7.3. Notice that the pentagonal numbers greater than 1 are represented by a square with a triangle on top of it.

Similarly, there are **hexagonal numbers, rectangular numbers,** and other figurate numbers.

In looking at the representations of the square numbers, it is easy to see that the first square number is $1 = 1^2$; the second is $4 = 2 \times 2 = 2^2$; the third is $9 = 3 \times 3 = 3^2$; and so on. Drawing pictures representing other square numbers, we are led to the generalization that the sixth square number is $6^2 = 6 \times 6 = 36$; the tenth square number is $10^2 = 10 \times 10 = 100$; and the nth square number is n^2. This explains why we read n^2 as "n squared".

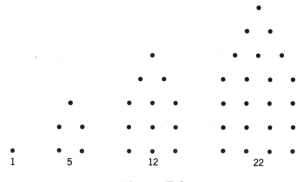

Figure 7.3

It is not difficult to discover a formula for finding the nth triangular number. If we observe the pictures of the representations of the triangular numbers, we see that the number of dots in the representation of the first triangular number is 1; of the second $1 + 2$; of the third $1 + 2 + 3$; and so on. The representation of each successive triangular number is thus obtained by adding another row of dots containing one more dot than the bottom row of the representation of the previous triangular number.

This leads us to discover that each triangular number after 1, which is the first figurate number of any form, is found by adding the next counting number to the sum that names the previous triangular number. Thus

1st	1
2nd	$1 + 2$
3rd	$1 + 2 + 3$
4th	$1 + 2 + 3 + 4$
	$\vdots$
kth	$1 + 2 + 3 + 4 + \cdots + k$

where k is a counting number.

This method of finding any desired triangular number entails a good deal of work if k is large. To simplify this task we shall develop a formula for finding the kth triangular number.

It is a well known fact that the area of a rectangular is given by the formula $A = lw$, where l is the length and w is the width. Look at Figure 7.4.

If we consider the shaded region formed by squares we find that

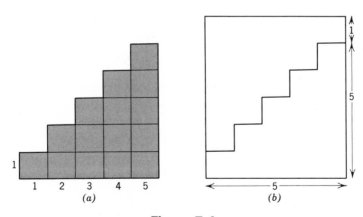

Figure 7.4

its area is equal to 15 square units; that is,

$$1 + 2 + 3 + 4 + 5 = 15.$$

If we take two such regions and fit them together as shown in Figure 7.4b, we obtain a rectangular region whose width is 5 and whose length is $5 + 1 = 6$, and whose area is $5 \times (5 + 1) = 30$. In other words, twice the sum of $1 + 2 + 3 + 4 + 5$ is equal to $5 \times (5 + 1)$.

We can use this argument to find the formula for the kth triangular number, which is given by the sum $1 + 2 + 3 + \cdots + k$. If we draw a figure like the one in Figure 7.4a but which has k square units in the bottom row, its total area will be $1 + 2 + 3 + \cdots + k$. Combining two such figures in the same way as before, we obtain a rectangular region whose length is $(k + 1)$ and whose width is k. The area of this rectangular region is $k(k + 1)$. This represents twice the sum of the counting numbers from 1 through k. Thus

$$2(1 + 2 + 3 + \cdots + k) = k(k + 1),$$

and the sum of the first k natural numbers is

$$1 + 2 + 3 + \cdots + k = \frac{k(k + 1)}{2}.$$

THEOREM 7.1. **The kth triangular number is given by**

$$\frac{k(k + 1)}{2}$$

where k is a natural number.

EXAMPLE i. Find the seventeenth square number.
Solution: The seventeenth square number is

$$17^2 = 17 \times 17 = 289.$$

EXAMPLE ii. Find the twentieth triangular number.
Solution: Using the formula

$$\frac{k(k + 1)}{2}$$

with k replaced by 20 we have

$$\frac{20(20 + 1)}{2} = \frac{20 \times 21}{2} = 210$$

EXAMPLE iii.   What is the sum of the first two hundred natural numbers?

Solution: We are asked to find

$$1 + 2 + \cdots + 200.$$

That is, we are looking for the 200th triangular number.
Using the formula

$$\frac{k(k + 1)}{2}$$

and replacing k by 200 we have

$$\frac{200(200 + 1)}{2} = \frac{200 \times 201}{2} = 20,100$$

7.2.   Subsets of the Set of Whole Numbers

If we have a set we can separate it into two or more disjoint subsets in many ways. For example, if we consider the set of persons in a mathematics class, we may separate this set into two disjoint subsets: the subset of women and the subset of men; or we may separate it into three disjoint subsets: the subset of persons over 21 years of age, the subset of persons 21 years of age and the subset of persons less than 21 years of age.

Similarly we may separate the set of whole numbers into disjoint subsets in various ways. We are going to study some of these special subsets of whole numbers.

All whole numbers may be classified as either **even** or **odd**. Even numbers are those numbers which have a remainder of 0 when divided by 2. Even numbers, therefore, are multiples of 2.

DEFINITION 7.1.   *A whole number is* **even** *if it is a multiple of 2.*

Every even number may be represented by $2n$ where n is a whole number. Thus

$$0 = 2 \times 0$$
$$4 = 2 \times 2$$
$$12 = 2 \times 6$$
$$58 = 2 \times 29$$

When a whole number is divided by 2, the remainder is either 0 or 1. Since even numbers are multiples of 2, the remainder when an even number is divided by 2 is 0. All whole numbers that have a remainder 1 when divided by 2 are called **odd** numbers.

DEFINITION 7.2. *A whole number is* **odd** *if it has a remainder of 1 when divided by 2.*

All odd numbers may be represented by $2n + 1$ where n is a whole number. Thus

$$1 = (2 \times 0) + 1$$
$$5 = (2 \times 2) + 1$$
$$17 = (2 \times 8) + 1$$
$$39 = (2 \times 19) + 1$$

7.3. Properties of Even and Odd Numbers

We can convince ourselves by observing examples such as

$$2 + 2 = 4 = 2 \times 2$$
$$4 + 6 = 10 = 2 \times 5$$
$$8 + 16 = 24 = 2 \times 12$$
$$24 + 36 = 60 = 2 \times 30$$

that the sum of two even numbers is always an even number.

Similarly, we see that the product of two even numbers is an even number. Observe that:

$$12 \times 6 = 72 \ = 2 \times 36$$
$$8 \times 4 = 32 \ = 2 \times 16$$
$$42 \times 6 = 252 = 2 \times 126$$

It is quite easy to prove that the sum and the product of two even numbers are even numbers.

THEOREM 7.2.   The sum of two even numbers is an even number.
Proof: Let two even numbers be represented by $2p$ and $2s$, where p and s are whole numbers. Then

$$2p + 2s = 2(p + s)$$

by the distributive property. Since p and s are whole numbers $p + s$ is a whole number by the closure property of addition. Let $p + s = n$. Then

$$2p + 2s = 2(p + s) = 2n$$

which is an even number.

THEOREM 7.3.   **The product of two even numbers is an even number.**
Proof: Let two even numbers be represented by $2p$ and $2s$ where p and s are whole numbers. Then

$$\begin{aligned}
2p \cdot 2s &= 2 \times (p \times 2 \times s) \\
&= 2 \times (2 \times p \times s) \\
&= 2 \times (2ps)
\end{aligned}$$

by the associative and commutative properties of multiplication. Since 2, p, and s are whole numbers their product is a whole number by the closure property of multiplication. Let $2ps = n$. Then

$$2p \cdot 2s = 2(2ps) = 2n$$

which is an even number.

In a similar fashion we can prove the following theorems.

THEOREM 7.4.   **The product of two odd numbers is an odd number.**
THEOREM 7.5.   **The sum of two odd numbers is an even number.**
THEOREM 7.6.   **The sum of an odd number and an even number is an odd number.**
THEOREM 7.7.   **The product of an odd number and an even number is an even number.**

The addition and multiplication tables below (Table 7.1) show in symbolic form the results of the theorems above. In the tables, E represents an even number and O represents an odd number.

Table 7.1

+	E	O
E	E	O
O	O	E

×	E	O
E	E	E
O	E	O

EXAMPLE i. Using Table 7.1 show that $O + (E \times O) + (O \times O)$ represents an even number.

Solution: $\quad\quad O + (E \times O) + (O \times O) = (O + E) + O$
$$= O + O$$
$$= E$$

EXAMPLE ii. Is the product of three odd numbers and an even number an odd number or an even number?

Solution: We have

$$O \times O \times O \times E = (O \times O) \times (O \times E)$$
$$= O \times E$$
$$= E$$

The product is an even number.

Exercise 7.1

1. Draw figures, using dots of the first ten triangular numbers.
2. Draw figures, using dots of the first seven pentagonal numbers.
3. Find the following square numbers.

   a. 4th   d. 15th
   b. 7th   e. 25th
   c. 12th   f. 100th

4. Find the following triangular numbers.

   a. 15th   d. 54th
   b. 27th   e. 150th
   c. 36th   f. 225th

5. Write the following even numbers in the form $2n$. In each case give the value of n.

   a. 18   d. 256   g. 8000
   b. 56   e. 500   h. 14,322
   c. 114   f. 1852   i. 768,398

6. Write the following odd numbers in the form $2n + 1$. In each case give the value of n.

   a. 9   d. 99   g. 5365
   b. 15   e. 113   h. 12,117
   c. 27   f. 279   i. 427,101

7. List the elements of the set of even numbers greater than 12 and less than 50.
8. List the elements of the set of odd numbers greater than 83 and less than 99.
9. Use Table 7.1. Tell whether the following are odd or even numbers.
    a. $E \times (O + O)$
    b. $(E + O) \times (E + E)$
    c. $(E \times O) + (O \times O)$
    d. $O \times (E + E) \times (O + E)$
10. Which of the following always represent even numbers? Which always represent odd numbers (n represents a natural number)?
    a. n d. $2n - 1$
    b. $2n$ e. $2n + 2$
    c. $2n + 1$ f. $n + 1$
11. Is $1001_{(two)}$ even or odd?
12. Is $12_{(three)}$ even or odd?
13. What number is a factor of every natural number?
14. What whole number greater than 1 is a factor of every even natural number?
15. Which of the following sets are closed under the operation of addition? multiplication?
    a. The set of even numbers.
    b. The set of odd numbers.
16. Which of the following are true statements?
    a. Adding 1 to any odd number gives an even number.
    b. Adding 1 to any even number gives an even number.
    c. The product of four even numbers is an even number.
    d. The product of two even numbers and two odd numbers is an odd number.
    e. Subtracting one from an odd number gives an even number.
    f. An even number always has a factor 2.
    g. The quotient of an odd number divided by one is an odd number.
    h. The quotient of an even number divided by itself is an odd number.
17. Prove Theorem 7.4.
18. Prove Theorem 7.5.
19. Prove Theorem 7.6.
20. Prove Theorem 7.7.

7.4 Prime Numbers and Composite Numbers

We have just shown that the set of whole numbers may be separated into two disjoint sets, the set of even numbers and the set of odd numbers.

Let us now omit zero from the set of whole numbers and consider the set of natural numbers:

$$\{1, 2, 3, \ldots\}.$$

Let us separate this set of natural numbers into three disjoint sets:

(1) The set A containing those natural numbers that have exactly one divisor.
(2) The set P containing those natural numbers that have exactly two divisors.
(3) The set C containing those natural numbers that have more than two divisors.

The set A contains only one element, the number 1. The set P contains those numbers that are divisible only by 1 and themselves. Such numbers are called **prime numbers** or **primes.**

$$P = \{2, 3, 5, 7, 11, 13, 17, 19, 23, \ldots\}$$

All the other natural numbers are in set C:

$$C = \{4, 6, 8, 9, 10, 12, 14, 15, 16, \ldots\}$$

The numbers in set C are called **composite numbers.**

Notice that the number 2 belongs to P because the only divisors of 2 are 1 and 2. The next number that belongs to P is 3 because it has only 1 and 3 as divisors. The number 4 belongs to set C because it has divisors 1, 2, and 4.

There are infinitely many prime numbers and infinitely many composite numbers. The only even prime is 2. Why?

DEFINITION 7.3. *A whole number $p > 1$ is a **prime number** if it has only two whole number divisions, 1 and itself.*

DEFINITION 7.4. *A whole number $p > 1$ is a **composite number** if it has more than two whole number divisors.*

A characteristic property of a composite number, m, consists of the possibility of representing it as the product of two factors, a and b,

$$m = ab$$

each of which is greater than 1. For example

$$4 = 2 \times 2 \qquad 10 = 2 \times 5$$
$$8 = 2 \times 4 \qquad 27 = 3 \times 9$$
$$9 = 3 \times 3 \qquad 56 = 7 \times 8$$

Such a representation is impossible for a prime number.

7.5. The Sieve of Eratosthenes

About 2200 years ago a Greek mathematician and scientist devised a scheme, called the **Sieve of Eratosthenes,** for finding all the prime numbers among the natural numbers less than some particular natural number. His method sieved out all the numbers that are not primes and left only the primes. Let us use his method to find all the prime numbers less than 100. We write down all the natural numbers from 2 to the number n which is to be tested (100 in our case). Two is the least prime, and the multiples of 2:

$$4, 6, 8, 10, \ldots$$

occur in the list of numbers at intervals of two following 2. Thus we scratch from the list every second number after 2, all of which are composite numbers because they have divisor 2 in addition to themselves and 1.

Now 3, the next number not scratched out, is a prime. Again, multiples of 3 occur in the list at intervals of three following 3; so we scratch out every third number after 3.

The next number not scratched out is 5 which is a prime. Again, multiples of 5 occur in the list at intervals of five following 5. We scratch out every fifth number after five.

Continuing in this fashion we find all the prime numbers less than n. Table 7.2 shows the completed sieve for $n = 100$. The primes less than 100 have been circled in the table.

In performing the procedure for finding all of the primes less than 100, we observe that it is not necessary to strike out multiples of 11 because these numbers were scratched out in previous steps. This was also true of multiples of 13, 17, 19, and all primes greater than 7.

In searching for the primes less than 100 the largest prime necessary to use in the sieving process was 7. We ask ourselves if this was a coincidence.

Table 7.2

Sieve of Eratosthenes for $n = 100$

If we wish to use the sieve method to find all of the primes less than, for example 200, would 7 be the largest prime necessary to use or would it be some other greater prime?

We are always sure to reach the end of the sieve process when we have crossed out the multiples of the prime p, where p is the greatest prime such that $p^2 \leqslant n$. This follows because if $n = ab$ is a composite number then at least one of the factors, say a, must be such that $a^2 \leqslant n$ and the other factor, b, is such that $b^2 \geqslant n$. That is, $a \leqslant \sqrt{n}$ ($\sqrt{n}$ is read: "the square root of n." If $\sqrt{n} = a$ then $a^2 = n$; thus $\sqrt{25} = 5$ because $5^2 = 25$). If both a and b were greater than $\sqrt{n}$, we would have $n = ab > \sqrt{n}\sqrt{n} = n$, an obvious contradiction. Hence if n is not crossed out when the proper multiples of p (and of all smaller primes) have been eliminated, then n must be a prime.

This gives a practical test to ascertain whether a given number is a prime number. It suffices to divide it by the primes less than or equal to its square root. If one divisor succeeds without a remainder, then the number is composite, otherwise it is a prime.

EXAMPLE i.   Is 101 a prime number?
Solution: Since $\sqrt{101}$ is between 10 and 11 ($10^2 = 100$ and $11^2 = 121$) we need test only primes not exceeding 10. These primes are 2, 3, 5, and 7.

Dividing 101 by each of these in turn we find that none is a divisor of 101. Hence 101 is a prime number.

EXAMPLE ii.   Is 681 a prime number?

Solution: Since $26^2 = 676$ and $27^2 = 729$, $\sqrt{681}$ is between 26 and 27. We test those primes not exceeding 26, to see whether or not they are divisors of 681. These primes are 2, 3, 5, 7, 11, 13, 17, 19 and 23. Since 3 divides 681 without a remainder, 3 is a divisor of 681 and 681 is not a prime.

7.6.   Factors

In the expression $3 \times 5 = 15$, 3 and 5 are called **factors** and 15 is called the **product.** When we write 3×5 as another name for 15, we say that we are writing 15 in **factored form.** In general if $n = ab$, a and b are factors of n and ab is a factored form of n.

Prime numbers may be written in factored form in only one way except for the order in which the factors are written. If p is a prime, then the only factored form of p is $1 \times p$. Thus

$$7 = 1 \times 7$$
$$11 = 1 \times 11$$
$$101 = 1 \times 101$$

Composite numbers may be written in factored form in more than one way. Some factored forms of 24 are:

$$1 \times 24$$
$$2 \times 12$$
$$3 \times 8$$
$$2 \times 2 \times 2 \times 3$$

It is observed that 1 and the number itself are always factors of a given number.

It is possible to find all the whole number factors of a given number by writing it as the product of two whole number factors in as many ways as possible (disregarding the order of the factors). Suppose we wish to find all the whole number factors of 48. Since 48 is a composite number it may be written as the product of two whole numbers one of which is less than or equal to $\sqrt{48}$. Since $\sqrt{48}$ is between 6 and 7 ($6^2 = 36$ and $7^2 = 49$) one factor in every factored form of 48 must be less than or equal to 6. All the possible factored forms of 48 having two whole number factors are (again disregarding the order of the factors):

$$1 \times 48$$
$$2 \times 24$$
$$3 \times 16$$
$$4 \times 12$$
$$6 \times 8$$

All the whole number factors of 48 are 1, 2, 3, 4, 6, 8, 12, 16, 24, and 48.

EXAMPLE i.   Find all the whole number factors of 81.
Solution: We write 81 as the product of two whole number factors in as many ways as possible (disgarding the order of the factors). Since $9^2 = 81$, $\sqrt{81} = 9$ and at least one of the two whole number factors of 81 must be less than or equal to 9.

$$1 \times 81$$
$$3 \times 27$$
$$9 \times 9$$

The whole number factors of 81 are 1, 3, 9, 27, and 81.

EXAMPLE ii.   Find all the whole number factors of 500.
Solution: We write 500 as the product of two whole number factors in as many ways as possible (disregarding the order of the factors). Since $\sqrt{500}$ is between 22 and 23 ($22^2 = 484$ and $23^2 = 529$) one factor must be less than or equal to 22:

$$1 \times 500$$
$$2 \times 250$$
$$4 \times 125$$
$$5 \times 100$$
$$10 \times 50$$
$$20 \times 25$$

The whole number factors of 500 are 1, 2, 4, 5, 10, 20, 25, 50, 100, 125, 250, and 500.

Another method of factoring a natural number is by constructing a **factor tree.** A factor tree is a means of showing how a natural number is built from small factors. The construction of a factor tree for 48 is shown below. We start by showing 48 as the product of two factors:

By extending the drawing, 48 is pictured as $2 \times 3 \times 8$:

In a further extension of this drawing, 48 is pictured as $2 \times 3 \times 2 \times 4$:

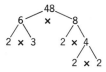

Another extension shows 48 written as the product of prime factors.

When a composite number is written as the product of prime factors we say that it is **factored completely.**

7.7.   The Fundamental Theorem of Arithmetic

A prime number can be expressed as the product of natural numbers in only one way (disregarding the order in which the factors are written), namely the product of itself and 1. Thus

$$7 = 7 \times 1$$
$$11 = 11 \times 1$$
$$23 = 23 \times 1$$
$$101 = 101 \times 1$$

A composite number has more than one factored form. For example, some factored forms of 48 are:

$$1 \times 48$$
$$2 \times 24$$
$$3 \times 16$$
$$4 \times 12$$
$$6 \times 8$$
$$2 \times 3 \times 8$$
$$2 \times 3 \times 4 \times 2$$
$$2 \times 2 \times 2 \times 2 \times 3$$

The factored form $2 \times 2 \times 2 \times 2 \times 3 = 2^4 \times 3$ is called the **complete factorization** or **prime factorization** of 48. It expresses 48 as the product of primes.

Every composite number can be factored, that is, it can be written as the product of at least two factors each of which is less than the original number. If one or more of these factors is composite, it can be written as the product of still smaller factors. This process cannot go on indefinitely since the factors are getting smaller and the least natural number is 1. Eventually we must come to a factored form each of whose factors is a prime. For example:

$$144 = 6 \times 24$$
$$= 2 \times 3 \times 24$$
$$= 2 \times 3 \times 3 \times 8$$
$$= 2 \times 3 \times 3 \times 4 \times 2$$
$$= 2 \times 3 \times 3 \times 2 \times 2 \times 2$$
$$144 = 12 \times 12$$
$$= 2 \times 6 \times 12$$
$$= 2 \times 2 \times 3 \times 12$$
$$= 2 \times 2 \times 3 \times 3 \times 4$$
$$= 2 \times 2 \times 3 \times 3 \times 2 \times 2$$
$$144 = 8 \times 18$$
$$= 2 \times 4 \times 18$$
$$= 2 \times 2 \times 2 \times 18$$
$$= 2 \times 2 \times 2 \times 2 \times 9$$
$$= 2 \times 2 \times 2 \times 2 \times 3 \times 3$$

Notice that although in each case above we started with a different factored form of 144, the complete factorization is the same except for the order in which the prime factors are written.

This example leads to the statement of the **Fundamental Theorem of Arithmetic.**

THEOREM 7.8. THE FUNDAMENTAL THEOREM OF ARITHMETIC. Every composite number can be written as a product of prime numbers in one and only one way except for the order in which the prime factors are written.

We shall accept the Fundamental Theorem of Arithmetic without proof.

Exercise 7.2

1. Define a prime number.
2. List the even primes.
3. What is the union of the set of prime numbers and the set of composite numbers?
4. What is the intersection of the set of prime numbers and the set of composite numbers?
5. Use the Sieve of Eratosthenes to find all the primes less than 200.
6. Which of the following are prime numbers?
   a. 127   c. 319   e. 649
   b. 313   d. 403   f.   1321
7. Make a factor tree factoring each of the following into prime factors.
   a. 72   b. 68   c. 188   d. 436
8. In 1742 a mathematician named Goldbach conjectured that every even number greater than 4 is the sum of two odd primes. To this day, nobody has proved or disproved Goldbach's Conjecture. Write the following even numbers as the sum of two odd prime numbers.
   a. 14   c. 28   e. 144
   b. 18   d. 68   f.   268
9. When a whole number is divided by 3, the possible remainders are 0, 1, and 2. Let A be the set of whole numbers that have a remainder 0 when divided by 3. Let B be the set of whole numbers that have a remainder 1 when divided by 3. Let C be the set of whole numbers that have a remainder 2 when divided by 3.
   a. What are the four least members of A?
   b. What are the four least members of B?

c. What are the five least members of C?

d. Are A, B, and C disjoint or overlapping sets?

e. Use several numerical examples to demonstrate the truth that the sum of a member of B and a member of C is a member of A.

f. Use several numerical examples to demonstrate the truth that the sum of a member of A and a member of B is a member of B.

g. Use several numerical examples to demonstrate the truth that the product of any two members of A is a member of A.

10. Which of the following have a factor 3?

a. 374   c. 861   e. 8043

b. 627   d. 7061   f. 87,765

11. List all of the whole number factors of the following.

a. 12   c. 18   e. 36

b. 17   d. 29   f. 45

12. Two odd primes whose difference is 2 are called **twin primes.** For example, 3 and 5 are twin primes since $5 - 3 = 2$. Give five examples of pairs of twin primes.

13. Every prime number of the form $4n + 1$, n a natural number, may be represented as the sum of two squares. For example $5 = (4 \times 1) + 1$ and $5 = 2^2 + 1^2$. Write the following primes as the sum of two squares.

a. 17   c. 13   e. 89

b. 41   d. 97   f. 61

14. State the Fundamental Theorem of Arithmetic.

15. What are the ones digits of the numerals of prime numbers greater than 5? Decide on your answer by looking at the primes less than 200 found in Problem 5.

7.8. Rules of Divisibility

Before discussing methods for factoring composite numbers completely, we shall discuss rules for divisibility by 2, 3, 5, 9, and 10. The justification of these rules depend upon the properties of the whole numbers, the place-value system of numeration used to name the numbers, and the following theorem which we shall accept without proof.

THEOREM 7.5. Divisibility Property of a Sum. For all whole numbers a, b, and c, $c \neq 0$, if a and b are divisible by c, then their sum, $a + b$ is divisible by c.

The numeral for every whole number n with $(k + 1)$ digits can be written in expanded notation as

$$n = a \cdot 10^k + b \cdot 10^{k-1} + c \cdot 10^{k-2} + \cdots$$
$$+ m \cdot 10^3 + h \cdot 10^2 + t \cdot 10^1 + u \cdot 10^0$$

where each of $a, b, c, \ldots, m, h, t,$ and u is one of the digits $0, 1, 2, \ldots, 9$. For example

$$9{,}683 = 9 \cdot 10^3 + 6 \cdot 10^2 + 8 \cdot 10^1 + 3 \cdot 10^0$$
$$236{,}872 = 2 \cdot 10^5 + 3 \cdot 10^4 + 6 \cdot 10^3 + 8 \cdot 10^2 + 7 \cdot 10^1 + 2 \cdot 10^0.$$

With this in mind, rules for divisibility by 2, 3, 5, 9, and 10 are easily derived when the numbers in question are written in base ten.

Divisibility by 2

Let us consider any whole number n. Then

$n = \cdots + m \cdot 10^3 + h \cdot 10^2 + t \cdot 10^1 + u \cdot 10^0$	expanded notation
$= [\cdots + m \cdot 10^3 + h \cdot 10^2 + t \cdot 10^1] + u$	associative property of addition and $10^0 = 1$
$= 10[\cdots + m \cdot 10^2 + h \cdot 10 + t] + u$	distributive property

Since 10 is divisible by 2, the first addend, $10[\cdots + m \cdot 10^2 + h \cdot 10 + t]$, is divisible by 2. If u is divisible by 2, then

$$n = 10[\cdots + m \cdot 10^2 + h \cdot 10 + t] + u$$

is divisible by 2 by Theorem 7.5. Conversely, if n is divisible by 2, then

$$u = n - 10[\cdots + m \cdot 10^2 + h \cdot 10 + t]$$

must be divisible by 2 by the distributive property of multiplication over subtraction.

We see then that a number is divisible by 2 if and only if its units digit is divisible by 2.

Since the units digit of a numeral must be one of the digits $0, 1, 2, \ldots, 9$, and of these only 0, 2, 4, 6 and 8 are divisible by 2, it follows that a number is divisible by 2 if and only if the units digit of its numeral is 0, 2, 4, 6 or 8.

THEOREM 7.6. A whole number is divisible by 2 if and only if the units digit of its numeral is 0, 2, 4, 6, or 8.

Divisibility by 5

A similar argument can be offered for divisibility by 5. It is readily seen that 5 divides

$$10(\cdots + m \cdot 10^2 + h \cdot 10 + t).$$

Therefore, it is only necessary for 5 to divide the units digit of its numeral for a number to be divisible by 5. The only possible units digits are 0, 1, 2, 3, ..., 9. Of these only 0 and 5 are divisible by 5. Hence a number is divisible by 5 if and only if the units digit of its numeral is 0 or 5.

THEOREM 7.7. A number is divisible by 5 if and only if the units digit of its numeral is 0 or 5.

Divisibility by 10

Divisibility by 10 is a consequence of the rules for divisibility by 2 and 5, since $10 = 2 \times 5$. If a number is divisible by 10, it must be divisible by both 2 and 5. To be divisible by 2 the units digit of the numeral of the number must be 0, 2, 4, 6, or 8. To be divisible by 5, the units digit must be 0 or 5. Since 0 is the only number common to these two sets, a number is divisible by 10 if and only if the units digit of its numeral is 0.

THEOREM 7.8. A number is divisible by 10 if and only if the unit digit of its numeral is 0.

Divisibility by 3

Demonstrating divisibility by 3 requires the rearrangement of the expanded form of the numeral of the number:

$$
\begin{aligned}
\cdots + m \cdot 10^3 &+ h \cdot 10^2 + t \cdot 10^1 + u \cdot 10^0 \\
&= \cdots + 1000m + 100h + 10t + u \\
&= \cdots + (999 + 1)m + (99 + 1)h + (9 + 1)t + u \\
&= \cdots + 999m + m + 99h + h + 9t + t + u \\
&= (\cdots + 999m + 99h + 9t) + (\cdots + m + h + t + u)
\end{aligned}
$$

Using the commutative and associative properties of addition it is possible to regroup the expanded form of a numeral as shown above. We now have

$$n = (\cdots + 999m + 99h + 9t) + (\cdots + m + h + t + u)$$

Since 3 divides

$$(\cdots + 999m + 99h + 9t) = 3(\cdots + 333m + 33h + 3t)$$

we see that if $(\cdots + m + h + t + u)$ is divisible by 3, then n is divisible by 3 by Theorem 7.5. Conversely, if n is divisible by 3, then $(\cdots + m + h + t + u)$ must be divisible by 3 by the distributive property of multiplication over subtraction. But $(\cdots + m + h + t + u)$ is the sum of the digits of the numeral of the number. Hence a number is divisible by 3 if and only if the sum of the digits in its numeral is divisible by 3.

THEOREM 7.9. **A number is divisible by 3 if and only if the sum of the digits in its numeral is divisible by 3.**

Divisibility by 9

An argument similar to the above for divisibility by 3 can be used to find a rule for divisibility by 9. By returning to the expression $(\cdots + 999m + 99h + 9t) + (\cdots + m + h + t + u)$ discussed in divisibility by 3, we see that $(\cdots + 999m + 99h + 9t) = 9(\cdots + 111m + 11h + t)$ has a factor 9. If $(\cdots + m + h + t + u)$ is divisible by 9, then the number n is divisible by 9 by Theorem 7.5. Conversely, if n is divisible by 9, then $(\cdots + m + h + t + u)$ must be divisible by 9 by the distributive property of multiplication over subtraction. Again $(\cdots + m + h + t + u)$ is the sum of the digits of the numeral of the number. Hence we conclude that a number is divisible by 9 if and only if the sum of the digits of its numeral is divisible by 9.

THEOREM 7.10. **A number is divisible by 9 if and only if the sum of the digits of its numeral is divisible by 9.**

EXAMPLE i.   Is 683,424 divisible by (a) 2; (b) 3; (c) 5?
Solution: (a) 683,424 is divisible by 2 since the units digit of 683,424 is 4.
(b) Since

$$6 + 8 + 3 + 4 + 2 + 4 = 27$$

and 27 is divisible by 3 683,424 is divisible by 3.
(c) 683,424 is not divisible by 5 since the units digit is 4 not 0 or 5.

EXAMPLE ii.   Is 683,424 divisible by 6?
Solution: If a number is divisible by 6, it must be divisible by both 2 and 3,

because $6 = 2 \times 3$. Since 683,424 is divisible by both 2 and 3 (see Example i) it is divisible by 6.

All of the theorems above were proved when numbers were named by numerals in the decimal system of numeration. Similar theorems can be found for numbers named in other bases. For example, a number named by an octal numeral is divisible by eight if and only if the units digit of its octal numeral is 0.

7.9. Complete Factorization

A composite number may be factored into prime factors, that is it may be factored completely, by drawing a factor tree. This method is fine for small numbers, but it is a rather long, tedious process for large numbers.

There is a more systematic way of factoring a composite number completely. This is a method of successive divisions and is called the **consecutive primes method.** We shall illustrate this method by an example. Let us factor 144 completely; that is, let us find the complete factorization of 144. We begin with the least prime, 2, and decide whether or not it is a factor of 144. We see by inspection that 144 is divisible by 2, hence

$$144 = 2 \times 72$$

Since $72 = 2 \times 36$, 2 is a factor of 72 and

$$144 = 2 \times 2 \times 36$$

Observing that 2 is also a factor of 36 since $36 = 2 \times 18$, we have

$$144 = 2 \times 2 \times 2 \times 18$$

Again, 2 is a factor of 18 and we have

$$144 = 2 \times 2 \times 2 \times 2 \times 9$$

Since 2 is not a factor of 9, we try the next prime, 3. We see that

$$144 = 2 \times 2 \times 2 \times 2 \times 3 \times 3 = 2^4 \times 3^2$$

Since all the factors in this expression are primes, we have factored 144 completely. The essential results of this method may be written in this shortened form:

$$
\begin{array}{r}
2)\overline{144} \\
2)\overline{72} \\
2)\overline{36} \\
2)\overline{18} \\
3)\overline{9} \\
\overline{3}
\end{array}
\qquad 144 = 2 \times 2 \times 2 \times 2 \times 3 \times 3 = 2^4 \times 3^2
$$

EXAMPLE i.   Use the consecutive primes method to find the complete factorization of 1500.

Solution:
$$
\begin{array}{r}
2)\overline{1500} \\
2)\overline{750} \\
3)\overline{375} \\
5)\overline{125} \\
5)\overline{25} \\
\overline{5}
\end{array}
$$

The complete factorization of 1500 is $2^2 \times 3 \times 5^3$.

EXAMPLE ii.   Use the consecutive primes method to find the complete factorization of 405.

Solution:
$$
\begin{array}{r}
3)\overline{405} \\
3)\overline{135} \\
3)\overline{45} \\
3)\overline{15} \\
\overline{5}
\end{array}
$$

The complete factorization of 405 is $3^4 \times 5$.

EXAMPLE iii.   Use the consecutive primes method to find the complete factorization of 198.

Solution:
$$
\begin{array}{r}
2)\overline{198} \\
3)\overline{99} \\
3)\overline{33} \\
\overline{11}
\end{array}
$$

The complete factorization of 198 is $2 \times 3^2 \times 11$.

Exercise 7.3

1. Find all the prime factors of each of the following.
    a. 288   c. 216
    b. 365   d. 404

2. Factor each of the following completely.
    a. 64   e. 3333
    b. 128   f. 1001
    c. 484   g. 4650
    d. 245   h. 21,700

3. Use a factor tree to factor each of the following completely.
    a. 36   f. 125
    b. 56   g. 700
    c. 84   h. 832
    d. 100   i. 999
    e. 112   j. 886

4. Which of the following are divisible by 2?
    a. 1076   f. 83,211
    b. 2076   g. 20,931
    c. 58,731   h. 72,642
    d. 27,399   i. 909,191
    e. 80,275   j. 607,329

5. Which of the numbers in Problem 4 are divisible by 3?
6. Which of the numbers in Problem 4 are divisible by 9?
7. Which of the numbers in Problem 4 are divisible by 6?
8. Determine in several different ways the complete factorization of each of the following. In each case verify the Fundamental Theorem of Arithmetic.
    a. 72   c. 56
    b. 60   d. 96

9. Name the greatest primes that divide the following.
    a. 421   d. 4235
    b. 299   e. 5246
    c. 176   f. 8088

10. Which of the following are true statements?
    a. If a number is divisible by 4, it is divisible by 2.
    b. If a number is not divisible by 3, then it is not divisible by 9.

---

c. If the numeral of a number has units digit 6, the number is divisible by 6.

d. If a number is divisible by 3 and 4, it is divisible by 7.

11. Given the numeral $346xy$, where x and y are each one of the digits $0, 1, 2, \ldots, 9$. Find all the values of x and y such that $346xy$ is divisible by

    a. 3   c. 6

    b. 2   d. 9

12. Let $n = (2 \cdot 3 \cdot 5 \cdot 7 \ldots 101) + 1$, where $(2 \cdot 3 \cdot 5 \cdot 7 \ldots 101)$ is the product of all the primes less than or equal to 101.

    a. What is the remainder when n is divided by 2?

    b. What is the remainder when n is divided by 3?

    c. What is the remainder when n is divided by 5?

    d. What is the remainder when n is divided by 101?

    e. What is the remainder when n is divided by any prime number less than or equal to 101?

    f. If n is a prime number is it greater than 101?

    g. If n is not a prime number and has a prime factor p, is p greater than 101? Why?

13. Name two values of s for which 7^3 will divide $7^2 s$.

14. What is the least value of s for which 7^4 will divide $7^2 \cdot s$?

15. What is the remainder when you divide each of the following by 9?

    a. 10   d. 10^4

    b. 10^2 e. 10^{50}

    c. 10^3 f. 10^n, n a whole number

16. Find a rule for divisibility by 15.

17. What is the rule for divisibility by 9 in the decimal system of numeration? Discover a similar rule for divisibility by 11 in the duodecimal system.

18. Is it true that if the base-three numeral of a number ends in 2, the number is even? Give a numerical example to substantiate your answer.

19. Discover a rule for divisibility by 7 when a number is named by an octal numeral.

20. Discover a rule for divisibility by 6 when a number is named by a duodecimal numeral.

7.10. Greatest Common Factor

Let a and b be two whole numbers. If c is a factor of both a and b it is called a **common factor** of a and b. Among the common factors of a and b there is a greatest one which is divisible by all the other common factors, and is called the **greatest common factor** of a and b.

All whole numbers are multiples of 1; hence 1 is a common factor of all whole numbers. When we look for common factors of several numbers we look for those common factors greater than 1. When two numbers have only one common factor, 1, we say that they are **relatively prime.**

Suppose we wish to find the common factors of 12 and 30. The set of all factors of 12 is

$$\{1, 2, 3, 4, 6, 12\}$$

The set of all factors of 30 is

$$\{1, 2, 3, 5, 6, 10, 15, 30\}$$

The set of common factors of 12 and 30 is the set of elements common to these two sets, that is the intersection of the two sets. Since

$$\{1, 2, 3, 4, 6, 12\} \cap \{1, 2, 3, 5, 6, 10, 15, 30\} = \{1, 2, 3, 6\}$$

the common factors of 12 and 30 are 1, 2, 3, and 6. The greatest common factor of 12 and 30 is the largest of the common factors, 6. Observe that all the other common factors of 12 and 30, namely 1, 2, 3, are factors of (that is divide) the greatest common factor, 6.

Writing the set of all factors of a number is sometimes troublesome, particularly if the number has many factors. An easier way to find the greatest common factor of several numbers is to use their complete factorizations. Suppose we want to find the greatest common factor of 72 and 90. Finding the complete factorizations of both of these numbers we have:

$$72 = 2^3 \times 3^2 = (2 \times 3 \times 3) \times 2 \times 2$$
$$90 = 2 \times 3^2 \times 5 = (2 \times 3 \times 3) \times 5$$

Notice that each number has 2 and 3^2 as common factors. Hence the greatest common factor of 72 and 90 is $2 \times 3^2 = 18$.

The greatest common factor is also called the **greatest common divisor** which we abbreviate **GCD.** Symbolically we write

$$(72, 90) = 18$$

We read this symbol: "The greatest common divisor of 72 and 90 is 18."

EXAMPLE i. Find the GCD of 48 and 136.
Solution: Factoring 48 and 136 completely we have

$$48 = 2^4 \times 3$$
$$136 = 2^3 \times 17$$

Then $(48, 136) = 2^3 = 8$.

EXAMPLE ii. Find the GCD of 121, 297, and 88.
Solution: Factoring 121, 297, and 88 completely we have

$$121 = 11^2$$
$$297 = 3^3 \times 11$$
$$88 = 2^3 \times 11$$

The GCD of 121, 297, and 88 is 11.

7.11. Euclid's Algorithm

A method of finding the GCD of two whole numbers was developed by Euclid and is called **Euclid's algorithm.** Euclid's algorithm is based on the division algorithm for whole numbers (Chapter 5, Section 5.5): If a and b are any two whole numbers, $b \neq 0$, unique whole numbers q and r can be found such that $a = bq + r$, where $0 \leqslant r < b$.

We shall demonstrate Euclid's algorithm by means of an example. Suppose we wish to find $(96, 26)$. Since

$$
\begin{array}{r}
3 \\
26\overline{)96} \\
78 \\
\hline
18
\end{array}
$$

we know that

$$96 = (26 \times 3) + 18 \quad \text{or} \quad 96 - (26 \times 3) = 18.$$

We see that any number that divides both 96 and 26 must also divide 18 by the distributive property of multiplication over subtraction. This means that we can reduce the problem to one of finding $(26, 18)$. Now

$$18 \overline{)26} \atop{\displaystyle \frac{18}{8}}$$

so

$$26 = (18 \times 1) + 8 \qquad \text{or} \qquad 26 - (18 \times 1) = 8.$$

Thus any number that divides both 26 and 18 also divides 8. So the problem is reduced to finding $(18, 8)$. Now

$$8 \overline{)18} \atop{\displaystyle \frac{16}{2}}$$

and, hence

$$18 = (8 \times 2) + 2 \qquad \text{or} \qquad 18 - (8 \times 2) = 2.$$

Any whole number that divides 18 and 8 must also divide 2, so the problem is reduced to finding $(8, 2)$. Now

$$2 \overline{)8} \atop{\displaystyle \frac{8}{0}}$$

so

$$8 = 2 \times 4$$

That is, 2 divides 8 and there is no greater divisor of 2 and 8. Therefore $(8, 2) = 2$. But

$$(96, 26) = (26, 18) = (18, 8) = (8, 2)$$

as shown in the steps above. Hence

$$(96, 26) = 2$$

The above work is usually shortened as follows:

$$26 \overline{)96} \atop{\displaystyle \frac{78}{\;}}$$
$$18 \overline{)26}$$
$$\frac{18}{\;}$$
$$8 \overline{)18}$$
$$\frac{16}{\;}$$
$$2 \overline{)8}$$
$$\frac{8}{0}$$

The GCD of the two given numbers is the last non-zero remainder in the division process (2 in this case).

EXAMPLE i. Find (544, 391).

Solution:

$$
\begin{array}{r}
1 \\
391\overline{)544} \\
391
\end{array}
$$

$$
\begin{array}{r}
2 \\
153\overline{)391} \\
306
\end{array}
$$

$$
\begin{array}{r}
1 \\
85\overline{)153} \\
85
\end{array}
$$

$$
\begin{array}{r}
1 \\
68\overline{)85} \\
68
\end{array}
$$

$$
\begin{array}{r}
4 \\
17\overline{)68} \\
68 \\
\hline
0
\end{array}
$$

Hence

$$(544, 391) = 17.$$

EXAMPLE ii. Find (360, 121).

Solution:

$$
\begin{array}{r}
2 \\
121\overline{)360} \\
242
\end{array}
$$

$$
\begin{array}{r}
1 \\
118\overline{)121} \\
118
\end{array}
$$

$$
\begin{array}{r}
39 \\
3\overline{)118} \\
117
\end{array}
$$

$$
\begin{array}{r}
3 \\
1\overline{)3} \\
3 \\
\hline
0
\end{array}
$$

Hence (360, 121) = 1, that is, 360 and 121 are relatively prime.

7.12. Least Common Multiple

A **multiple** of a given whole number is the product of that number and another whole number. Thus 24 is a multiple of 3 since $24 = 3 \times 8$; 36 is also a multiple of 3 since $36 = 3 \times 12$. A number has many multiples.

The set of the multiples of 3 is

$$\{0, 3, 6, 9, 12, 15, 18, 21, 24, \ldots\}$$

Since $0 = 0 \times a$ for any number a, *0 is a multiple of every number.*

Any number that is a multiple of two or more numbers is called a **common multiple** of these numbers. Thus 24 is a common multiple of 3 and 4 since $24 = 3 \times 8$ and $24 = 4 \times 6$. Two numbers have many common multiples. The set of multiples of 3 is shown above. The set of multiples of 4 is

$$\{0, 4, 8, 12, 16, 20, 24, 28, \ldots\}$$

The set of common multiples of 3 and 4 is the set of elements common to these two sets (the set of multiples of 3 and the set of multiples of 4), that is the intersection of the two sets. Since

$$\{0, 3, 6, 9, 12, \ldots\} \cap \{0, 4, 8, 12, 16, 20, \ldots\} = \{0, 12, 24, 36, \ldots\},$$

the common multiples of 3 and 4 are the multiples of 12.

The **least common multiple,** denoted by **LCM,** of two or more numbers is the least *natural* number that is a multiple of each of the numbers. Although zero is a multiple of every whole number, it is excluded when determining the least common multiple of several numbers.

Let us determine the least common multiple of 8 and 12. The set of non-zero multiples of 8 is

$$\{8, 16, 24, 32, 40, 48, \ldots\}$$

The set of non-zero multiples of 12 is

$$\{12, 24, 36, 48, 60, 72, \ldots\}$$

The common multiples of 8 and 12 are the elements in the intersection of these two sets:

$$\{24, 48, 72, 96, \ldots\}$$

The least common multiple of 8 and 12 is the least element, 24, in this set.

We can use the complete factorization of numbers to find least common multiple of several numbers. Suppose we wish to find the LCM of 144 and 64. We find the complete factorization of these two numbers:

$$144 = 2^4 \times 3^2$$
$$64 = 2^6$$

Any number which is a multiple of 64 must have a factor 2^6, hence the least common multiple of 64 and 144 must have a factor 2^6. Any number that is a multiple of 144 must have factors 2^4 and 3^2. Since 2^4 is a factor of 2^6 ($2^6 = 2^4 \times 2^2$), the LCM of 64 and 144 is $2^6 \times 3^2 = 576$.

To find the LCM of several numbers we factor each of the numbers completely. We then form the product of the prime numbers which are factors of any of the numbers. We use each prime number as a factor the greatest number of times it appears in the complete factorization of any one of the numbers.

EXAMPLE i. Find the LCM of 24 and 32.
Solution: First we find the complete factorization of 24 and 36:

$$24 = 2^3 \times 3$$
$$36 = 2^2 \times 3^2$$

The LCM of 24 and 36 is

$$2^3 \times 3^2 = 72$$

EXAMPLE ii. Find the LCM of 12, 16, and 15.
Solution: We find the complete factorization of the three numbers:

$$12 = 2^2 \times 3$$
$$16 = 2^4$$
$$15 = 3 \times 5$$

The LCM is

$$2^4 \times 3 \times 5 = 240.$$

Exercise 7.4

1. Use sets of divisors to find the GCD of the following.
   a. 16 and 56   c. 66 and 99
   b. 65 and 135   d. 36 and 78
2. List all the common divisors of 84 and 196.
3. List all the common divisors of 144 and 84.
4. Find the GCD of the following pairs of numbers.
   a. 45; 75   d. 72; 175
   b. 21; 77   e. 36; 108
   c. 84; 198   f. 144; 196

5. Use Euclid's algorithm to find the GCD of the following pairs of numbers.
   a. (97, 21)   c. (7232, 1806)
   b. (504, 142)   d. (80301, 972)
6. Find the GCD of the following sets of numbers by the prime factorization method.
   a. 48; 198   d. 36; 148; 356
   b. 92; 175   e. 723; 72; 81
   c. 144; 356   f. 144; 356; 360
7. Find the GCD.
   a. (0, 6)   c. (0, 18)
   b. (0, 9)   d. (0, n); n a natural number.
8. Which of the following pairs of numbers are relatively prime?
   a. (16, 9)   d. (70, 105)
   b. (17, 87)   e. (2187, 36)
   c. (64, 137)   f. (846, 937)
9. What is the GCD of p and q if both are prime numbers?
10. What is the GCD of two relatively prime numbers?
11. Find the LCM of the following pairs of numbers.
    a. 4; 18   d. 36; 54
    b. 6; 26   e. 75; 125
    c. 36; 42   f. 303; 33
12. Find the LCM of the following sets of numbers.
    a. 16; 36; 72   c. 36; 56; 72
    b. 42; 90; 135   d. 108; 84; 15
13. What is the LCM of p and q if both are prime numbers?
14. Write zero as a multiple of 5; of 8; of the natural number n.

8.1. Informal and Formal Geometry

Before we begin discussion of this chapter, we must be sure that we understand the difference between formal and informal geometry. **Formal** geometry is developed as a deductive system. That is, certain terms are taken as undefined; certain statements, called **axioms** or **postulates,** about these undefined terms are accepted as true without proof; other words are defined in terms of the undefined words; and statements, called **theorems,** are proved by the process of logical reasoning. The objects studied in geometry are abstract and have no existence in the physical world.

Informal geometry calls attention to what may be thought of as geometric properties of familiar objects. For example, we develop mathematical ideas of points, lines, planes, curves, angles, and space by studying some of the models of these ideas in the physical world around us. Informal geometry is studied through observation, intuition, experimentation, and reasoning by induction. Informal geometry is sometimes called **intuitive geometry.** The approach of this chapter is largely informal.

Topics from Geometry

VIII

8.2.   Undefined Terms*

Every definition must eventually depend upon ideas and words which have not been defined. It is impossible to define every term we use in any subject. If we attempt to define every term we become involved in what is known as **circular reasoning.** For example, if we look up the meaning of a word in the dictionary, we find this word defined in terms of other words. If we continue looking up each of these words in the dictionary, we soon find that we are going around in a circle. That is, one of the words which we are looking up is defined in terms of the original word whose meaning we are seeking.

For example, suppose we use the dictionary to find the meaning of "to investigate." We may find the following:

To avoid circular reasoning there are some words in any deductive system that we do not attempt to define. We accept them as undefined. In geometry some of these terms are point, line, and space.

8.3.   Points, Lines, and Space.

A **point** may be thought of as a precise location. It cannot be seen or felt; it has no size. A point may be represented by a dot on paper or as the end of a sharply pointed pencil. These representations are merely attempts to symbolize the idealized geometric entity we call a point. The dot which we use to represent a point, no matter how small, covers many locations.

A point is a fixed location. If we mark a dot to fix a particular location and then erase the dot, the location still remains. The point has not disappeared, only the representation of it has been erased. Just as the symbol "4" is not the number four, this dot · is not a point but merely a pictorial representation of a point.

* This topic was discussed in Chapter 2, but because of its importance it is reviewed here.

We do not define **space** but it may be thought of as the set of all points. The concept of line is also undefined but we may think of a line as a set of points which is a subset of space.

We symbolize points by dots, space by our three-dimensional universe and draw "paths" to represent lines. These symbols must not be confused with the abstract geometric concepts of point, line, and space.

Suppose we represent two points in space by the two dots in Figure 8.1. It is customary to use capital letters to name points in space. We refer to the particular points in Figure 8.1 as "point A" and "point B."

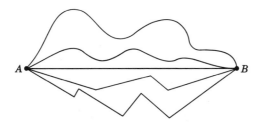

Figure 8.1

Now let us trace a path from point A to point B. We may trace infinitely many such paths. Several are shown in Figure 8.1. All such paths are infinite sets of points called **curves.** One of the paths from A to B is a "straight line." If we use our fingers to hold a piece of string, the various positions in which the string falls represent paths between the two points represented by our fingers. The special path that is represented by the string pulled taut is called a **line segment.** Figure 8.2 shows a representa-

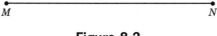

Figure 8.2

tion of a line segment consisting of the points M and N and all the points between them. The symbol used for this line segment is $\overline{MN}$ or $\overline{NM}$. The points M and N are called the **endpoints** of line segment $\overline{MN}$. A line segment is a set of points consisting of two points called endpoints and all the points between them. A line segment is independent of its representation; if the printed path in Figure 8.2 were removed, the line segment would remain since it is a set of locations (points).

A line may be thought of as an extension in both directions of a line segment. It is symbolized by $\overleftrightarrow{AB}$, where A and B are any two points on the line. We represent a line as shown in Figure 8.3. The arrows on the ends are to suggest that the line extends indefinitely in both directions. A line may be named by any two points on it. The line represented in Figure 8.3 may be named $\overleftrightarrow{AB}$ or $\overleftrightarrow{XY}$ or $\overleftrightarrow{AX}$ or $\overleftrightarrow{BY}$.

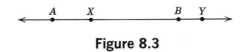

Figure 8.3

8.4. Planes

Let us now consider another subset of the set of points called space. This subset is called a **plane.** Again, we do not define a plane but describe its relation to lines and points.

Any flat surface such as the top of a table, the floor, a sheet of paper, or a wall suggests the idea of a mathematical plane. Like the line, a plane is unlimited in extent. That is, any flat surface used to represent a plane is only a representation of a portion of the plane. The drawings in Figure 8.4 are representations of portions of planes. We refer to a plane by a single letter, as plane P (Figure 8.4a), or by naming three or more points of the plane as $ABCD$ (Figure 8.4b).

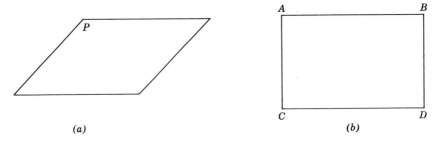

(a)   (b)

Figure 8.4

8.5. Properties of Lines and Planes

We now intuitively discover some properties of lines and planes. From the previous discussion we see that:

Property 1. Through any two points in space there is exactly one line, or simply, two points in space determine one line.

Suppose two points are in a plane as represented in Figure 8.5. These two points determine one line by Property 1. This line certainly lies entirely in the plane. We conclude:

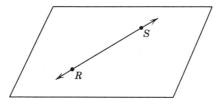

Figure 8.5

Property 2. If a line contains two different points of a plane it lies in the plane.

Consider a line through two points. How many planes contain this line? Think of two points on the spine of a book. Each page of the book repre-

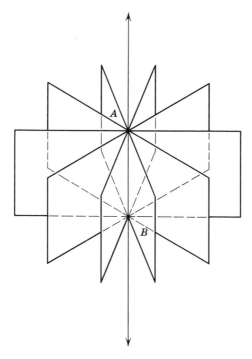

Figure 8.6

sents a different plane passing through the two points. As another example, think of the hinges of a door as two points. Since the door, which represents a portion of a plane through these two points, can swing freely, we again see that many planes contain the two points. These examples suggest:

Property 3. A line lies on infinitely many planes, or, infinitely many planes contain a line.

Three points may lie on one line, as points A, B, and C in Figure 8.7a. Such points are called **collinear** points. If three points are not collinear, as points R, S, and T in Figure 8.7b, they determine three lines as shown in the figure. Such points are called **noncollinear** points.

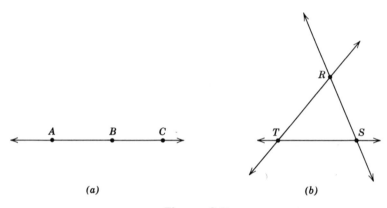

(a)   (b)

Figure 8.7

Let us consider the closed box pictured in Figure 8.8. The diagram suggests that besides A and B and all the points on the line $\overleftrightarrow{AB}$ determined by points A and B, the plane which we have labeled K (suggested by the top of the box) also contains point C. Point C is not on $\overleftrightarrow{AB}$. There is no plane other than K that contains all three points A, B, and C. This suggests:

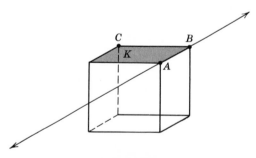

Figure 8.8

Property 4.   Through three noncollinear points there is exactly one plane, or three noncollinear points determine a plane.

Because of Property 4, a three-legged stool always sits firmly on the ground, whereas a four-legged table may wobble. Because of this property, photographers use tripods to support their cameras.

Now let us consider two lines in space. If two lines are **coplanar,** that is, they lie on the same plane, they either intersect as shown in Figure 8.9a or have no points in common as shown in Figure 8.9b. Two coplanar lines are called **parallel** lines if they have no points in common. Two different lines in a plane are called **intersecting** lines if they have common points. It is intuitively obvious that if two different coplanar lines intersect, they intersect in a single point.

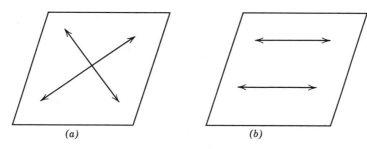

(a)   (b)

Figure 8.9

Property 5.   If two different lines intersect, they intersect in exactly one point.

Let us consider the diagram in Figure 8.10. Plane K contains both of

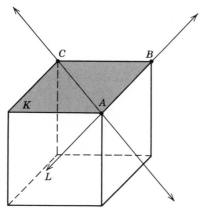

Figure 8.10

the intersecting lines $\overleftrightarrow{AB}$ and $\overleftrightarrow{AC}$. No other plane contains both of these lines. We see then that:

Property 6. Two intersecting lines determine a plane.

Now let us consider the diagram in Figure 8.11. We see that lines $\overleftrightarrow{AB}$ and $\overleftrightarrow{CD}$ are parallel. They both lie in plane R. No other plane contains both of these lines. We conclude:

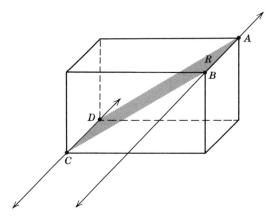

Figure 8.11

Property 7. Two parallel lines determine a plane.

Look at the diagram in Figure 8.12. Lines $\overleftrightarrow{AB}$ and $\overleftrightarrow{EF}$ are neither parallel nor intersecting. They are **noncoplanar,** that is, they do not lie in the same plane. Such lines are called **skew lines.**

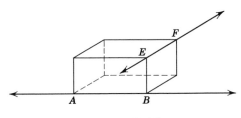

Figure 8.12

If a line and a plane intersect, their intersection is either a point or the entire line as suggested by Figure 8.13.

Property 8. If a line and a plane intersect their intersection is either one point or the entire line.

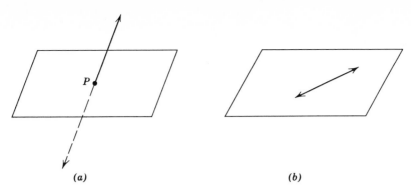

(a)   (b)

Figure 8.13

Of course, a line and a plane may not intersect at all. In this case we say that the line and the plane are **parallel.** In Figure 8.14 plane P and line $\overset{\leftrightarrow}{AB}$ are parallel.

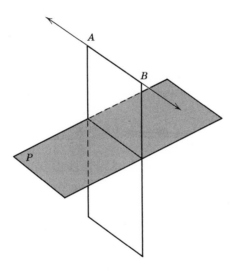

Figure 8.14

Now let us consider two planes. They may intersect as shown in Figure 8.15a or they may have no points in common as shown in Figure 8.15b. When two planes intersect their intersection is a line. When two planes have no points in common they are said to be **parallel.**

Property 9. If two planes intersect, their intersection is a line.

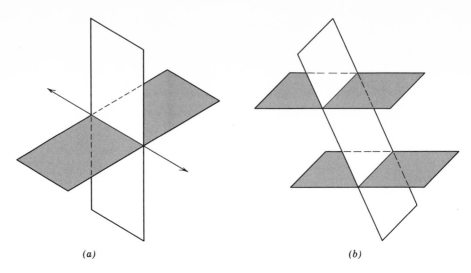

(a)   (b)

Figure 8.15

8.6. Separation of Space

Think of a wall as representing a portion of a plane. The wall separates space into three disjoint sets of points: the set of points in the wall, the set of points in front of the wall, and the set of points behind the wall. We say that a plane separates space into two **half-spaces.** In our model, the set of points in front of the wall forms a half-space and the set of points behind the wall forms the other half-space. The points in the wall are in neither half-space. The separating plane (the wall in our model) is called the **boundary** of each half-space.

Observe Figure 8.16. In the diagram, B is on one side of plane P and C is on the other side. We say that B and C are in different or **opposite** half-spaces. Notice that the line segment $\overline{BC}$ with B and C as endpoints contains point D of P. In general, if two points lie in different half-spaces, the line segment with the two points as endpoints contains a point of the boundary. Points A and B are on the same side of P and are in the same half-space. Notice that the line segment with A and B as end points does not contain a point of P. This is true in general, if two points are in the same half-space, the line segment connecting them does not contain a point of the boundary. Point D is in neither half-space; it lies on the boundary P.

A line separates a plane into three disjoint sets of points: the set

---

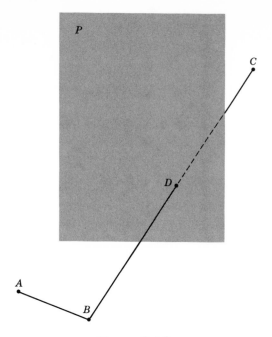

Figure 8.16

of points of the line, and the two sets of points on either side of the line. The two sets of points on either side of the line are called **half-planes.** The line is called the **boundary.** In Figure 8.17, $\overleftrightarrow{AB}$ separates plane P into two half-planes, the half-plane containing F and the half-plane containing C. Notice that $\overline{FC}$ contains point D of the boundary. We say that F and C are in different or **opposite** half-planes. In general, if two points lie in different half-planes, the line segment with the two points as endpoints contains a point of the boundary. Points F and E are in the same half-plane because $\overline{FE}$ does not contain a point of the boundary.

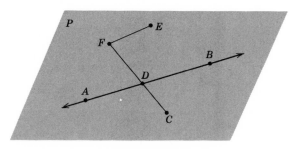

Figure 8.17

A point separates a line into three disjoint sets of points: the set consisting of the point itself, and the two sets of points on either side of the separation point. These two disjoint sets of points are called **half-lines** and the separation point is called the **boundary.** In Figure 8.18, point P separates line $\overleftrightarrow{AB}$ into two half-lines, the half-line to the left of P containing point A and the half-line to the right of P containing point B.

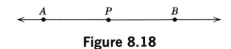

Figure 8.18

Exercise 8.1

1. On the map at the right, name the point which represents the intersection of the following streets.
    a. Kingshighway and Walsh
    b. Hampton and Itaska
    c. Gravois and Kingshighway
    d. Goethe and Murdock
    e. Watson and Delor
    f. Jamieson and Walsh
    g. Watson and Goethe
    h. Main and Delor
    i. Main and Watson

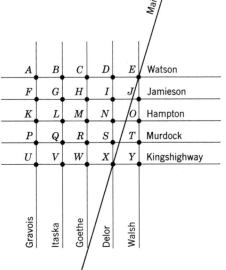

2. Use the figure at the right. The plane determined by $\overleftrightarrow{EH}$ and $\overleftrightarrow{HB}$ is named plane $ABHE$. Name the planes determined by
    a. $\overleftrightarrow{EF}$ and $\overleftrightarrow{HG}$
    b. $\overleftrightarrow{HB}$ and $\overleftrightarrow{BC}$
    c. $\overleftrightarrow{FG}$ and $\overleftrightarrow{EF}$

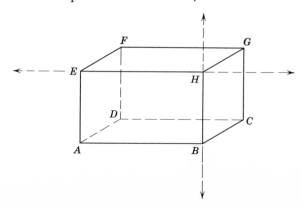

3. Use the figure at the right. Are the following statements true or false? Give reasons for your answer.
   a. Points B and E are in the same half-plane.
   b. Points B and D are in the same half-plane.
   c. Points D and C are in opposite half-planes.
   d. Point A is in the same half-plane as point C.

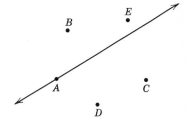

4. Use the figure at the right.
   a. Name three lines parallel to $\overleftrightarrow{CD}$.
   b. Name a plane parallel to plane $ABFE$.
   c. Name two lines in plane $ABCD$ parallel to $\overleftrightarrow{FG}$.
   d. Name the intersection of plane $ABCD$ and plane $BCGF$.
   e. Name the intersection of $\overleftrightarrow{BC}$, $\overleftrightarrow{AB}$, and $\overleftrightarrow{BD}$.

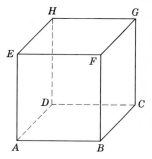

5. How many lines can be drawn through four points, a pair at a time, if the points are all on the same plane but no three of them are on the same line?
6. What is the intersection of two planes?
7. What is the intersection of a plane and a line not in the plane and not parallel to the plane?
8. Draw a diagram representing two coplanar lines whose intersection is
   a. ϕ b. one point
9. Draw a diagram representing two line segments whose intersection is
   a. ϕ b. one point   c. a line segment
10. How many lines are determined by two points?
11. How many lines are determined by three non-collinear points?
12. Which of the following are physical models of a portion of a plane?
    a. ball   c. table top   e. chalkboard
    b. apple   d. pencil   f. a sheet of plywood
13. Which of the following are true statements?
    a. A line has two endpoints.
    b. A line segment has two endpoints.
    c. The intersection of a line and a plane is the empty set if the line and the plane are parallel.

---

d. Two points in space determine infinitely many lines.
e. Two points in space are contained in exactly one plane.
f. A line is an infinite set of points.
g. A plane is an infinite set of points.
h. A line segment is a finite set of points.
d. Two points in space determine infinitely many lines.
14. Describe the intersection set of the following curves.

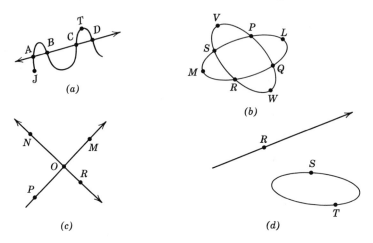

8.7. RAYS

A **ray** is the union of a half-line and its boundary. Figure 8.19 represents a ray that has endpoint P and that contains point S.

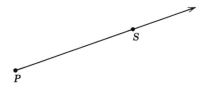

Figure 8.19

We use the symbol $\overrightarrow{PS}$ to denote this ray. Notice that the first letter in the symbol $\overrightarrow{PS}$ names the endpoint or **origin** of the ray; the second letter names any point of the ray.

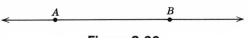

Figure 8.20

Let us consider rays $\overrightarrow{AB}$ and $\overrightarrow{BA}$ on line $\overleftrightarrow{AB}$ (Figure 8.20). Note that $\overrightarrow{AB}$ is not the same ray as $\overrightarrow{BA}$. We see that

$$\overrightarrow{AB} \cup \overrightarrow{BA} = \overleftrightarrow{AB}$$
$$\overrightarrow{AB} \cap \overrightarrow{BA} = \overline{AB}.$$

From our knowledge of a line, we can assert that a line contains infinitely many rays since any point on the line may serve as the endpoint of a ray. Since there are only two directions on a line from a fixed point, there are only two distinct rays on a line with the fixed point as the common endpoint.

In a plane there is an unlimited number of rays with a common endpoint as shown in Figure 8.21.

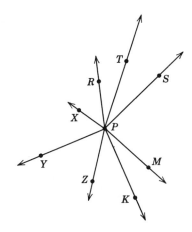

Figure 8.21

8.8. Angles

An **angle** is the union of two rays that have a common endpoint. The common endpoint of the two rays is called the **vertex** of the angle. The two rays forming the angle are called its **sides.** Angles are named by three points of the angle including the endpoint and a point on each side. The symbol for

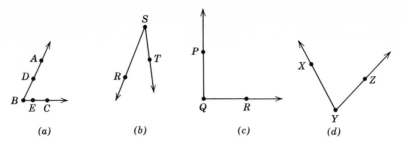

Figure 8.22

angle is $\angle$. The angles represented in Figure 8.22 are: (a) $\angle ABC$ or $\angle CBA$; (b) $\angle RST$ or $\angle TSR$; (c) $\angle PQR$ or $\angle RQP$; (d) $\angle XYZ$ or $\angle ZYX$. Angles may have many names if each represents the same set of points. For example in Figure 8.22, $\angle ABC$ may also be named $\angle CBA$ or $\angle DBE$ or $\angle EBD$.

An angle separates a plane into three disjoint sets of points: the set of points in the angle itself, and two sets of points called the **interior** and the **exterior** of the angle. In Figure 8.23, a portion of the interior of $\angle XYZ$ is shown using vertical shading and a portion of the exterior of $\angle XYZ$ is shown using horizontal shading.

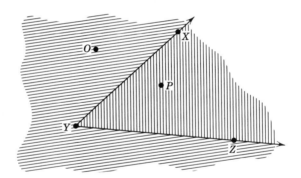

Figure 8.23

Point 0 lies in the exterior of $\angle XYZ$ and point P lies in the interior. We can determine the interior of an angle by using the separation property of a plane and a line. In Figure 8.24 $\angle XYZ$ is determined by rays $\overrightarrow{YX}$ and $\overrightarrow{YZ}$. These rays determine lines $\overleftrightarrow{YX}$ and $\overleftrightarrow{YZ}$ respectively. Line $\overleftrightarrow{YX}$ separates the plane into two half-planes, and Z lies in one of these. This half-plane is shaded by vertical line segments in the figure. Line $\overleftrightarrow{YZ}$ separates the plane into two half-planes, and X lies in one of these. This half-plane is shaded

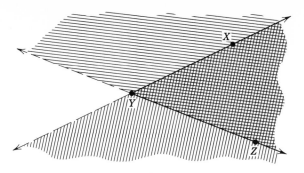

Figure 8.24

by horizontal line segments in the figure. That portion of the plane which is cross-hatched is the interior of ∠ *XYZ*.

Exercise 8.2

1. Use the figure at the right to answer the following.
   a. What is the endpoint of $\overrightarrow{PS}$?
   b. What is the endpoint of $\overleftrightarrow{RT}$?
   c. What is $\overrightarrow{PR} \cup \overrightarrow{PT}$?
   d. What is $\overrightarrow{PS} \cap \overleftrightarrow{RT}$?
   e. What is $\overleftrightarrow{QS} \cup \overrightarrow{RS}$?
   f. What is $\overrightarrow{QP} \cap \overrightarrow{RQ}$?

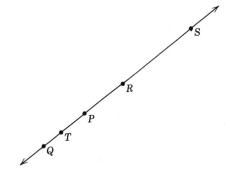

2. How many endpoints does a ray have?
3. Draw a diagram representing two rays whose union is a line.
4. Draw a diagram representing two rays whose intersection is a ray.
5. Use the figure at the right for the following.
   a. Name four angles.
   b. Name three points in the interior of ∠ *CPB*.
   c. Name two angles that have $\overrightarrow{PC}$ as a common side.
   d. Name the sides of ∠ *BPD*.
   e. Name the vertex of ∠ *BPC*.

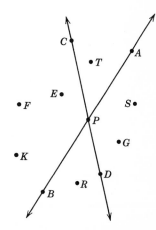

6. Consider the diagram below.

    a. What is $\overline{PQ} \cup \overline{QS}$?
    b. What is $\overline{PR} \cap \overline{QS}$?
    c. What is $\overrightarrow{PS} \cap \overrightarrow{RQ}$?

7. Use the diagram at the right to decide which of the following are true statements.
    a. $\overleftrightarrow{PQ}$ is a set of points.
    b. Point R separates $\overleftrightarrow{PQ}$ into two rays.
    c. Point P lies on $\overleftrightarrow{RQ}$.
    d. P is an endpoint of $\overline{PQ}$.
    e. $\overrightarrow{RQ} \cup \overrightarrow{RP} = \overleftrightarrow{PQ}$.
    f. $\overrightarrow{PQ} \cap \overrightarrow{RP} = \overline{PR}$.

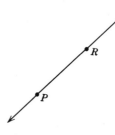

8. Define a ray.
9. Define an angle.
10. Consider the diagram below.

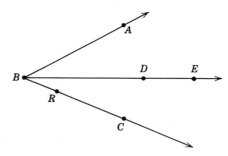

    a. What is the vertex of $\angle ABC$?
    b. Name a line segment lying wholly in the interior of $\angle ABC$.
    c. Name a ray lying wholly in the exterior of $\angle ABD$.

11. In the diagram below $\overleftrightarrow{AB}$ and $\overleftrightarrow{CD}$ are parallel lines and $\overleftrightarrow{BC}$ and $\overleftrightarrow{AD}$ are parallel lines.
    a. Name the angle whose sides are $\overrightarrow{AB}$ and $\overrightarrow{AD}$.
    b. Name the angle whose sides are $\overrightarrow{BJ}$ and $\overrightarrow{BL}$.
    c. What is another name of $\angle BAD$?
    d. What is $\overrightarrow{AP} \cup \overrightarrow{AD}$?

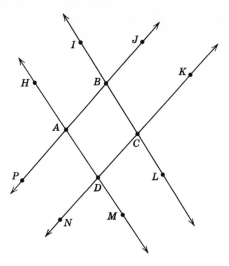

e. Name three points in the interior of $\angle ADC$.
f. Name four points in the exterior of $\angle BCD$.
g. Name two rays lying entirely in the interior of $\angle ADC$.
h. Name two rays lying entirely in the interior of $\angle BAM$.

12. Use the diagram at the right. What is

a. $\overline{AF} \cup \overline{FD}$?

b. $\overrightarrow{DE} \cup \overrightarrow{DF}$?

c. $\overline{AB} \cap \overrightarrow{CB}$?

d. $\overleftrightarrow{AF} \cap \overline{FD}$?

e. $\overline{AB} \cup \overrightarrow{BC}$?

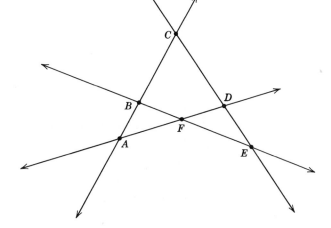

8.9. Simple Closed Curves

A **curve** is a set of points which may be represented by a path that can be traced without lifting the pencil from the paper. The drawings in Figure 8.25

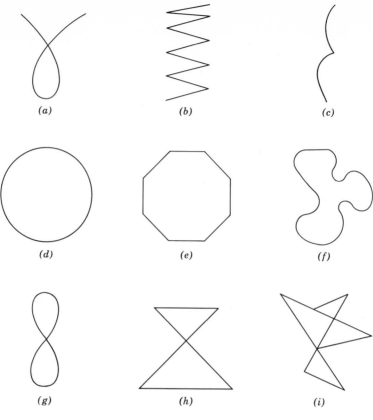

(a) (b) (c)

(d) (e) (f)

(g) (h) (i)

Figure 8.25

represent curves. Figures 8.25a, 8.25b, and 8.25c represent **open curves;** the other pictures represent **closed curves.** In tracing a picture of a closed curve it is possible to begin tracing at any point and return to the starting point without lifting the pencil from the paper. Figures 8.25d, 8.25e, and 8.25f represent simple closed curves. A **simple closed curve** is a set of coplanar points represented by a path that begins and ends at the same point and does not intersect itself.

A simple closed curve separates the plane into three disjoint sets of points: the set of points inside the curve, called the **interior** of the curve, the set of points outside the curve, called the **exterior** of the curve, and the set of points of the curve. The interior of any simple closed curve is called a **region.** The curve is called the **boundary** of the region. The union of the region and its boundary is called a **closed region.**

8.10. Polygons

A **polygon** is a simple closed curve which is the union of three or more coplanar line segments each of which intersects exactly two of the other segments one at each endpoint. The segments that form a polygon are called the **sides** of the polygon. Each endpoint of a side is called a **vertex** of the polygon.

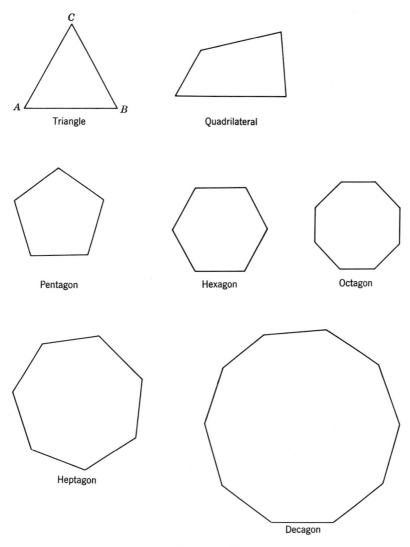

Figure 8.26

Polygons are identified by naming the vertices. For example, we refer to the polygon in Figure 8.26 whose vertices are A, B, and C as polygon ABC or triangle ABC.

Polygons having three, four, five, six, seven, eight, and ten sides are called, respectively, **triangles, quadrilaterals, pentagons, hexagons, heptagons, octagons,** and **decagons.**

8.11. Triangles

Triangles are polygons with three sides. We name a triangle by its vertices. The symbol for a triangle is $\triangle$. In Figure 8.27 we have pictured $\triangle ABC$. $\triangle ABC$ may also be named $\triangle BAC$ or $\triangle CBA$.

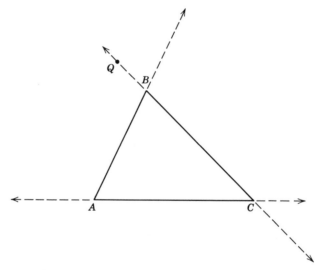

Figure 8.27

A triangle determines three angles called the **angles** of the triangle. Although $\angle ABC$ is an angle of $\triangle ABC$, not all of the points of $\angle ACB$ are points of the triangle. (Figure 8.27) Notice that Q is a point of $\angle ACB$ but not a point of $\triangle ABC$. Similar statements can be made about the other angles of the triangle.

A triangle is a simple closed curve and hence separates the points of the plane into three disjoint sets of points: the set of points of the triangle, the set of points in the interior of the triangle, and the set of points in the exterior of the triangle.

Exercise 8.3

1. Which of the following are simple closed curves?
   a. triangle
   b. ray
   c. line
   d. line segment
   e. hexagon
   f. angle
   g. polygon
   h. decagon
2. Define a polygon.
3. Consider the numerals 0, 1, 2, 3, 4, 5, 6, 7, 8, and 9. Which are simple closed curves?
4. Which of the diagrams below represent simple closed curves?

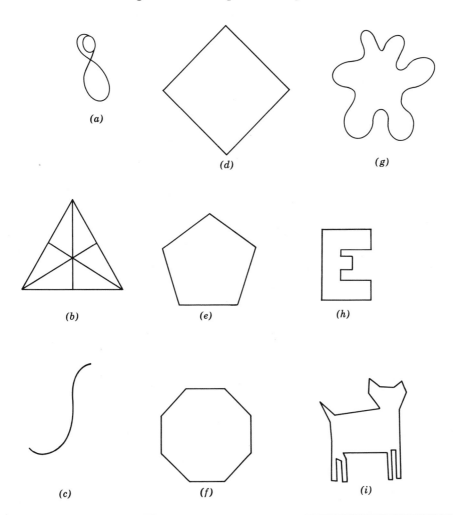

(a)

(d)

(g)

(b)

(e)

(h)

(c)

(f)

(i)

5. In $\triangle RST$ pictured at the right, name
   a. the vertices
   b. the sides
   c. the angles

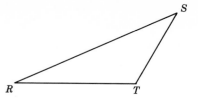

6. Which of the points in the diagram at the right are
   a. on the triangle?
   b. in the interior of the triangle?
   c. in the exterior of the triangle?

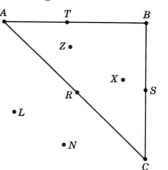

7. Use the diagram below to answer the following.

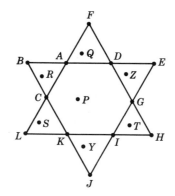

   a. Which points are in the interior of $\triangle BEJ$?
   b. Which points are in the interior of $\triangle FLH$?
   c. Which points are in the exterior of $\triangle BEJ$?
   d. Which points are in the exterior of $\triangle FLH$?
   e. What is the intersection set of $\triangle BEJ$ and $\triangle FLH$?

8. Draw a diagram (if possible) of two triangles whose intersection consists of

   a. one point   c. three points   e. five points
   b. two points   d. four points   f. six points

---

9. Draw a diagram (if possible) of a triangle and a line whose intersection is
   a. the empty set
   b. one point
   c. two points
10. Draw a diagram (if possible) of a triangle and an angle whose intersection is
    a. φ
    b. two points
    c. four points
11. Which of the simple closed curves pictured below are polygons?

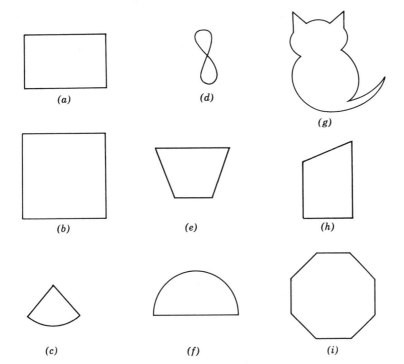

(a)

(d)

(g)

(b)

(e)

(h)

(c)

(f)

(i)

12. Which of the polygons pictured below are quadrilaterals?

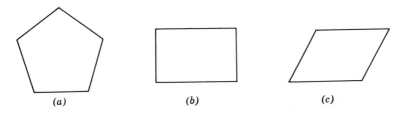

(a)

(b)

(c)

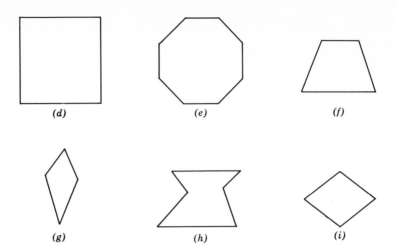

(d)   (e)   (f)

(g)   (h)   (i)

13. Define a triangle.
14. Define a hexagon.
15. Define a quadrilateral.
16. How many vertices does a decagon have?
17. Use the diagram below to answer the following.

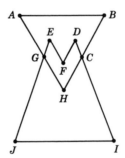

   a. What is the intersection of $\triangle ABH$ and simple closed curve CDF EGH?
   b. What points are in the interior of $\triangle ABH$?
   c. What points are in the exterior of $\triangle ABH$?

8.12. Congruence

Congruence is a complex idea with many consequences in geometry. It applies to geometric figures of all kinds. We shall confine ourselves to an intuitive idea of congruence, that is, if one geometric figure is an exact replica of another, we shall say that the two figures are **congruent.**

The best way to determine whether or not two line segments are congruent is by using a compass. If the points of a compass are separated and made to cover the endpoints of the first segment as shown in Figure 8.28, and then, on being transferred to the second segment, cover the endpoints of this segment, the two segments are congruent.

Figure 8.28

The line segments represented in Figure 8.29 are congruent. We use the symbol ≅ for congruent and write $\overline{AB} \cong \overline{CD}$.

Note that we do *not* write $\overline{AB} = \overline{CD}$, because equality means "names the same thing as" and obviously $\overline{AB}$ and $\overline{CD}$ are not names for the same line segment.

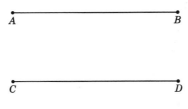

Figure 8.29

8.13. Circles

A **circle** is a simple closed curve having a point, O, in its interior such that if P and Q are any two points of the circle, then $\overline{OP} \cong \overline{OQ}$. The point O is called the **center** of the circle. The center is *not* a point of the circle. We usually refer to a circle by the name of its center. Thus the circle in Figure 8.30 is called circle O.

A line segment with one endpoint at the center and the other endpoint on the circle is called a **radius** (plural: radii) of the circle. In Figure 8.30, $\overline{OP}$, $\overline{OQ}$, $\overline{OT}$, and $\overline{OS}$ are all radii of circle O. *All radii of a given circle are congruent.*

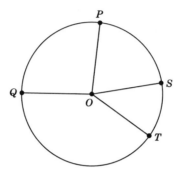

Figure 8.30

A **chord** of a circle is a line segment whose endpoints lie on the circle. In Figure 8.31, $\overline{AB}$, $\overline{AK}$, and $\overline{CD}$ are chords of circle O'. The **diameter** of a circle is a chord which contains the center of the circle. In Figure 8.31, $\overline{AK}$ is a diameter of the circle. *All diameters of a given circle are congruent.*

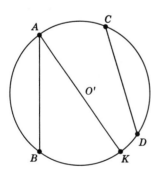

Figure 8.31

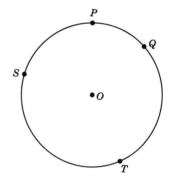

Figure 8.32

A portion of a circle is called an **arc**. Any two points of a circle separate the circle into two arcs. For example, P and T in Figure 8.32 separate the circle into the arc containing Q and the arc containing S. These arcs are denoted by $\overparen{PQT}$ and $\overparen{PST}$.

Since the circle is a simple closed curve, it separates the plane into three disjoint sets of points: the set of points of the circle, the set of points of the interior of the circle and the set of points of the exterior of the circle.

8.14. Congruent Angles

If two angles, $\angle ABC$ and $\angle DEF$, are given, we can take as a representation of $\angle ABC$ a tracing, say $\angle A'B'C'$, and place the tracing so that ray $\overrightarrow{B'C'}$ falls along $\overrightarrow{EF}$ with $\overrightarrow{B'A'}$ falling in the same half-plane as $\overrightarrow{ED}$ and with B' falling on E. Now if $\overrightarrow{B'A'}$ falls along $\overrightarrow{ED}$ we say that $\angle ABC$ is congruent to $\angle DEF$ and write

$$\angle ABC \cong \angle DEF$$

We can construct an angle congruent to a given angle using a straight edge and a compass as follows.

Angle to be copied	New angle	Directions
		1. Draw a ray $\overrightarrow{DE}$
		2. Place the point of the compass on B. Draw an arc that meets the sides of $\angle ABC$ at P and Q. Without changing the compass setting, put the point of the compass on D and draw a similar arc which cuts $\overrightarrow{DE}$ at R.
		3. Place the compass point on P and the compass pencil point on Q to get the distance. Do not change the compass setting. Place the compass point on R and make a mark S as shown.
		4. Draw $\overrightarrow{DS}$. $\angle FDE \cong \angle ABC$

8.15. Congruent Triangles

In Figure 8.33, the two triangles represented are congruent. If $\triangle DEF$ were traced on a piece of paper and the paper cut along the sides of the triangle, the paper pattern would represent the triangle and its interior. The paper pattern of $\triangle DEF$, when placed on $\triangle ABC$, would fit exactly. If vertex D were placed on vertex A of $\triangle ABC$, with $\overline{DF}$ along $\overline{AC}$, vertex F would fall on vertex C and vertex E would fall on vertex B. We say that vertex A of

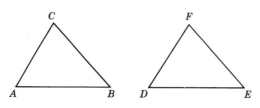

Figure 8.33

$\triangle ABC$ corresponds to vertex D of $\triangle DEF$. Similarly vertex E corresponds to vertex B and vertex F corresponds to vertex C.

    In these two congruent triangles there are three pairs of congruent line segments and three pairs of congruent angles:

$$\begin{aligned} \overline{AB} &\cong \overline{DE} & \angle CAB &\cong \angle FDE \\ \overline{BC} &\cong \overline{EF} & \angle ABC &\cong \angle DEF \\ \overline{AC} &\cong \overline{DF} & \angle BCA &\cong \angle EFD \end{aligned}$$

Since $\overline{AB} \cong \overline{DE}$, we say that $\overline{AB}$ and $\overline{DE}$ are **corresponding** sides of the two congruent triangles. Similarly $\overline{BC}$ and $\overline{EF}$ are corresponding sides and $\overline{AC}$ and $\overline{DF}$ are corresponding sides. Since $\angle CAB \cong \angle FDE$, we say that $\angle CAB$ and $\angle FDE$ are **corresponding angles** of congruent triangles. Similarly $\angle ABC$ and $\angle DEF$ are corresponding angles. Name another pair of corresponding angles.

To denote that $\triangle ABC$ is congruent to $\triangle DEF$ we write

$$\triangle ABC \cong \triangle DEF$$

Care must be taken in writing this statement. We must be sure that corresponding letters are names for corresponding vertices. In this instance, it would be incorrect to write $\triangle ABC \cong \triangle FED$, since vertex A does not correspond to vertex F.

In formal geometry the following theorems are either proved or given as axioms. We shall accept them without proof.

THEOREM 8.1. **Two triangles are congruent if three sides of one triangle are congruent respectively to three sides of the other triangle.**

THEOREM 8.2. **Two triangles are congruent if two sides and the included angle of one triangle are congruent respectively to two sides and the included angle of the other triangle.**

THEOREM 8.3. **Two triangles are congruent if two angles and the side which lies between them of one triangle are congruent respectively to two angles and the side which lies between them of the other triangle.**

There are several ways of constructing a triangle congruent to a given triangle. Suppose we wish to construct a triangle congruent to $\triangle ABC$ in Figure 8.34. We shall discuss three methods of constructing a triangle congruent to a given triangle.

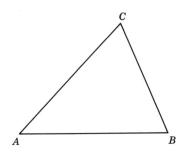

Figure 8.34

Method 1.

 (1) Draw a line. On this line lay off a segment $\overline{A'B'}$ congruent to $\overline{AB}$.

 (2) Open the legs of a compass so that the opening is exactly the same length as $\overline{AC}$. With this opening a radius, draw an arc of a circle with center at A'.

 (3) Now open the legs of a compass so that the opening is exactly the same length as $\overline{BC}$. With this opening as radius draw an arc of a circle with center at B'.

 (4) These two arcs will intersect in a point. Label this point C'.

 (5) Draw $\overline{A'C'}$ and $\overline{B'C'}$.

 (6) $\triangle ABC \cong \triangle A'B'C'$ (Figure 8.35) Why?

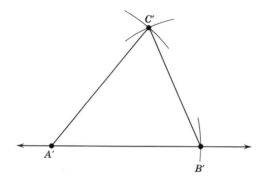

Figure 8.35

Method 2.

 (1) Draw a line. On this line lay off segment $\overline{A'B'}$ congruent to $\overline{AB}$.

 (2) Construct an angle with vertex at A' congruent to $\angle CAB$. Figure 8.36 shows the constructed angle, $\angle KA'B'$, which is congruent to $\angle ABC$.

 (3) On $\overrightarrow{A'K}$ construct $\overline{A'C'} \cong \overline{AC}$.

 (4) Draw $\overline{B'C'}$ making $\triangle A'B'C'$. Since $\triangle A'B'C'$ has two sides and an included angle congruent respectively to two sides and an included angle of $\triangle ABC$, $\triangle ABC \cong \triangle A'B'C'$.

Method 3.

 (1) Same as in Method 2, Step 1.

 (2) Same as in Method 2, Step 2.

 (3) Construct an angle with vertex at B' and congruent to $\angle CBA$.

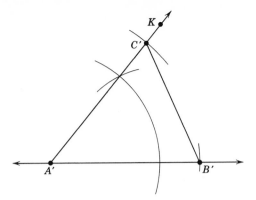

Figure 8.36

Figure 8.37 shows the constructed angle, $\angle LB'A'$, which is congruent to $\triangle CBA$.

(4) $\overrightarrow{A'K}$ and $\overrightarrow{B'L}$ will intersect in a point. Label this point C'. We now have $\triangle A'B'C'$ which is congruent to $\triangle ABC$ because two angles and the side which lies between them are congruent respectively to the corresponding angles and the side that lies between them of $\triangle ABC$.

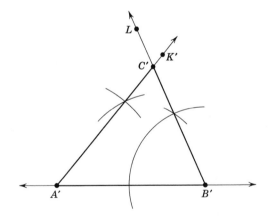

Figure 8.37

Exercise 8.4

1. Construct line segments congruent to the line segments represented below.

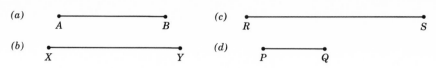

2. Construct angles congruent to the angles below.

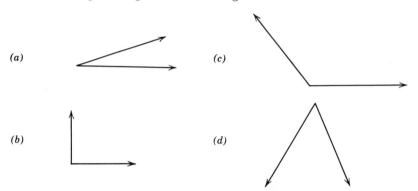

3. Using Method 1 of Section 8.15 construct triangles congruent to the triangles below.

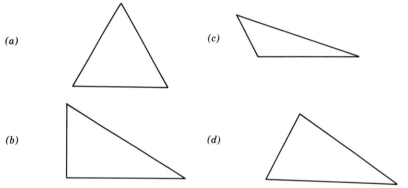

4. Using Method 2 of Section 8.15 construct triangles congruent to the triangles below.

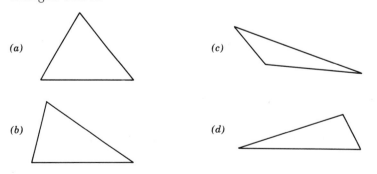

5. Using Method 3 of Section 8.15 construct triangles congruent to the triangles below.

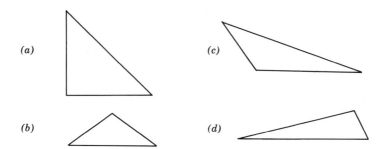

(a)

(c)

(b)

(d)

6. Using Method 1 of Section 8.15 construct a triangle with sides congruent to the line segments below.

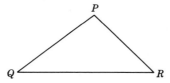

7. Construct an angle congruent to ∠*PQR* of △*PQR* below.

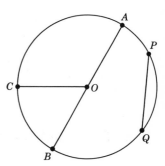

8. Use the figure at the right.
   a. Name a diameter of the circle.
   b. Name the center of the circle.
   c. Name four points of the circle.
   d. Name a point of the interior of the circle.

9. Complete the following statements.
   a. If $\overline{AB} \cong \overline{CD}$ and $\overline{AB} \cong \overline{PQ}$, then ____ .
   b. If $\overline{AB} \cong \overline{CD}$ and $\overline{CD} \cong \overline{EF}$, then ____ .

10. Draw two circles in such a way that they have the same point as center. Such circles are called **concentric circles.**
11. Draw two circles whose intersection is the empty set.
12. Draw two circles whose intersection is a set containing exactly one point.
13. Draw two circles whose intersection is a set containing exactly two points.
14. List the elements of the set which is the intersection of the two curves in the diagram at the right.

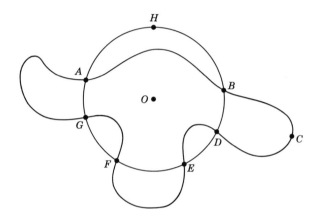

15. Which of the following are true statements?
    a. A diameter is a chord of a circle.
    b. A radius is a chord of a circle.
    c. Every chord of a circle contains exactly three points of the interior of the circle.
    d. No chord of a circle passes through the center of the circle.
    e. All the diameters of a circle intersect in one point.
    f. Two congruent circles have congruent radii.

8.16. Solid Geometric Figures

We have discussed some plane geometric figures such as polygons and circles. These figures lie entirely in a plane. We are now going to discuss some common geometric solids. Any one of a variety of physical objects may be used to explain the idea of a **solid.** We may talk about children's alphabet blocks, ice cream cones, and inflated rubber balloons. (Figure 8.38.) Of course, geometric solids are not composed of wood, ice cream, or rubber; they are infinite sets of points. When we think of the set of points of space

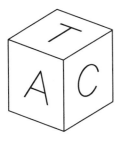

Figure 8.38

occupied by a balloon, then we have a good intuitive idea of a geometric solid called a **sphere.** When we think of the set of points of space occupied by an ice cream cone we have a good intuitive idea of a geometric solid called a **cone.** When we think of the set of points of space occupied by an alphabet block we have a good idea of a geometric solid called a **cube.**

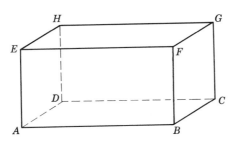

Figure 8.39

Consider an ordinary closed box (Figure 8.39). This figure represents a **rectangular prism.** When we look at the diagram in Figure 8.39 we see a closed figure whose boundary consists of six rectangles and their interiors. The **bases** of this prism are the rectangles *ABCD* and *EFGH.* The bases of a prism lie in parallel planes and are congruent. The other four rectangles, *ADHE, BFCG, ABFE,* and *DCGH,* are called the **lateral faces** of the prism. The lateral faces together with the bases of a rectangular prism are called its **faces.**

The eight vertices of the faces are called the **vertices** of the prism. The sides of the faces are called the **edges** of the prism. Any of the four edges of the prism not lying in a base may be called an **altitude** of the prism. Notice that all of the altitudes of a prism are parallel to each other.

A rectangular prism is an example of a **simple closed surface.** A simple closed surface separates the points of space into three disjoint sets of points: (1) the set of points of the surface itself; (2) the set of points interior to the surface; and (3) the set of points exterior to the surface. One must pass through the simple closed surface to get from an interior point to an exterior point.

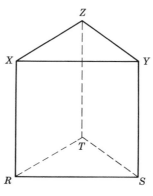

Figure 8.40

Figure 8.40 shows another kind of prism—a **triangular prism.** Its **bases,** which lie in parallel planes and are congruent, are triangles *RST* and *XYZ.* Its **lateral faces** are rectangles *XYSR, STZY,* and *XRTZ.* The **altitudes** are $\overline{XR}, \overline{YS},$ and $\overline{ZT},$ and are all congruent and parallel to each other.

The figures shown in Figure 8.41 are also prisms. There are many kinds of prisms. A **prism** is a simple closed surface consisting of two polygonal regions bounded by congruent polygons which lie in parallel planes and the regions bounded by the line segments whose endpoints are determined by corresponding vertices of the bases.

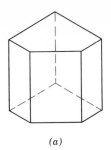 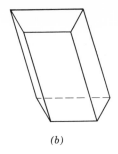 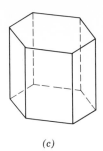

(a)   (b)   (c)

Figure 8.41

A **pyramid** consists of one polygonal region and a number of triangular regions together with their boundaries. Two pyramids are shown in Figure 8.42. The polynomial region is called the **base** of the pyramid. The base of the pyramid in Figure 8.42a is triangle ABC; the base of the pyramid in Figure 8.42b is the square $WXYZ$. The point at the top of a pyramid is

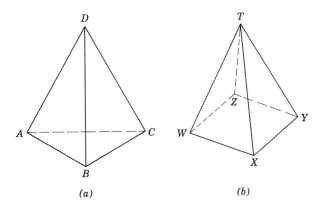

(a)   (b)

Figure 8.42

called its **vertex.** The vertex in Figure 8.42a is point D; the vertex in Figure 8.42b is point T. The triangular regions of a pyramid are called its **lateral faces.** The sides of the triangles which bound the lateral faces are called the **lateral edges** of the pyramid.

A **cylinder** is defined in a manner very similar to the way we defined a prism, except that the bases are regions bounded by simple closed curves instead of polygons. Thus, in a very general sense, a prism may be

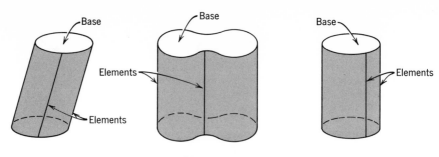

Figure 8.43

considered a special case of a cylinder. A line segment which connects two corresponding points in the curves bounding the bases is called an **element** of the cylinder. (Figure 8.43.)

A **cone** is a simple closed surface which is a set of points consisting of a plane region bounded by a simple closed curve, a point called the **vertex** not in the plane of the curve, and all the line segments of which one endpoint is the vertex and the other endpoint is on the given curve. (Figure 8.44.)

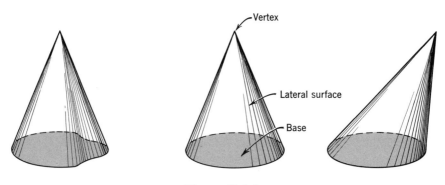

Figure 8.44

The simple closed curve and its interior is called the **base** of the cone. The point is called the **vertex.** The union of all the line segments is called the **lateral surface** of the cone.

A **sphere** is a simple closed surface having a point O in its interior and such that if A and B are any two points on the surface $\overline{OA} \cong \overline{OB}$.

The point O is called the **center** of the sphere. Line segments $\overline{OA}$ and $\overline{OB}$ are called **radii** of the sphere. The line segment $\overline{AK}$ is called a **diameter** of the sphere.

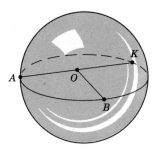

Figure 8.45

Exercise 8.5

1. Name the solids pictured below.

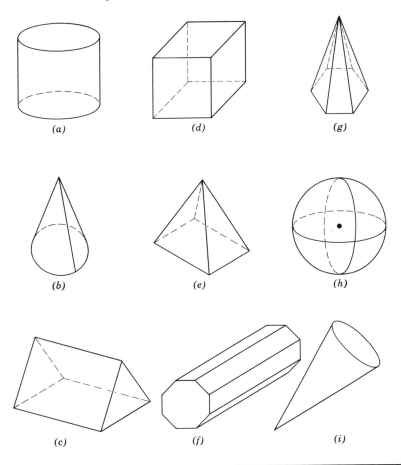

(a)

(d)

(g)

(b)

(e)

(h)

(c)

(f)

(i)

2. Sketch a prism with a pentagon base.
3. Sketch a cone with a circle as a base.
4. Sketch a sphere.
5. Sketch a pyramid with a base that is a pentagon.

9.1. The Word "Fraction"

The term "fraction" is commonly used in three distinct ways. One usage of the word fraction is reflected in expressions such as "Add the fractions $\frac{1}{2}$ and $\frac{1}{3}$"; "If we multiply $\frac{3}{4}$ and $\frac{1}{6}$ the product is $\frac{1}{8}$." Since addition and multiplication are operations performed on numbers, when we use the term "fraction" in this way, we are thinking of fractions as numbers. We see then, that the word "fraction" as used in this manner means fractions as numbers.

We also hear expressions like these: "What is the numerator of the fraction $\frac{5}{6}$?" "Do the fractions $\frac{1}{3}$ and $\frac{3}{6}$ have a common denominator?" and "Which fraction has the smaller numerator, $\frac{2}{3}$ or $\frac{1}{2}$?" Here we are not using the term "fraction" to mean number. In this usage the word "fraction" means a numerator-denominator pair, that is, an ordered pair of numbers.

The third usage of the term "fraction" is encountered in expressions like: "Write 1.5 as a fraction"; "What numeral is written above the bar in the fraction $\frac{5}{8}$?" Here we are using the term "fraction" as a symbol.

Fractions

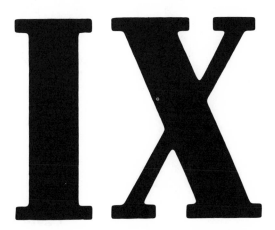

The three common usages of the term "fraction" are then:

(1) Fractions as numbers
(2) Fractions as numerator-denominator pairs, that is, as ordered pairs of numbers.
(3) Fractions as symbols.

In this chapter we shall be concerned with fractions as symbols for number pairs. In Chapter 10 we shall be concerned with fractions as numbers.

We shall utilize all three usages of the term "fraction" in this book, relying on the content to reveal which of the meanings we are using.

9.2. Fractions as Symbols for Number Pairs

Let us consider the unit regions shown in Figure 9.1. Each region has been divided into a number of congruent subregions and in each case a number of these subregions have been shaded.

Table 9.1 gives the total number of congruent subregions of each unit region, the number of shaded subregions of each region, the number pair associated with the number of shaded subregions in each case, and the symbol used to denote this number pair.

With Figure 9.1a we associate the number pair $(6, 9)$ and denote it by the symbol $\frac{6}{9}$; with Figure 9.1b we associate the number pair $(1, 8)$ and denote it by the symbol $\frac{1}{8}$; and so forth.

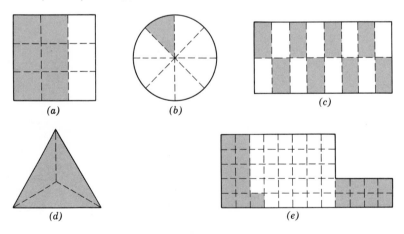

(a)   (b)   (c)

(d)   (e)

Figure 9.1

Table 9.1

Region	(a)	(b)	(c)	(d)	(e)
Number of Shaded Regions	6	1	8	3	19
Number of Congruent Subregions	9	8	16	3	48
Number Pair	$(6, 9)$	$(1, 8)$	$(8, 16)$	$(3, 3)$	$(19, 48)$
Symbol Denoting Number Pair	$\frac{6}{9}$	$\frac{1}{8}$	$\frac{8}{16}$	$\frac{3}{3}$	$\frac{19}{48}$

The symbols denoting these number pairs, that is the symbols $\frac{6}{9}$, $\frac{1}{8}$, $\frac{8}{16}$, $\frac{3}{3}$, and $\frac{19}{48}$ are called **common fractions** or, simply, **fractions.** Here we are using the term fraction to mean a written symbol for a number pair.

EXAMPLE i. What fractional part of each region in Figure 9.1 is not shaded?
Solution: (a) $\frac{3}{9}$; (b) $\frac{7}{8}$; (c) $\frac{8}{16}$; (d) $\frac{0}{3}$; since all of the congruent subregions in Figure 9.1d are shaded the number of unshaded subregions is 0; (e) $\frac{29}{48}$.

EXAMPLE ii. How many dots in the array below are black? How many dots are in the array? What fractional part of the dots are black?

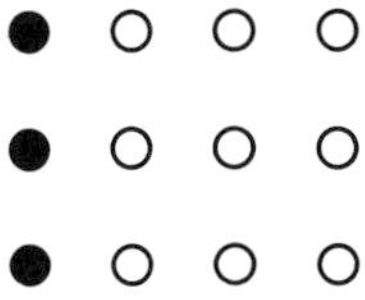

Figure 9.2

Solution: There are 12 dots in the array; 3 are black; $\frac{3}{12}$ of the dots are black.

EXAMPLE iii. What fractional part of the bar in Figure 9.3 is shaded?
Solution: There are 5 congruent parts and 3 are shaded. The fractional part shaded is $\frac{3}{5}$.

Figure 9.3

EXAMPLE iv. What fraction denotes the shaded portion of the triangular region in Figure 9.4?

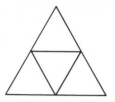

Figure 9.4

Solution: There are 4 congruent parts; the number of shaded parts is 0. The fraction denoting the shaded portion of the figure is $\frac{0}{4}$.

EXAMPLE v. What fraction denotes the shaded part of the circular region in Figure 9.5?

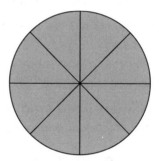

Figure 9.5

Solution: There are 8 congruent subregions and all 8 are shaded. The fraction denoting the shaded portion of the figure is $\frac{8}{8}$.

We note there that $\frac{0}{5}$ is a meaningful fraction since this fraction denotes the number pair (0, 5), meaning that we are considering none of five congruent parts of a unit region. On the other hand, $\frac{5}{0}$ for example, is not a meaningful fraction. We never consider a congruent region to consist of less than one (that is the whole) unit region. We summarize by noting that a number pair (a, b), a and b whole numbers and $b \neq 0$, may be denoted by the fraction $\frac{a}{b}$.

9.3. Parts of a Fraction

We defined a fraction as the symbol for a number pair. A fraction, for example $\frac{3}{5}$, has three parts: the numeral above the bar, the numeral below the bar, and the horizontal bar. The number named by the numeral above the bar is called the **numerator** of the fraction; the number named by the numeral below the bar is called the **denominator.** The numerator of $\frac{3}{5}$ is 3; the denominator is 5.

Observe that a fraction is defined here as a symbol whereas the numerator and denominator are defined as numbers. This is done because we wish to talk about operations involving numerators and denominators, and our operations are defined only for numbers, not for numerals.

Exercise 9.1

1. Use the figures shown to complete the table below.

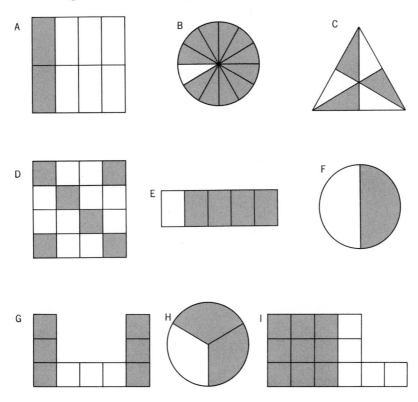

Region		Number Pair	Fraction of the Region Shaded
a.	A		
b.	B		
c.	C		
d.	D		
e.	E		
f.	F		
g.	G		
h.	H		
i.	I		

2. What fractional part of each bar is shaded?

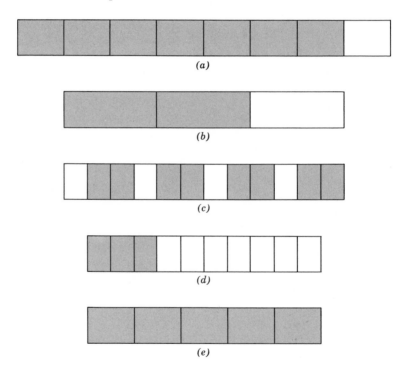

(a)

(b)

(c)

(d)

(e)

3. Name the numerator and denominator of each of the following fractions.

a. $\frac{1}{2}$ d. $\frac{5}{8}$ g. $\frac{12}{79}$

b. $\frac{3}{4}$ e. $\frac{1}{3}$ h. $\frac{18}{18}$

c. $\frac{3}{5}$ f. $\frac{9}{16}$ i. $\frac{42}{100}$

4. A box contains 48 cans of dog food; 18 cans are taken from the box and put on a shelf. What fractional part of the cans have been placed on the shelf?

5. There are 12 horses in a corral; 5 are rented to a group of riders. What fractional part of the horses is left in the corral?

6. A shipment of 350 dresses was received by a store in December. In January 40 of the dresses had not been sold. What fractional part of the dresses was sold?

7. A can contains 46 ounces of tomatoe juice. Sixteen ounces of the tomatoe juice is poured into a pitcher. What fractional part of the tomatoe juice is still in the can?

8. A large parcel of land is subdivided into 15 lots of the same size and shape. Eight of the lots have been sold. What fractional part of the land has been sold?

9. Complete the table below.

	Numerator	Denominator	Fraction
a.	3	9	
b.	5	12	
c.		16	$\frac{3}{16}$
d.	12		$\frac{12}{100}$
e.	0		$\frac{0}{16}$
f.		100	$\frac{72}{100}$
g.		8	$\frac{0}{8}$
h.	7	7	

10. Write a fraction for each problem below.
    a. The numerator is 13 and the denominator is 25.
    b. The numerator is 6 and the denominator is 5 more than the numerator.
    c. The denominator is 36 and the numerator is half of the denominator.
    d. The denominator is 64 and the numerator is 16 less than the denominator.
    e. The numerator is 250 and the denominator is four times the numerator.

9.4. Fractions Whose Numerators Are Equal to or Greater than the Denominator

Up to this point we have considered only those fractions whose denominators are less than or equal to the denominator. If we divide a unit region into eight congruent parts and consider all of them, the fractional part considered is denoted by $\frac{8}{8}$. What does the fraction $\frac{16}{16}$ mean? We can interpret the fraction $\frac{16}{16}$ to denote that a unit region has been divided into 16 congruent subregions and that all of them are considered.

What does a fraction such as $\frac{5}{4}$ denote? A good way to understand what such a fraction denotes is to consider a ruler that is marked into one-fourth inches. Let us compare the bar in Figure 9.6a with the inch. We

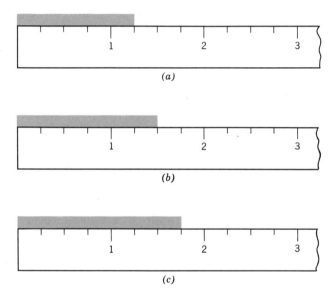

Figure 9.6

use the fraction $\frac{5}{4}$ to compare the length of the bar in Figure 9.6a to the length of an inch. Similarly, we use the fraction $\frac{6}{4}$ to compare the bar in Figure 9.6b with the inch. What fraction do we use to compare the bar in Figure 9.6c with the inch?

Fractions such as $\frac{8}{8}$, $\frac{4}{4}$, and $\frac{16}{16}$ have numerators equal to their denominators. Fractions such as $\frac{5}{4}$, $\frac{6}{4}$, and $\frac{7}{4}$ have numerators that are greater than their denominators. Fractions that have numerators equal to or greater than their denominators are sometimes called **improper fractions.**

9.5. Equivalent Fractions

Let us consider two congruent square regions (Figure 9.7). The square region in Figure 9.7a is divided into two congruent subregions; the square region in Figure 9.7b is divided into four congruent subregions. The fraction denoting the shaded portion of Figure 9.7a is $\frac{1}{2}$; the fraction denoting

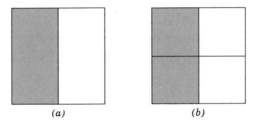

(a)   *(b)*

Figure 9.7

the shaded portion of Figure 9.7b is $\frac{2}{4}$. Notice that in both cases the same portion of the region is shaded. We say that $\frac{1}{2}$ and $\frac{2}{4}$ are **equivalent fractions** and write $\frac{1}{2} \rightarrow \frac{2}{4}$ (notice we do not use an equal sign). In Figure 9.8 we see that $\frac{3}{4}$ and $\frac{6}{8}$ are equivalent fractions.

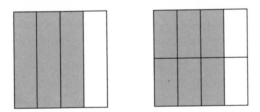

Figure 9.8

Notice that $\frac{1}{2}$ and $\frac{2}{4}$ are equivalent fractions and that $1 \times 4 = 2 \times 2$. Similarly $\frac{3}{4}$ and $\frac{6}{8}$ are equivalent fractions and $3 \times 8 = 4 \times 6$. We now give a formal definition for equivalent fractions.

DEFINITION 9.1. *The fractions $\frac{a}{b}$ and $\frac{c}{d}$ are **equivalent** if and only if $ad = bc$.*

Every fraction is equivalent to infinitely many other fractions. All of the fractions in each set below are equivalent to every other fraction in the set:

(1) $\{\frac{1}{2}, \frac{2}{4}, \frac{3}{6}, \frac{4}{8}, \frac{5}{10}, \frac{6}{12}, \frac{7}{14}, \ldots\}$

(2) $\{\frac{2}{3}, \frac{4}{6}, \frac{6}{9}, \frac{8}{12}, \frac{10}{15}, \ldots\}$

(3) $\{\frac{5}{6}, \frac{10}{12}, \frac{15}{18}, \frac{20}{24}, \ldots\}$

(4) $\{\frac{9}{2}, \frac{18}{4}, \frac{27}{6}, \frac{36}{8}, \ldots\}$

(5) $\{\frac{0}{1}, \frac{0}{2}, \frac{0}{3}, \frac{0}{4}, \frac{0}{5}, \ldots\}$

(6) $\{\frac{1}{1}, \frac{2}{2}, \frac{3}{3}, \frac{4}{4}, \frac{5}{5}, \ldots\}$

Let us consider Set 2 above. Observe

$$\frac{2}{3} \qquad \frac{4}{6} \qquad \frac{6}{9} \qquad \frac{8}{12} \qquad \frac{10}{15} \quad \cdots$$

$$\downarrow \qquad \downarrow \qquad \downarrow \qquad \downarrow \qquad \downarrow$$

$$\frac{2 \times 1}{3 \times 1} \qquad \frac{2 \times 2}{3 \times 2} \qquad \frac{2 \times 3}{3 \times 3} \qquad \frac{2 \times 4}{3 \times 4} \qquad \frac{2 \times 5}{3 \times 5}$$

The above pattern gives a method of obtaining infinitely many other fractions equivalent to a given fraction. Notice that multiplying numerator and denominator of a fraction by a natural number produces a fraction equivalent to the given fraction.

Basic Property 1. **The fraction $\frac{a}{b}$ is equivalent to the fraction $\frac{an}{bn}$, n a natural number.**

Now let us look at $\frac{5}{10}$ and $\frac{8}{16}$, both elements of Set 1. Notice that the numerator and denominator of $\frac{5}{10}$ have 1 and 5 as common factors. Now let us divide numerator and denominator of $\frac{5}{10}$ by their common factor 5:

$$\frac{5 \div 5}{10 \div 5} \rightarrow \frac{1}{2}$$

Notice that $\frac{1}{2}$ is equivalent to $\frac{5}{10}$. The common divisors of 8 and 16 are 1, 2, 4, and 8. Let us divide numerator and denominator of $\frac{8}{16}$ by their common factor 4:

$$\frac{8 \div 4}{16 \div 4} \rightarrow \frac{2}{4}$$

Notice that $\frac{8}{16}$ and $\frac{2}{4}$ are equivalent fractions. In general if the numerator and denominator of a fraction are divided by one of their common factors, the resulting fraction is equivalent to the given fraction.

Basic Property 2. **If n, a natural number, is a factor of a and b, then $\frac{a}{b}$ is equivalent to $\frac{a \div n}{b \div n}$.**

Examining the definition of equivalent fractions and sets of equivalent fractions we observe that:

(1) The definition of equivalent fractions partitions the set of all fractions into classes of equivalent fractions called **equivalence classes.**
(2) Any two fractions in a given equivalence class are equivalent.
(3) No fraction in one equivalence class is equivalent to a fraction in another equivalence class.
(4) There are infinitely many fractions in each equivalence class.
(5) Every fraction is in one and only one equivalence class.

EXAMPLE i. Find the next three fractions in the equivalence class

$$\left\{ \frac{2}{5}, \frac{4}{10}, \frac{6}{15}, \frac{8}{20}, \cdots \right\}$$

Solution: $\frac{10}{25}, \frac{12}{30}, \frac{14}{35}$.

EXAMPLE ii. Find the fraction with denominator 48 belonging to the equivalence class

$$\left\{ \frac{3}{8}, \frac{6}{16}, \frac{9}{24}, \cdots \right\}$$

Solution: Since $8 \times 6 = 48$ we find

$$\frac{3 \times 6}{8 \times 6} \rightarrow \frac{18}{48}$$

9.6. Simplest Form of a Fraction

Let us look at the fractions in the equivalence class

$$\left\{ \frac{1}{2}, \frac{2}{4}, \frac{3}{6}, \cdots \right\}$$

Among the fractions in the equivalence class above, only the fraction $\frac{1}{2}$ has numerator and denominator relatively prime, that is $(1, 2) = 1$. When the numerator and denominator of a fraction are relatively prime the fraction is said to be in **simplest form** or in **lowest terms.**

DEFINITION 9.2. *A fraction whose numerator and denominator are relatively prime is said to be in* **simplest form.**

To decide whether or not a given fraction is in simplest form, we must look for common factors of the numerator and denominator. The simplest way to do this is to factor both numerator and denominator into their prime factors. For example, suppose we wish to find the simplest form of $\frac{216}{576}$. Factoring the numerator and denominator into their prime factors we have

$$216 = 2 \times 2 \times 2 \times 3 \times 3 \times 3 = 2^3 \times 3^3$$
$$576 = 2 \times 2 \times 2 \times 2 \times 2 \times 2 \times 3 \times 3 = 2^6 \times 3^2$$

The highest common factor of 216 and 576 is

$$2 \times 2 \times 2 \times 3 \times 3 = 2^3 \times 3^2 = 72.$$

Using Basic Property 2 and dividing both the numerator and denominator of $\frac{216}{576}$ by their highest common factor, 72, we obtain

$$\frac{216 \div 72}{576 \div 72} \rightarrow \frac{3}{8}$$

Since $(3, 8) = 1$, $\frac{3}{8}$ is in simplest form.

We can always find a fraction equivalent to a given fraction and in simplest form by dividing numerator and denominator of the given fraction by their highest common factor. Finding the simplest form of a fraction is called **reducing** it to simplest form.

EXAMPLE i.   Reduce $\frac{8}{12}$ to simplest form.

Solution: The highest common factor of 8 and 12 is 4. Dividing numerator and denominator of $\frac{8}{12}$ by 4 we obtain

$$\frac{8 \div 4}{12 \div 4} \rightarrow \frac{2}{3}$$

EXAMPLE ii.   Reduce $\frac{49}{56}$ to simplest form.
Solution: The highest common factor of 49 and 56 is 7.

$$\frac{49 \div 7}{56 \div 7} \rightarrow \frac{7}{8}$$

Given any fraction we can build the equivalence class containing that fraction. If we are given the simplest form fraction in the class, we multiply numerator and denominator of this fraction by each of the natural numbers greater than 1 in turn. Each fraction thus obtained is equivalent to the given fraction. Thus to build the equivalence fraction containing $\frac{1}{2}$, which is in simplest form, we obtain fractions equivalent to $\frac{1}{2}$ by multiplying the numerator and denominator of $\frac{1}{2}$ by 2, 3, 4, and so on. Thus

$$\frac{1 \times 2}{2 \times 2} \rightarrow \frac{2}{4} \qquad \frac{1 \times 3}{2 \times 3} \rightarrow \frac{3}{6} \qquad \frac{1 \times 4}{2 \times 4} \rightarrow \frac{4}{8} \qquad \frac{1 \times 5}{2 \times 5} \rightarrow \frac{5}{10} \qquad \cdots$$

The equivalence class containing $\frac{1}{2}$ is

$$\left\{ \frac{1}{2}, \frac{2}{4}, \frac{3}{6}, \frac{4}{8}, \frac{5}{10}, \cdots \right\}$$

If the given fraction is not in simplest form we reduce it to simplest form and proceed as above.

EXAMPLE iii.   Build the equivalence class that contain $\frac{12}{18}$.
Solution: Since $(12, 18) \neq 1$, the given fraction is not in simplest form. We first reduce it to simplest form:

$$\frac{12 \div 6}{18 \div 6} \rightarrow \frac{2}{3}$$

The equivalence class containing $\frac{2}{3}$ is

$$\left\{ \frac{2}{3}, \frac{4}{6}, \frac{6}{9}, \frac{8}{12}, \cdots \right\}$$

Exercise 9.2

1. Give the next three fractions in each equivalence class.
    a. $\left\{ \frac{1}{5}, \frac{2}{10}, \frac{3}{15}, \cdots \right\}$
    b. $\left\{ \frac{5}{6}, \frac{10}{12}, \frac{15}{18}, \cdots \right\}$
    c. $\left\{ \frac{1}{4}, \frac{2}{8}, \frac{3}{12}, \cdots \right\}$
    d. $\left\{ \frac{1}{16}, \frac{2}{32}, \frac{3}{48}, \cdots \right\}$
    e. $\left\{ \frac{5}{7}, \frac{10}{14}, \frac{15}{21}, \cdots \right\}$
2. Write a fraction equivalent to the given fraction and having the denominator indicated in parentheses.
    a. $\frac{3}{4}$ (16)   e. $\frac{4}{3}$ (132)
    b. $\frac{5}{12}$ (72)   f. $\frac{8}{7}$ (49)
    c. $\frac{7}{14}$ (42)   g. $\frac{3}{100}$ (5000)
    d. $\frac{12}{9}$ (108)   h. $\frac{9}{12}$ (132)

3. Give the equivalence class containing each fraction.

   a. $\frac{5}{6}$ e. $\frac{15}{48}$

   b. $\frac{7}{8}$ f. $\frac{12}{16}$

   c. $\frac{12}{14}$ g. $\frac{24}{36}$

   d. $\frac{8}{32}$ h. $\frac{15}{3}$

4. Write the following in simplest form.

   a. $\frac{10}{12}$ d. $\frac{36}{148}$

   b. $\frac{9}{27}$ e. $\frac{196}{224}$

   c. $\frac{15}{65}$ f. $\frac{169}{130}$

5. Write the following in simplest form.

   a. $\frac{125}{175}$ d. $\frac{96}{144}$

   b. $\frac{121}{132}$ e. $\frac{119}{187}$

   c. $\frac{45}{100}$ f. $\frac{60}{225}$

6. Which of the following pairs of fractions belong to the same equivalence class?

   a. $\frac{9}{16}, \frac{3}{4}$ d. $\frac{16}{136}, \frac{68}{578}$

   b. $\frac{14}{5}, \frac{28}{10}$ e. $\frac{24}{100}, \frac{54}{225}$

   c. $\frac{9}{72}, \frac{72}{192}$ f. $\frac{84}{168}, \frac{75}{150}$

7. Find the whole number values of n that make the following true statements.

   a. $\frac{3}{5}$ is equivalent to $\frac{n}{125}$

   b. $\frac{2}{3}$ is equivalent to $\frac{n}{24}$

   c. $\frac{9}{16}$ is equivalent to $\frac{108}{n}$

   d. $\frac{7}{8}$ is equivalent to $\frac{105}{n}$

   e. $\frac{5}{12}$ is equivalent to $\frac{480}{n}$

   f. $\frac{13}{15}$ is equivalent to $\frac{169}{n}$

8. Fill in the blanks to make true statements.

   a. $\frac{4}{6}$ is equivalent to $\frac{6}{9}$ because $4 \times$ _____ = _____ $\times$ 6.

   b. $\frac{16}{10}$ is equivalent to $\frac{8}{5}$ because $16 \times$ _____ = $10 \times$ _____.

   c. $\frac{75}{100}$ is equivalent to $\frac{3}{4}$ because _____ $\times$ _____ = 100×3.

   d. $\frac{12}{24}$ is equivalent to _____ because $12 \times 48 = 24 \times 24$.

   e. $\frac{3}{8}$ is equivalent to _____ because $3 \times 32 = 8 \times 12$.

9. Give the simplest form fraction for each set.

   a. $\left\{ \frac{9}{15}, \frac{21}{35}, \frac{24}{40}, \frac{30}{50} \right\}$

   b. $\left\{ \frac{5}{50}, \frac{7}{70}, \frac{100}{1000}, \frac{40}{400} \right\}$

   c. $\left\{ \frac{16}{6}, \frac{32}{12}, \frac{80}{30}, \frac{64}{24} \right\}$

   d. $\left\{ \frac{24}{27}, \frac{88}{99}, \frac{32}{36}, \frac{56}{63} \right\}$

10. What number would you divide numerator and denominator by to reduce the following to simplest form?

a. $\frac{63}{72}$ d. $\frac{135}{432}$

b. $\frac{60}{75}$ e. $\frac{288}{648}$

c. $\frac{100}{350}$ f. $\frac{560}{160}$

11. Give the equivalence class containing each of the following.

a. $\frac{2}{3}$ d. $\frac{84}{48}$

b. $\frac{5}{25}$ e. $\frac{21}{700}$

c. $\frac{36}{16}$ f. $\frac{18}{135}$

12. Give the missing fractions in each set.

a. $\{\frac{1}{3}, \frac{2}{6}, \frac{3}{9},$ ———— $, \frac{4}{12}, \frac{5}{15}, \ldots\}$

b. $\{\frac{4}{5}, \frac{8}{10}, \frac{12}{15}, \frac{16}{20},$ ———— $, \frac{24}{30},$ ———— $,$ ———— $, \ldots\}$

c. $\{$ ———— $,$ ———— $,$ ———— $, \frac{36}{40}, \frac{45}{50}, \ldots\}$

d. $\{$ ———— $, \frac{16}{10}, \frac{24}{15},$ ———— $,$ ———— $, \frac{48}{30}, \ldots\}$

13. Complete the following.

a. A serving of one-sixth of a pizza provides the same amount of pizza as two- ———— of the pizza.

b. If a team won 8 of 16 games played, then the team won eight-sixteenths or one- ———— of the games played.

c. A girl spends 6 out of every 9 dollars that she earns for clothes. She spends six-ninths or ———— -thirds of her earnings for clothes.

d. If 18 of the 20 students in a mathematics class are men, then eighteen-twentieths or ———— -tenths of the class are men.

14. Which fraction in each set below is not equivalent to the others in the set?

a. $\{\frac{5}{6}, \frac{20}{24}, \frac{25}{36}, \frac{40}{48}\}$

b. $\{\frac{14}{16}, \frac{35}{40}, \frac{56}{63}, \frac{70}{80}\}$

c. $\{\frac{49}{112}, \frac{14}{32}, \frac{35}{80}, \frac{70}{150}\}$

d. $\{\frac{27}{6}, \frac{36}{8}, \frac{99}{22}, \frac{107}{24}\}$

15. Find the highest common factor of the numerator and denominator of the following.

a. $\frac{18}{36}$ d. $\frac{21}{112}$

b. $\frac{48}{108}$ e. $\frac{248}{388}$

c. $\frac{105}{60}$ f. $\frac{700}{475}$

16. Name a fraction with denominator 100 in the same equivalence class as the following.

---

a. $\frac{3}{5}$ d. $\frac{9}{25}$

b. $\frac{7}{4}$ e. $\frac{11}{25}$

c. $\frac{5}{2}$ f. $\frac{12}{5}$

17. Name a fraction with denominator 144 in the same equivalence class as the following.

a. $\frac{2}{3}$ d. $\frac{9}{16}$

b. $\frac{5}{16}$ e. $\frac{11}{12}$

c. $\frac{7}{72}$ f. $\frac{1}{36}$

18. If both the numerator and denominator of a fraction are even, can the fraction be in simplest form? Why?

19. The highest common factor of the numerator and denominator of a fraction is 1. Is the fraction in simplest form?

20. If the numerator and denominator of a fraction are two different prime numbers is the fraction in simplest form? Why?

10.1. Numbers Named by Fractions

We recall that we developed the concept of cardinal numbers by thinking about classes of equivalent sets. Thus we associated the cardinal number three with the class of all sets that are equivalent to the set $\{1, 2, 3\}$. Just as we developed the concept of the cardinal numbers from classes of equivalent sets, we develop the concept of **non-negative rational numbers,** which we shall, in this chapter, call **rational numbers,** from classes of equivalent fractions.

Let us consider the equivalence class

$$\left\{\frac{1}{2}, \frac{2}{4}, \frac{3}{6}, \frac{4}{8}, \cdots\right\}$$

With this set we associate one rational number. Every fraction in this equivalence class is a symbol for this rational number. Thus we see that *every rational number has many names.* For example $\frac{1}{2}$, $\frac{4}{8}$, $\frac{6}{12}$, and $\frac{80}{160}$ are names for the same rational number.

Non-Negative
Rational Numbers

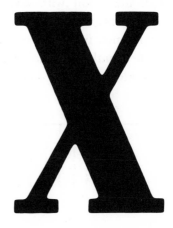

Although a rational number has many names, the simplest form fraction is the most commonly used name for a rational number. Thus the rational number associated with the equivalence class above is usually named $\frac{1}{2}$.

We recall that each whole number is associated with one and only one point on the number line. A rational number is also associated with one and only one point on the number line. As in the case of the whole numbers, this point is called the **graph** of the rational number; the rational number is called the **coordinate** of the point on the number line. Figure 10.1 shows the graphs of the rational numbers $\frac{1}{4}$, $\frac{1}{2}$, and $\frac{3}{4}$.

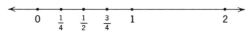

Figure 10.1

Suppose we wish to locate the point on the number line that is the graph of $\frac{5}{16}$. We divide that part of the number line between 0 and 1 into sixteen congruent sublengths. The point at the right hand end of the fifth part represents $\frac{5}{16}$, that is, is the graph of this rational number (Figure 10.2).

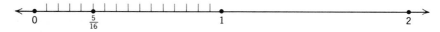

Figure 10.2

Although we think of exactly one number for a given point on the number line and exactly one point for a given rational number, we may label this point with any fraction from the set of equivalent fractions naming this number. Thus the point associated with $\frac{1}{4}$ may be labeled with any fraction from the set $\{\frac{1}{4}, \frac{2}{8}, \frac{3}{12}, \frac{4}{16}, \ldots\}$. Figure 10.3 shows some points on the number line labeled by several fractions in the equivalence class associated with the coordinate of that point.

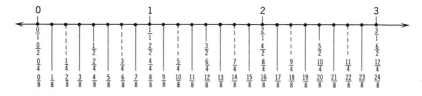

Figure 10.3

We now formally define a rational number.

DEFINITION 10.1. *A **non-negative rational number** is an ordered pair,* (a, b), *of whole numbers,* $b \neq 0$, *and is denoted by* $\frac{a}{b}$.

Using set-builder notation we denote the set, $R^\#$, of non-negative rational numbers by

$$R^\# = \{\tfrac{a}{b} \mid a \text{ and } b \text{ are whole numbers and } b \neq 0\}$$

10.2. Equality of Rational Numbers

The fact that every rational number has many names leads us to define equality of two rational numbers. Since $\frac{1}{2}$ and $\frac{2}{4}$ are in the same equivalence class, that is they are equivalent fractions, the symbols "$\frac{1}{2}$" and "$\frac{2}{4}$" are names for the same rational number. We write

$$\frac{1}{2} = \frac{2}{4}$$

to indicate that $\frac{1}{2}$ and $\frac{2}{4}$ are names for the same rational number. In general, we say that

$$\frac{a}{b} = \frac{c}{d}$$

means that $\frac{a}{b}$ and $\frac{c}{d}$ are names for the same rational number. Thus

$$\frac{5}{6} = \frac{15}{18}$$

$$\frac{4}{6} = \frac{16}{24}$$

$$\frac{9}{3} = \frac{3}{1}$$

Note that

$$\frac{5}{6} = \frac{15}{18} \text{ and } 5 \times 18 = 6 \times 15$$

$$\frac{4}{6} = \frac{16}{24} \text{ and } 4 \times 24 = 6 \times 16$$

$$\frac{9}{3} = \frac{3}{1} \text{ and } 9 \times 1 = 3 \times 3$$

In general, two symbols name the same rational number if and only if they are equivalent fractions.

DEFINITION 10.2. *Two rational numbers $\frac{a}{b}$ and $\frac{c}{d}$ are **equal**, denoted by $\frac{a}{b} = \frac{c}{d}$ if and only if $\frac{a}{b}$ and $\frac{c}{d}$ are equivalent fractions that is $ad = bc$.*

EXAMPLE i. Is $\frac{30}{45}$ equal to $\frac{22}{33}$?

Solution: Using Definition 10.2 we find that $\frac{30}{45}$ and $\frac{22}{33}$ are equivalent fractions since

$$30 \times 33 = 990$$
$$45 \times 22 = 990$$

Hence $\frac{30}{45} = \frac{22}{33}$.

EXAMPLE ii. Find the solution set of $\frac{2}{3} = \frac{n}{24}$. The domain is the set of whole numbers.

Solution: Using equivalent fractions we find

$$2 \times 24 = 3n$$
$$48 = 3n$$

The solution set of this equation is $\{16\}$.

10.3. Ordering the Rational Numbers

If we wish to order two rational numbers such as $\frac{1}{8}$ and $\frac{3}{8}$, we conclude that $\frac{1}{8} < \frac{3}{8}$ by considering the physical interpretation from which the fractions $\frac{1}{8}$ and $\frac{3}{8}$ arise. Similarly we see that $\frac{5}{8} < \frac{8}{8}$, $\frac{4}{8} < \frac{16}{8}$, and $\frac{15}{8} < \frac{24}{8}$. We conclude that rational numbers named by fractions with the same denominator present themselves in a very definite order. For example

$$\frac{0}{8}, \frac{1}{8}, \frac{2}{8}, \frac{3}{8}, \ldots, \frac{8}{8}, \frac{9}{8}, \ldots$$

Thus rational numbers named by fractions with the same denominator are ordered by the numerators of these fractions. We see that each succeeding rational number in the sequence above is greater than the preceding one:

$$\frac{0}{8} < \frac{1}{8} < \frac{2}{8} < \frac{3}{8} < \cdots < \frac{8}{8} < \frac{9}{8} < \cdots$$

because $0 < 1 < 2 < 3 < \cdots < 8 < 9 < \cdots$

DEFINITION 10.3. *If $\frac{a}{b}$ and $\frac{c}{b}$ are two rational numbers, $\frac{a}{b} < \frac{c}{b}$ if and only if $a < c$, and $\frac{a}{b} > \frac{c}{b}$ if and only if $a > c$.*

Using Definition 10.3 it is possible to order any two rotational numbers. All that is necessary is to choose fractions from the equivalence class of each rational number so that they have the same denominator. For example, let us order $\frac{2}{3}$ and $\frac{5}{8}$. The fractions $\frac{2}{3}$ and $\frac{16}{24}$ are in the same equivalence class, so the rational numbers are equal:

$$\frac{2}{3} = \frac{16}{24}$$

Similarly

$$\frac{5}{8} = \frac{15}{24}$$

By Definition 10.3

$$\frac{2}{3} > \frac{5}{8} \quad \text{since} \quad \frac{16}{24} > \frac{15}{24}$$

Observe the following:

$$\frac{2}{3} > \frac{1}{2} \quad \text{and } 2 \times 2 > 3 \times 1$$

$$\frac{10}{3} > \frac{5}{2} \quad \text{and } 10 \times 2 > 3 \times 5$$

$$\frac{5}{6} = \frac{10}{12} \quad \text{and } 5 \times 12 = 6 \times 10$$

$$\frac{1}{3} < \frac{5}{8} \quad \text{and } 1 \times 8 < 3 \times 5$$

$$\frac{5}{6} < \frac{9}{7} \quad \text{and } 5 \times 7 < 6 \times 9$$

The above examples lead us to the following generalization.

DEFINITION 10.4. *If $\frac{a}{b}$ and $\frac{c}{d}$ are two nonnegative rational numbers, then*

(1) $\frac{a}{b} < \frac{c}{d}$ if and only if $ad < bc$
(2) $\frac{a}{b} = \frac{c}{d}$ if and only if $ad = bc$
(3) $\frac{a}{b} > \frac{c}{d}$ if and only if $ad > bc$.

Just as the order of whole numbers shows up clearly by the position of their graphs on the number line, the order of the rational numbers is also shown on the number line. The graph of the whole number 8 is to the right of the graph of the whole number 3 and $8 > 3$. Similarly the graph of the rational number $\frac{3}{4}$ is to the right of the graph of the rational number $\frac{1}{2}$, and correspondingly, $\frac{3}{4} > \frac{1}{2}$.

EXAMPLE i. Is the graph of $\frac{7}{8}$ to the right or to the left of the graph of $\frac{4}{5}$?

Solution: We see that $4 \times 8 < 5 \times 7$, hence $\frac{4}{5} < \frac{7}{8}$; so the graph of $\frac{7}{8}$ is to the right of the graph of $\frac{4}{5}$.

EXAMPLE ii. Find the solution set of $\frac{7}{8} > \frac{3}{n}$. The domain is the set of whole numbers.

Solution: Using Definition 10.4 we find

$$7n > 24$$

We now solve this inequality. We see that n must be a natural number (n cannot be 0; why?) greater than 3. The solution set is $\{4, 5, 6, 7, \ldots\}$.

EXAMPLE iii. Find the solution set of $\frac{n}{3} < \frac{9}{16}$. The domain is the set of whole numbers.

Solution: Using Definition 10.4 we find

$$16n < 27$$

The solution set of this inequality is $\{0, 1\}$.

Exercise 10.1

1. Replace * with the symbols $<$, $=$ or $>$ to make true statements.
   a. $\frac{2}{3} * \frac{1}{3}$
   f. $\frac{28}{3} * \frac{30}{4}$
   b. $\frac{7}{16} * \frac{12}{16}$
   g. $\frac{8}{11} * \frac{9}{21}$
   c. $\frac{3}{4} * \frac{1}{2}$
   h. $\frac{130}{169} * \frac{10}{13}$
   d. $\frac{2}{3} * \frac{6}{9}$
   i. $\frac{68}{96} * \frac{48}{56}$
   e. $\frac{17}{20} * \frac{15}{17}$
   j. $\frac{121}{125} * \frac{364}{488}$

2. Order the following rational numbers from the least to the greatest:
   $\frac{2}{3}, \frac{1}{6}, \frac{3}{16}, \frac{5}{8}, \frac{1}{5}, \frac{0}{9}$.

3. Draw a number line. Mark the point representing $\frac{1}{2}$ by X, the point representing $\frac{3}{4}$ by Y, the point representing $\frac{5}{8}$ by Z, and the point representing $\frac{1}{4}$ by T.

4. Find the solution sets. The domain is the set of whole numbers.
    a. $\frac{3}{12} = \frac{n}{4}$ c. $\frac{7}{5} = \frac{n}{125}$
    b. $\frac{5}{6} = \frac{15}{n}$ d. $\frac{3}{16} = \frac{24}{n}$
5. Find the solution sets. The domain is the set of whole numbers.
    a. $\frac{3}{4} < \frac{5}{n}$ c. $\frac{7}{9} > \frac{2}{n}$
    b. $\frac{1}{n} > \frac{1}{3}$ d. $\frac{n}{3} < \frac{8}{11}$
6. Find the solution sets. The domain is the set of whole numbers.
    a. $\frac{n}{4} > \frac{3}{8}$ c. $\frac{n}{3} < \frac{5}{8}$
    b. $\frac{2}{3} < \frac{5}{n}$ d. $\frac{2}{n} > \frac{3}{7}$
7. Tell which fraction in each given pair names the greater number. If the fractions name the same number, so state.
    a. $\frac{1}{2}, \frac{1}{3}$ e. $\frac{7}{8}, \frac{21}{24}$
    b. $\frac{5}{6}, \frac{3}{8}$ f. $\frac{7}{10}, \frac{75}{100}$
    c. $\frac{0}{7}, \frac{5}{3}$ g. $\frac{3}{16}, \frac{5}{17}$
    d. $\frac{5}{10}, \frac{51}{100}$ h. $\frac{18}{12}, \frac{9}{6}$
8. A baseball team won $\frac{13}{15}$ of the games it played. Another team won $\frac{49}{55}$ of the games it played. Which team had the better record?
9. Phil Jones answered $\frac{15}{17}$ of the questions on a test correctly. Jay Thomas answered $\frac{48}{57}$ of the questions correctly. Which boy had the higher score on the test?
10. David Spencer spends $\frac{6}{9}$ of his income and saves the rest. Mike Pilgrim spends $\frac{15}{16}$ of his income and saves the rest. Which boy saves the greater part of his income?

10.4. Addition of Rational Numbers

We now define the operation of addition of rational numbers named by common fractions with the same denominator.

DEFINITION 10.5. *If $\frac{a}{b}$ and $\frac{c}{b}$ are two rational numbers, their* **sum,** *denoted by $\frac{a}{b} + \frac{c}{b}$, is $\frac{a+c}{b}$.*

We see from Definition 10.5 that when we add two rational numbers named by common fractions with the same denominator we add the numerators and use the same denominator.

EXAMPLE i. Find the sum $\frac{2}{8} + \frac{3}{8}$.
Solution: Using Definition 10.5 we obtain

$$\frac{2}{8} + \frac{3}{8} = \frac{2+3}{8} = \frac{5}{8}.$$

We can show the addition of two rational numbers, for example $\frac{2}{8} + \frac{3}{8}$, on the number line (Figure 10.4).

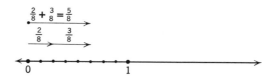

Figure 10.4

If two rational numbers named by common fractions with different denominators are to be added, they are first renamed by fractions with the same denominator, called a **common denominator**. They are then added using Definition 10.5.

EXAMPLE ii.   Find the sum $\frac{3}{4} + \frac{2}{3}$.

Solution: We first rename $\frac{3}{4}$ and $\frac{2}{3}$ by fractions with a common denominator:

$$\frac{3}{4} = \frac{3 \times 3}{4 \times 3} = \frac{9}{12}$$

$$\frac{2}{3} = \frac{2 \times 4}{3 \times 4} = \frac{8}{12}$$

Using Definition 10.5 we obtain

$$\frac{3}{4} + \frac{2}{3} = \frac{9}{12} + \frac{8}{12}$$

$$= \frac{9 + 8}{12}$$

$$= \frac{17}{12}$$

Now let us develop a rule for finding the sum of any two rational numbers, $\frac{a}{b}$ and $\frac{c}{d}$. In order to use Definition 10.5 we must rename these two rational numbers by fractions with a common denominator. Let us multiply numerator and denominator of $\frac{a}{b}$ by d and numerator and denominator of $\frac{c}{d}$ by b:

$$\frac{a}{b} = \frac{a \times d}{b \times d} = \frac{ad}{bd}$$

$$\frac{c}{d} = \frac{c \times b}{d \times b} = \frac{cb}{db}$$

Using Definition 10.5 we have

$$\frac{a}{b} + \frac{c}{d} = \frac{ad}{bd} + \frac{cb}{db}$$

$$= \frac{ad}{bd} + \frac{bc}{bd} \qquad \text{Why?}$$

$$= \frac{ad + bc}{bd}$$

DEFINITION 10.6. *If $\frac{a}{b} + \frac{c}{d}$ are two rational numbers, their* **sum,** *$\frac{a}{b} + \frac{c}{d}$, is $\frac{ad + bc}{bd}$.*

10.5. Subtraction of Rational Numbers

We define subtraction of rational numbers as the inverse of addition of rational numbers. Thus $\frac{7}{3} - \frac{2}{3} = \frac{5}{3}$ because $\frac{2}{3} + \frac{5}{3} = \frac{7}{3}$.

DEFINITION 10.7. *If $\frac{a}{b}$ and $\frac{c}{b}$, $a \geq c$ are two rational numbers, their* **difference** *denoted by $\frac{a}{b} - \frac{c}{b}$ is $\frac{a - c}{b}$.*

We can show the difference of two rational numbers, for example $\frac{7}{3} - \frac{2}{3}$, on the number line (Figure 10.5).

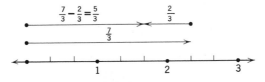

Figure 10.5

If two rational numbers named by fractions with different denominators are to be subtracted, in order to use Definition 10.7 we must first rename them by fractions with a common denominator. Suppose we wish to find the difference $\frac{a}{b} - \frac{c}{d}$. We change $\frac{a}{b}$ to an equivalent fraction by multiplying numerator and denominator by d; we change $\frac{c}{d}$ to an equivalent fraction by multiplying numerator and denominator by b. Then

$$\frac{a}{b} = \frac{a \times d}{b \times d} = \frac{ad}{bd}$$

$$\frac{c}{d} = \frac{c \times b}{d \times b} = \frac{cb}{db} = \frac{bc}{bd}$$

Now, using Definition 10.7, we obtain

$$\frac{a}{b} - \frac{c}{d} = \frac{ad}{bd} - \frac{bc}{bd} = \frac{ad - bc}{bd}$$

DEFINITION 10.8. *If $\frac{a}{b}$ and $\frac{c}{d}$ are rational numbers, $ad \geqslant bc$, their* **difference** $\frac{a}{b} - \frac{c}{d}$ *is* $\frac{ad - bc}{bd}$.

EXAMPLE i. Subtract $\frac{3}{4} - \frac{1}{2}$.
Solution: Using Definition 10.8 we obtain

$$\frac{3}{4} - \frac{1}{2} = \frac{(3 \cdot 2) - (4 \cdot 1)}{4 \cdot 2}$$

$$= \frac{6 - 4}{8}$$

$$= \frac{2}{8}$$

$$= \frac{1}{4}$$

EXAMPLE ii. Subtract $\frac{13}{8} - \frac{3}{4}$. Write the fraction naming the difference in simplest form.
Solution: Using Definition 10.8 we have

$$\frac{13}{8} - \frac{3}{4} = \frac{(13 \cdot 4) - (3 \cdot 8)}{8 \cdot 4}$$

$$= \frac{52 - 24}{32}$$

$$= \frac{28}{32}$$

$$= \frac{7}{8}$$

10.6. Properties of Addition of Rational Numbers

We shall now show that the set of rational numbers and the operation of addition as defined in Section 10.4 possess the properties listed below. Since two or more rational numbers may be named by common fractions with the same denominator, we shall prove the following properties naming the rational numbers by fractions with the same denominator without losing any generality.

R#-1 **Closure Property of Addition.** If $\frac{a}{b}$ and $\frac{c}{d}$ are rational numbers, then $\frac{a}{b} + \frac{c}{b}$ is a rational number.

R#-2 **Commutative Property of Addition.** If $\frac{a}{b} + \frac{c}{b}$ are rational numbers, then $\frac{a}{b} + \frac{c}{b} = \frac{c}{b} + \frac{a}{b}$.

R#-3 **Associative Property of Addition.** If $\frac{a}{b}$, $\frac{c}{b}$ and $\frac{d}{b}$ are rational numbers, then $(\frac{a}{b} + \frac{c}{b}) + \frac{d}{b} = \frac{a}{b} + (\frac{c}{b} + \frac{d}{b})$.

R#-4 **Identity Element for Addition.** In the set of rational numbers there exists an element $\frac{0}{b}$ such that for every rational number $\frac{a}{b}$, $\frac{a}{b} + \frac{0}{b} = \frac{0}{b} + \frac{a}{b} = \frac{a}{b}$. The rational number $\frac{0}{b}$ is called the **identity element of addition** or the **additive identity.**

Proof of R#-1: By the definition of addition of rational numbers

$$\frac{a}{b} + \frac{c}{b} = \frac{a + c}{b}.$$

By Definition 10.1, a, b, and c are whole numbers, $b \neq 0$, hence $a + c$ is a whole number by the closure property of addition of whole numbers. Then $\frac{a + c}{b}$ is a rational number by Definition 10.1.

Proof of R#-2: By the definition of addition of rational numbers

$$\frac{a}{b} + \frac{c}{b} = \frac{a + c}{b}$$

$$= \frac{c + a}{b} \qquad \text{Commutative property of addition of whole numbers}$$

$$= \frac{c}{b} + \frac{a}{b} \qquad \text{Definition 10.5}$$

Proof of R#-3: By the definition of addition of rational numbers

$$\left(\frac{a}{b} + \frac{c}{b}\right) + \frac{d}{b} = \frac{a+c}{b} + \frac{d}{b}$$

$$= \frac{(a+c)+d}{b} \qquad \text{Definition 10.5}$$

$$= \frac{a+(c+d)}{b} \qquad \text{Associative property of addition of whole numbers}$$

$$= \frac{a}{b} + \frac{c+d}{b} \qquad \text{Definition 10.5}$$

$$= \frac{a}{b} + \left(\frac{c}{b} + \frac{d}{b}\right) \qquad \text{Definition 10.5}$$

Proof of R#-4: $\quad \frac{a}{b} + \frac{0}{b} = \frac{a+0}{b} \qquad \text{Definition 10.5}$

$$= \frac{a}{b} \qquad \text{Additive identity for whole numbers}$$

The proof that $\frac{0}{b} + \frac{a}{b} = \frac{a}{b}$ is left to the reader.

Exercise 10.2

1. Add. Name the sums in simplest form.
   a. $\frac{2}{8} + \frac{4}{8}$ e. $\frac{9}{16} + \frac{1}{16}$
   b. $\frac{5}{16} + \frac{7}{16}$ f. $\frac{7}{12} + \frac{3}{12}$
   c. $\frac{3}{5} + \frac{4}{5}$ g. $\frac{5}{24} + \frac{7}{24}$
   d. $\frac{3}{25} + \frac{12}{25}$ h. $\frac{13}{100} + \frac{12}{100}$

2. Add. Name the sums in simplest form.
   a. $\frac{1}{3} + \frac{1}{4}$ e. $\frac{5}{16} + \frac{1}{3}$
   b. $\frac{1}{3} + \frac{2}{5}$ f. $\frac{9}{25} + \frac{3}{5}$
   c. $\frac{3}{8} + \frac{2}{3}$ g. $\frac{3}{5} + \frac{3}{100}$
   d. $\frac{3}{4} + \frac{5}{8}$ h. $\frac{2}{3} + \frac{5}{8}$

3. Add. Name the sums in simplest form.
   a. $\frac{1}{2} + \frac{1}{4} + \frac{1}{8}$
   b. $\frac{3}{8} + \frac{1}{2} + \frac{1}{3}$
   c. $\frac{5}{16} + \frac{5}{8} + \frac{2}{3}$
   d. $\frac{5}{6} + \frac{3}{8} + \frac{3}{4}$

4. Add. Name the sums in simplest form.
   a. $\frac{2}{3} + \frac{3}{8} + \frac{5}{16}$
   b. $\frac{3}{4} + \frac{1}{2} + \frac{3}{5}$
   c. $\frac{5}{3} + \frac{2}{5} + \frac{1}{4}$
   d. $\frac{1}{2} + \frac{3}{4} + \frac{7}{16}$

5. Subtract. Name the differences in simplest form.
   a. $\frac{5}{8} - \frac{1}{8}$ c. $\frac{9}{25} - \frac{4}{25}$
   b. $\frac{3}{4} - \frac{1}{4}$ d. $\frac{9}{16} - \frac{5}{16}$

6. Subtract. Name the differences in simplest form.
   a. $\frac{3}{4} - \frac{1}{2}$ d. $\frac{5}{16} - \frac{1}{4}$
   b. $\frac{7}{8} - \frac{1}{12}$ e. $\frac{17}{12} - \frac{5}{8}$
   c. $\frac{1}{2} - \frac{3}{10}$ f. $\frac{15}{24} - \frac{3}{8}$

7. Find the solution sets. The domain is the set of rational numbers.
   a. $\frac{2}{3} + n = \frac{7}{3}$ c. $\frac{5}{8} + n = \frac{23}{24}$
   b. $\frac{5}{8} - n = \frac{1}{2}$ d. $\frac{19}{16} - n = \frac{3}{4}$

8. Find the solution sets. The domain is the set of rational numbers.
   a. $\frac{1}{2} + n = \frac{5}{8}$ c. $n - \frac{1}{3} = \frac{5}{8}$
   b. $\frac{3}{4} - n = \frac{1}{3}$ d. $n - \frac{2}{3} = \frac{9}{16}$

9. Perform the indicated operations.
   a. $(\frac{3}{4} - \frac{2}{3}) + \frac{1}{2}$ c. $\frac{25}{16} - (\frac{1}{2} + \frac{1}{8})$
   b. $(\frac{2}{3} + \frac{5}{8}) - \frac{3}{4}$ d. $\frac{31}{6} - (\frac{2}{3} - \frac{1}{2})$

10. The sum of two rational numbers is $\frac{9}{8}$. One of the addends is $\frac{15}{24}$. What is the other addend?

11. Justify each statement below by a property of addition of rational numbers.
    a. $\frac{5}{6} + \frac{2}{3} = \frac{2}{3} + \frac{5}{6}$
    b. $\frac{3}{4} + \frac{0}{9} = \frac{3}{4}$
    c. $\frac{2}{3} + (\frac{3}{4} + \frac{5}{6}) = (\frac{2}{3} + \frac{3}{4}) + \frac{5}{6}$
    d. $(\frac{2}{3} + \frac{3}{5}) + (\frac{7}{8} + \frac{3}{8}) = (\frac{3}{5} + \frac{2}{3}) + (\frac{7}{8} + \frac{3}{8})$
    e. $\frac{2}{3} + (\frac{3}{4} + \frac{1}{3}) = \frac{2}{3} + (\frac{1}{3} + \frac{3}{4})$

12. Which of the following are true statements?
    a. Addition of rational numbers is a commutative operation.
    b. The sum of $\frac{3}{4}$ and $\frac{1}{4}$ is a rational number.
    c. The set of non-negative rational numbers is closed under the operation of subtraction.
    d. The fractions $\frac{5}{6}$ and $\frac{10}{18}$ name two different rational numbers.

13. The **least common denominator** of the common denominators of several fractions is the least common multiple of the denominators of

the fractions. What is the least common denominator of the sets of fractions below?

a. $\frac{1}{2}, \frac{3}{4}$ e. $\frac{5}{8}, \frac{7}{6}, \frac{1}{10}$

b. $\frac{2}{3}, \frac{5}{6}$ f. $\frac{1}{16}, \frac{5}{12}, \frac{7}{30}$

c. $\frac{3}{4}, \frac{5}{8}$ g. $\frac{1}{2}, \frac{5}{6}, \frac{7}{27}$

d. $\frac{3}{8}, \frac{7}{12}$ h. $\frac{3}{100}, \frac{2}{25}, \frac{1}{15}$

14. Add. Use the least common denominator.

     a. $\frac{1}{2} + \frac{3}{4} + \frac{2}{3}$ d. $\frac{5}{16} + \frac{1}{14} + \frac{3}{7}$

     b. $\frac{2}{3} + \frac{1}{5} + \frac{5}{6}$ e. $\frac{7}{8} + \frac{2}{3} + \frac{9}{15}$

     c. $\frac{1}{8} + \frac{2}{3} + \frac{7}{6}$ f. $\frac{4}{5} + \frac{3}{100} + \frac{2}{3}$

15. During a missile launch, the countdown was delayed four times. The first delay lasted $\frac{1}{4}$ of an hour; the second, $\frac{1}{5}$ of an hour; the third, $\frac{2}{5}$ of an hour; and the fourth, $\frac{2}{3}$ of an hour. In all, how long was the missile launch delayed?

16. Mrs. Foster reported to her husband: "Last year we spent $\frac{1}{4}$ of our income to pay off the mortgage on the house, $\frac{2}{5}$ for food and recreation, and $\frac{1}{10}$ for clothing. We invested $\frac{1}{12}$ of our income in stocks and bonds. I just can't account for the way we spent the rest." For what part of their income was Mrs. Foster unable to account?

17. At the end of the first hour of trading on the New York Stock Exchange, the price of a certain stock had risen $\frac{3}{8}$ of a point. During the next hour the price of the same stock rose another $\frac{2}{3}$ of a point. The price remained the same until the last hour of trading when it dropped $\frac{1}{5}$ of a point. Did the stock have a net gain or a net loss in price for the day? How much was the net gain or loss?

18. During a recent rain storm it rained $\frac{1}{2}$ of an inch on Monday, $\frac{2}{3}$ of an inch on Tuesday, $\frac{3}{10}$ of an inch on Wednesday, and $\frac{3}{4}$ of an inch on Friday. During the rest of the week it was cloudy, but it did not rain. What was the total rainfall for the week?

19. Fred Felstar was required to write a research paper for a literature class. He completed the paper on Wednesday and began typing it on Thursday. He typed $\frac{4}{15}$ of the paper on Thursday and $\frac{5}{12}$ of it on Friday. He completed the typing on Sunday. What portion of the paper did he type on Sunday?

20. The Carltons were sorting the books in their library. They found that $\frac{2}{3}$ of the books were novels, $\frac{1}{16}$ of the books were technical books, $\frac{1}{24}$ of the books were reference books, and the rest were cookbooks. What portion of the Carltons' books were cookbooks?

10.7. Multiplication of Rational Numbers

We now define multiplication of two rational numbers.

DEFINITION 10.9. *If $\frac{a}{b}$ and $\frac{c}{d}$ are rational numbers, their* **product** *$\frac{a}{b} \cdot \frac{c}{d}$ is* $\frac{ac}{bd}$.

EXAMPLE i. Find the product $\frac{1}{2} \cdot \frac{2}{3}$.

Solution: Using Definition 10.9 we have

$$\frac{1}{2} \cdot \frac{2}{3} = \frac{1 \cdot 2}{2 \cdot 3} = \frac{2}{6}$$

Multiplication of rational numbers can be illustrated by using unit regions. Let us consider the product in Example i. First we take a unit square, that is, a square whose side is one unit long as shown in Figure 10.6a. Now we partition this unit square into two congruent subregions by horizontal line segments (Figure 10.6b). Next we partition the unit square into three congruent subregions by vertical line segments as shown in Figure 10.6c.

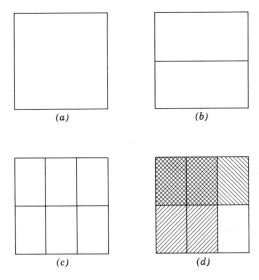

Figure 10.6

Observing this unit square we see that we have, with vertical and horizontal line segments, partitioned it into six congruent subregions. To show the product $\frac{1}{2} \times \frac{2}{3}$, we shade one-half of the unit region (shading in one direction) and then two-thirds of it (shading in another direction).

The double shading (cross-hatching) shows the set of congruent subregions which we define to correspond to the product $\frac{1}{2} \cdot \frac{2}{3}$ (Figure 10.6d). Of the six congruent subregions we have cross-hatched two, hence

$$\frac{1}{2} \cdot \frac{2}{3} = \frac{2}{6}.$$

Figure 10.7 illustrates the product $\frac{1}{2} \cdot \frac{3}{4}$ (Figure 10.7a) and the product $\frac{2}{3} \cdot \frac{3}{4}$ (Figure 10.7b).

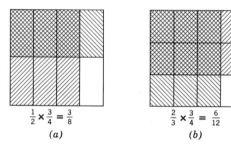

$$\frac{1}{2} \times \frac{3}{4} = \frac{3}{8} \qquad\qquad \frac{2}{3} \times \frac{3}{4} = \frac{6}{12}$$

(a) $\qquad\qquad\qquad\qquad$ (b)

Figure 10.7

The definition of multiplication of rational numbers has two very important special cases. If $c = 1$ and $d = 1$, we have

$$\frac{a}{b} \cdot \frac{c}{d} = \frac{a}{b} \cdot \frac{1}{1} = \frac{a}{b}$$

Similarly, if $a = 1$ and $b = 1$ we have

$$\frac{a}{b} \cdot \frac{c}{d} = \frac{1}{1} \cdot \frac{c}{d}$$

$$= \frac{c}{d}$$

We see that $\frac{1}{1}$ is the **identity element** for multiplication of rational numbers since

$$\frac{1}{1} \cdot \frac{a}{b} = \frac{a}{b} \cdot \frac{1}{1} = \frac{a}{b}$$

We conclude that the product of any rational number and $\frac{1}{1}$ is that rational number.

Notice that

$$\frac{2}{3} \cdot \frac{3}{2} = \frac{1}{1},$$

$$\frac{5}{6} \cdot \frac{6}{5} = \frac{1}{1}$$

$$\frac{9}{1} \cdot \frac{1}{9} = \frac{1}{1}$$

$$\frac{a}{b} \cdot \frac{b}{a} = \frac{1}{1}, \ a \neq 0, \ b \neq 0.$$

We see that for any rational number $\frac{a}{b}$ except $\frac{0}{b}$

$$\frac{a}{b} \cdot \frac{b}{a} = \frac{1}{1}$$

The two rational numbers $\frac{a}{b}$ and $\frac{b}{a}$, $a \neq 0$, are called **reciprocals** or **multiplicative inverses** of each other. The reciprocal of $\frac{3}{5}$ is $\frac{5}{3}$; the reciprocal of $\frac{7}{8}$ is $\frac{8}{7}$. Every rational number except $\frac{0}{b}$ has a multiplicative inverse.

DEFINITION 10.10. *If $\frac{a}{b}$ is a rational number, $a \neq 0$, the rational number $\frac{b}{a}$ is called the* **reciprocal** *or* **multiplicative inverse** *of $\frac{a}{b}$.*

10.8. Properties of Multiplication of Rational Numbers

The operation of multiplication of rational numbers has all of the properties of whole numbers plus the reciprocal property discussed in Section 10.7.

R#-5. **Closure Property of Multiplication.** If $\frac{a}{b}$ and $\frac{c}{d}$ are rational numbers, then their product $\frac{a}{b} \cdot \frac{c}{d}$ is a rational number.

R#-6. **Commutative Property of Multiplication.** If $\frac{a}{b}$ and $\frac{c}{d}$ are rational numbers, then $\frac{a}{b} \cdot \frac{c}{d} = \frac{c}{d} \cdot \frac{a}{b}$.

R#-7. **Associative Property of Multiplication.** If $\frac{a}{b}$, $\frac{c}{d}$, and $\frac{e}{f}$ are rational numbers, then $(\frac{a}{b} \cdot \frac{c}{d}) \cdot \frac{e}{f} = \frac{a}{b} \cdot (\frac{c}{d} \cdot \frac{e}{f})$.

R#-8. **Identity Element for Multiplication.** In the set of rational numbers there is an element $\frac{1}{1}$ such that for all rational numbers $\frac{a}{b}$, $\frac{a}{b} \cdot \frac{1}{1} = \frac{1}{1} \cdot \frac{a}{b} = \frac{a}{b}$. The number $\frac{1}{1}$ is called the **identity element for multiplication** or the **multiplicative identity**.

R#-9. **Distributive Property of Multiplication over Addition.** If $\frac{a}{b}$, $\frac{c}{d}$, and $\frac{e}{f}$ are rational numbers, then $\frac{a}{b} \cdot (\frac{c}{d} + \frac{e}{f}) = \frac{a}{b} \cdot \frac{c}{d} + \frac{a}{b} \cdot \frac{e}{f}$.

R#-10. Multiplicative Inverse Property (Reciprocals). Every rational number $\frac{a}{b}$, $a \neq 0$, has a **multiplicative inverse (reciprocal),** namely $\frac{b}{a}$, such that $\frac{a}{b} \cdot \frac{b}{a} = \frac{b}{a} \cdot \frac{a}{b} = \frac{1}{1}$.

We shall now prove the above properties.

Proof of R#-5: By Definition 10.9

$$\frac{a}{b} \cdot \frac{c}{d} = \frac{ac}{bd}$$

Since a and c are whole numbers their product, ac, is a whole number. Since b and d are whole numbers, $b \neq 0$ and $d \neq 0$, their product is a non-zero whole number. Hence $\frac{ac}{bd}$ is a rational number by Definition 10.1.

Proof of R#-6: By Definition 10.9

$$\frac{a}{b} \cdot \frac{c}{d} = \frac{ac}{bd}$$

$$= \frac{ca}{db} \qquad \text{Commutative property of mul-}$$
$$\text{tiplication of whole numbers}$$

$$= \frac{c}{d} \cdot \frac{a}{b} \qquad \text{Definition 10.9}$$

Proof of R#-7: $\left(\frac{a}{b} \cdot \frac{c}{d}\right) \cdot \frac{e}{f} = \frac{ac}{bd} \cdot \frac{e}{f} \qquad$ Definition 10.9

$$= \frac{(ac)e}{(bd)f} \qquad \text{Definition 10.9}$$

$$= \frac{a(ce)}{b(df)} \qquad \text{Associative property of multi-}$$
$$\text{plication of whole numbers}$$

$$= \frac{a}{b} \cdot \frac{ce}{df} \qquad \text{Definition 10.9}$$

$$= \frac{a}{b} \cdot \left(\frac{c}{d} \cdot \frac{e}{f}\right) \qquad \text{Definition 10.9}$$

Proof of R#-8: $\quad \frac{a}{b} \cdot \frac{1}{1} = \frac{a \cdot 1}{b \cdot 1} \qquad$ Definition 10.9

$$= \frac{a}{b} \qquad \text{Identity element for multipli-}$$
$$\text{cation of whole numbers.}$$

$$\frac{1}{1} \cdot \frac{a}{b} = \frac{1 \cdot a}{1 \cdot b} \qquad \text{Definition 10.9}$$

$$= \frac{a}{b} \qquad \text{Why?}$$

Proof of R#-9: Since rational numbers may be named by fractions with the same denominator, the distributive property may, without loss of generality, be stated

$$\frac{a}{b} \cdot \left(\frac{c}{d} + \frac{e}{d}\right) = \frac{a}{b} \cdot \frac{c}{d} + \frac{a}{b} \cdot \frac{e}{d}$$

We shall now prove the distributive property in the above form.

$$\frac{a}{b} \cdot \left(\frac{c}{d} + \frac{e}{d}\right) = \frac{a}{b} \cdot \frac{c + e}{d} \qquad \text{Definition 10.5}$$

$$= \frac{a(c + e)}{bd} \qquad \text{Definition 10.9}$$

$$= \frac{ac + ae}{bd} \qquad \text{Distributive property of whole numbers}$$

$$= \frac{ac}{bd} + \frac{ae}{bd} \qquad \text{Definition 10.5}$$

$$= \frac{a}{b} \cdot \frac{c}{d} + \frac{a}{b} \cdot \frac{e}{d} \qquad \text{Definition 10.9}$$

Proof of R#-10: $\quad \dfrac{a}{b} \cdot \dfrac{b}{a} = \dfrac{ab}{ba} \qquad$ Definition 10.9

$$= \frac{ab}{ab} \qquad \text{Commutative property of multiplication of whole numbers}$$

$$= \frac{1}{1} \qquad \text{Definition 10.2}$$

In a similar manner we can show that $\frac{b}{a} \cdot \frac{a}{b} = \frac{1}{1}$.

10.9. Division of Rational Numbers

The operation of division of rational numbers is defined as the inverse operation of multiplication of these numbers. Thus

$$\frac{2}{3} \div \frac{1}{4} = \frac{e}{f} \qquad \text{means} \qquad \frac{2}{3} = \frac{1}{4} \cdot \frac{e}{f}$$

In general

$$\frac{a}{b} \div \frac{c}{d} = \frac{e}{f} \qquad \text{means} \qquad \frac{c}{d} \cdot \frac{e}{f} = \frac{a}{b}$$

As in the whole number system we exclude division by the additive identity.

The property that every rational number $\frac{a}{b}$, $a \neq 0$, has a multiplicative inverse is basic to the definition of division of rational numbers.

Let us consider the division $\frac{2}{3} \div \frac{1}{4}$. We shall assume that this quotient exists and is a rational number $\frac{e}{f}$:

$$\frac{2}{3} \div \frac{1}{4} = \frac{e}{f}$$

We ask ourselves how can we determine $\frac{e}{f}$? We know that

$$\frac{2}{3} \div \frac{1}{4} = \frac{e}{f} \quad \text{means} \quad \frac{2}{3} = \frac{1}{4} \cdot \frac{e}{f}.$$

Since $\frac{1}{4}$ is not the additive identity it has a reciprocal, $\frac{4}{1}$. Multiplying both members of $\frac{2}{3} = \frac{1}{4} \cdot \frac{e}{f}$ by $\frac{4}{1}$ we obtain

$$\frac{4}{1} \cdot \frac{2}{3} = \frac{4}{1} \cdot \left(\frac{1}{4} \cdot \frac{e}{f} \right)$$

$$\frac{4}{1} \cdot \left(\frac{1}{4} \cdot \frac{e}{f} \right) = \frac{4}{1} \cdot \frac{2}{3} \qquad \text{Symmetric property of equality}$$

$$\left(\frac{4}{1} \cdot \frac{1}{4} \right) \cdot \frac{e}{f} = \frac{4}{1} \cdot \frac{2}{3} \qquad \text{Associative property of multiplication of rational numbers}$$

$$\frac{1}{1} \cdot \frac{e}{f} = \frac{4}{1} \cdot \frac{2}{3} \qquad \text{Multiplicative inverse property}$$

$$\frac{e}{f} = \frac{4}{1} \cdot \frac{2}{3} \qquad \text{Multiplicative identity for rational numbers}$$

$$\frac{e}{f} = \frac{8}{3} \qquad \text{Definition 10.9}$$

We have now shown that if there is a rational number $\frac{e}{f}$ such that $\frac{2}{3} \div \frac{1}{4} = \frac{e}{f}$, then $\frac{e}{f} = \frac{8}{3}$. We will now show that $\frac{8}{3}$ is truly that number which satisfies the sentence $\frac{2}{3} \div \frac{1}{4} = \frac{e}{f}$. Let us replace $\frac{e}{f}$ in this equation by $\frac{8}{3}$. We obtain

$$\frac{2}{3} \div \frac{1}{4} = \frac{8}{3}$$

which means

$$\frac{1}{4} \cdot \frac{8}{3} = \frac{2}{3}$$

But

$$\frac{1}{4} \cdot \frac{8}{3} = \frac{8}{12}$$

$$= \frac{2}{3}$$

As we look back at the steps in the division $\frac{2}{3} \div \frac{1}{4}$ we observe that the quotient $\frac{8}{3}$ was obtained by multiplying $\frac{2}{3}$ by the reciprocal of $\frac{1}{4}$, that is

$$\frac{2}{3} \div \frac{1}{4} = \frac{2}{3} \cdot \frac{4}{1}.$$

DEFINITION 10.11. *If $\frac{a}{b}$ and $\frac{c}{d}$ are rational numbers, $c \neq 0$, then the* **quotient,** $\frac{a}{b} \div \frac{c}{d}$, *is* $\frac{a}{b} \cdot \frac{d}{c}$.

Definition 10.11 tells us that *dividing by a rational number is the same as multiplying by its reciprocal.* Since every rational number except the additive identity has a reciprocal, we see that *the set of rational numbers is closed under the operation of division except by the additive identity.*

Exercise 10.3

1. Multiply. Name the products in simplest form.
   a. $\frac{1}{2} \cdot \frac{1}{3}$ d. $\frac{36}{17} \cdot \frac{51}{48}$
   b. $\frac{9}{25} \cdot \frac{5}{27}$ e. $\frac{19}{100} \cdot \frac{25}{7}$
   c. $\frac{3}{4} \cdot \frac{8}{15}$ f. $\frac{32}{25} \cdot \frac{5}{16}$
2. Multiply. Name the products in simplest form.
   a. $\frac{1}{4} \cdot \frac{2}{3} \cdot \frac{9}{16}$
   b. $\frac{5}{8} \cdot \frac{3}{5} \cdot \frac{4}{9}$
   c. $\frac{7}{8} \cdot \frac{5}{16} \cdot \frac{32}{25}$
   d. $\frac{3}{5} \cdot \frac{25}{51} \cdot \frac{9}{100}$
3. What are the multiplicative inverses of the following?
   a. $\frac{5}{6}$ e. $\frac{8}{7}$
   b. $\frac{3}{4}$ f. $\frac{1}{1}$
   c. $\frac{3}{8}$ g. $\frac{12}{7}$
   d. $\frac{19}{27}$ h. $\frac{2}{5}$
4. Write the multiplication sentences associated with each division sentence.

a. $36 \div n = 9$ e. $\frac{1}{2} \div \frac{3}{4} = n$

b. $n \div 4 = 8$ f. $n \div \frac{1}{2} = \frac{3}{5}$

c. $\frac{2}{3} \div n = \frac{1}{2}$ g. $2 \div 7 = n$

d. $n \div \frac{1}{4} = \frac{2}{3}$ h. $n \div 4 = 6$

5. Divide. Name the quotients in simplest form.

   a. $\frac{2}{3} \div \frac{4}{3}$ e. $\frac{5}{8} \div \frac{16}{100}$

   b. $\frac{3}{8} \div \frac{1}{4}$ f. $\frac{7}{12} \div \frac{1}{3}$

   c. $\frac{2}{5} \div \frac{16}{25}$ g. $\frac{5}{12} \div \frac{15}{48}$

   d. $\frac{9}{16} \div \frac{3}{4}$ h. $\frac{4}{3} \div \frac{16}{9}$

6. Find the solution sets. The domain is the set of rational numbers.

   a. $\frac{5}{8}n = \frac{1}{1}$ d. $\frac{3}{5}n = \frac{1}{1}$

   b. $\frac{2}{3}n = \frac{1}{1}$ e. $\frac{7}{8}n = \frac{1}{1}$

   c. $\frac{9}{16}n = \frac{1}{1}$ f. $\frac{9}{100}n = \frac{1}{1}$

7. Find the solution sets. The domain is the set of rational numbers.

   a. $\frac{2}{3} \cdot \frac{3}{4} = n$ d. $\frac{5}{12}n = \frac{5}{6}$

   b. $\frac{5}{6}n = \frac{2}{3}$ e. $\frac{3}{16}n = \frac{9}{4}$

   c. $\frac{7}{8}n = \frac{21}{4}$ f. $\frac{4}{3}n = \frac{7}{2}$

8. Justify each statement below by a property of the rational numbers and their operations.

   a. $\frac{2}{3} \cdot \frac{3}{5} = \frac{3}{5} \cdot \frac{2}{3}$

   b. $\frac{1}{2} \cdot (\frac{3}{4} \cdot \frac{2}{3}) = (\frac{1}{2} \cdot \frac{3}{4}) \cdot \frac{2}{3}$

   c. $\frac{1}{3} \cdot \frac{1}{1} = \frac{1}{3}$

9. Replace each * by an expression or a numeral that will make the resulting statement an application of the given property of rational numbers.

   a. $\frac{2}{3} \cdot \frac{3}{5} = *$; commutative property of multiplication

   b. $\frac{1}{3} \cdot \frac{1}{1} = *$; identity element of multiplication

   c. $\frac{1}{3}(\frac{2}{3} + \frac{5}{6}) = *$; distributive property

   d. $\frac{3}{2} \cdot (\frac{1}{3} \cdot \frac{2}{5}) = *$; associative property of multiplication

10. Construct some numerical examples to show that division of rational numbers is not a commutative operation.

11. Construct some numerical examples to show that division of rational numbers is not an associative operation.

12. Construct some numerical examples to show that multiplication of rational numbers distributes over subtraction of rational numbers.

13. Construct some numerical examples to show that division of rational numbers distributes over subtraction of rational numbers.

14. If $\frac{a}{b}$, $\frac{c}{d}$, and $\frac{e}{f}$ are non-negative rational numbers and $\frac{a}{b} < \frac{c}{d}$ is $\frac{a}{b} + \frac{e}{f}$ less than or greater than $\frac{c}{d} + \frac{e}{f}$?

15. Mr. Parsons owns $\frac{1}{4}$ of the stock of the Parsons Publication Co. His wife owns $\frac{5}{6}$ as much of the stock as her husband does.

   a. What part of the stock is owned by Mrs. Parsons?

   b. What part of the stock is owned by persons other than Mr. and Mrs. Parsons.

16. The product of two rational numbers is $\frac{1}{1}$. One of the factors is $\frac{12}{37}$. What is the other factor?

17. The product of two rational numbers is $\frac{1}{1}$. One of the factors is $\frac{37}{100}$. What is the other factor?

18. Give a reason for each step in the proof below.

   Prove: (Cancellation Property of Multiplication of Rational Numbers)

   For all rational numbers $\frac{a}{b}, \frac{c}{d}$, and $\frac{e}{f}$, $a \neq 0$, if $\frac{a}{b} \cdot \frac{c}{d} = \frac{a}{b} \cdot \frac{e}{f}$ then $\frac{c}{d} = \frac{e}{f}$.

   Proof. (a) Since $a \neq 0$, $\dfrac{a}{b}$ has a multiplicative inverse $\dfrac{b}{a}$

   (b) $\dfrac{a}{b} \cdot \dfrac{c}{d} = \dfrac{a}{b} \cdot \dfrac{e}{f}$

   (c) $\dfrac{b}{a} \cdot \left(\dfrac{a}{b} \cdot \dfrac{c}{d} \right) = \dfrac{b}{a} \cdot \left(\dfrac{a}{b} \cdot \dfrac{e}{f} \right)$

   (d) $\left(\dfrac{b}{a} \cdot \dfrac{a}{b} \right) \cdot \dfrac{c}{d} = \left(\dfrac{b}{a} \cdot \dfrac{a}{b} \right) \cdot \dfrac{e}{f}$

   (e) $\dfrac{1}{1} \cdot \dfrac{c}{d} = \dfrac{1}{1} \cdot \dfrac{e}{f}$

   (f) $\dfrac{c}{d} = \dfrac{e}{f}$

19. Find the solution sets. The domain is the set of rational numbers.

   a. $\frac{3}{4}n + \frac{1}{2} = \frac{5}{4}$ c. $\frac{3}{4} \div \frac{1}{2} = \frac{1}{3}n$

   b. $\frac{5}{6}n = \frac{5}{4} - \frac{1}{10}$ d. $\frac{5}{6} \div \frac{2}{3} = \frac{3}{5}n$

20. What is the solution set of $14n = 27$ if the domain is

   a. The set of whole numbers?

   b. The set of rational numbers?

10.10. The Whole Numbers as a Subset of the Rational Numbers

Let us consider the set of all rational numbers that may be named by fractions whose denominators are, 1. This set, F, is

$$F = \left\{ \frac{0}{1}, \frac{1}{1}, \frac{2}{1}, \frac{3}{1}, \frac{4}{1}, \cdots \right\}$$

The set, W, of whole numbers is

$$W = \{0, 1, 2, 3, 4, \ldots\}$$

We can set up a one-to-one correspondence between the elements of F and the elements of W as shown below:

$$
\begin{array}{cccccc}
\dfrac{0}{1} & \dfrac{1}{1} & \dfrac{2}{1} & \dfrac{3}{1} & \dfrac{4}{1} \cdots & \dfrac{n}{1} \cdots \\[8pt]
\updownarrow & \updownarrow & \updownarrow & \updownarrow & \updownarrow & \updownarrow \\[4pt]
0 & 1 & 2 & 3 & 4 \cdots & n \cdots
\end{array}
$$

This matching preserves both the order and the fundamental operations. For example

$$6 > 2 \qquad \text{and} \qquad \frac{6}{1} > \frac{2}{1}$$

$$6 + 2 = 8 \quad \text{and} \quad \frac{6}{1} + \frac{2}{1} = \frac{8}{1}$$

$$6 - 2 = 4 \quad \text{and} \quad \frac{6}{1} - \frac{2}{1} = \frac{4}{1}$$

$$6 \cdot 2 = 12 \quad \text{and} \quad \frac{6}{1} \cdot \frac{2}{1} = \frac{12}{1}$$

$$6 \div 2 = 3 \quad \text{and} \quad \frac{6}{1} \div \frac{2}{1} = \frac{3}{1}$$

We observe that so far as their order and the performance of the arithmetic operations are concerned, the rational numbers of F and the whole numbers behave exactly alike. Because of this similarity between the elements of F and the whole numbers, it is customary to use the symbols "0", "1", "2", ... to denote both the whole numbers and their corresponding elements in F. That is, the symbols "0", "1", "2", ..., are regarded as other names for $\frac{0}{1}, \frac{1}{1}, \frac{2}{1}, \ldots$, respectively. Similarly, the whole numbers may be renamed as convenient, for all purposes of arithmetic, by the fractions $\frac{0}{1}, \frac{1}{1}, \frac{2}{1}, \ldots$.

Because of the above consistency of order relations and the arithmetic operations between the whole numbers and the rational numbers of F, we say that *the system of whole numbers is embedded in the system of rational numbers, or that the system of whole numbers is a subsystem of the system of rational numbers.*

10.11. The Density Property of the Rational Numbers

We know that each member of the set of whole numbers has an immediate successor—the next larger whole number. The successor of 0 is 1; the successor of 1 is 2; the successor of 2 is 3; and so on. The successor of any whole number n is $n + 1$.

Every whole number n except 0 has an immediate predecessor, the next smaller whole number. The predecessor of 4 is 3; the predecessor of 18 is 17. In general, the predecessor of any whole number n, $n \neq 0$, is $n - 1$.

The rational numbers do not possess this property. A rational number has neither successor nor predecessor. Let us try to visualize, on the number line, the set of points corresponding to the whole numbers. On the number line there are no points which are graphs of whole numbers between the points corresponding to 0 and 1. Similarly, for any successive whole numbers n and $n + 1$, there are no points corresponding to whole numbers between the two points corresponding to n and $n + 1$. This means that on the number line there are wide gaps between the points which are graphs of the whole numbers.

Now let us try to visualize, on the number line, the set of points between the points corresponding to 0 and 1. Let us concentrate on those rational numbers which are greater than zero and less than one. Now let us visualize the points on the number line that are graphs of these numbers. We call points on the number line that correspond to rational numbers **rational points.** Figure 10.8 shows the rational points between 0 and 1 that correspond to the rational numbers named by fractions with denominators 2, 4, and 8.

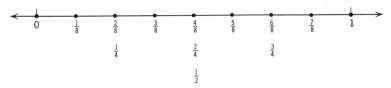

Figure 10.8

We now ask ourselves: How many rational points are there between two rational points, for example, between the points corresponding to $\frac{1}{8}$ and $\frac{1}{4}$? We know that

$$\frac{1}{8} = \frac{2}{16} \quad \text{and} \quad \frac{1}{4} = \frac{4}{16}$$

We see that there is at least one rational point between the points corresponding to $\frac{1}{8}$ and $\frac{1}{4}$, namely the point corresponding to $\frac{3}{16}$.

Since

$$\frac{1}{8} = \frac{4}{32} \qquad \text{and} \qquad \frac{3}{16} = \frac{6}{32}$$

the point that corresponds to $\frac{5}{32}$ is between the points corresponding to $\frac{1}{8}$ and $\frac{3}{16}$. We now have two rational points between the points corresponding to $\frac{1}{8}$ and $\frac{1}{4}$. However

$$\frac{1}{8} = \frac{8}{64} \qquad \text{and} \qquad \frac{5}{32} = \frac{10}{64}$$

and hence the point corresponding to $\frac{9}{64}$ is between the points corresponding to $\frac{1}{8}$ and $\frac{5}{32}$ and therefore between the points corresponding to $\frac{1}{8}$ and $\frac{1}{4}$.

Clearly this process could be carried on indefinitely, and would apply no matter what rational numbers were involved. We see then, that between any two rational points, there are infinitely many other rational points. Since each rational point corresponds to a rational number, we see that between any two rational numbers, no matter how close together they are, that is, no matter how small their difference, there is always a third rational number. Accordingly, between two rational numbers there must be infinitely many rational numbers. We describe this situation by saying that the set of rational numbers is **dense.**

One way to find a rational number between any two given rational numbers is to find the average of the two given numbers. For example, the average of $\frac{1}{8}$ and $\frac{1}{4}$ is

$$\left(\frac{1}{8} + \frac{1}{4} \right) \div 2 = \frac{3}{8} \div 2$$

$$= \frac{3}{16}$$

We know from the previous discussion that $\frac{3}{16}$ is between $\frac{1}{8}$ and $\frac{1}{4}$. In fact, the average of two rational numbers is a rational number midway between the two given numbers.

EXAMPLE i. Find a rational number midway between $\frac{2}{3}$ and $\frac{7}{8}$.
Solution: We find the average of $\frac{2}{3}$ and $\frac{7}{8}$:

$$\left(\frac{2}{3}+\frac{7}{8}\right) \div 2 = \left(\frac{2\cdot 8 + 3\cdot 7}{3\cdot 8}\right) \div 2$$

$$= \frac{37}{24} \div 2$$

$$= \frac{37}{24}\cdot\frac{1}{2}$$

$$= \frac{37}{48}$$

The rational number $\frac{37}{48}$ is midway between $\frac{2}{3}$ and $\frac{7}{8}$.

EXAMPLE ii. Find the coordinates of the rational points that separate the line segment with rational points corresponding to $\frac{1}{2}$ and $\frac{5}{6}$ as endpoints into four congruent parts.

Solution: Find the average of the two given rational numbers:

$$\left(\frac{1}{2}+\frac{5}{6}\right) \div 2 = \frac{8}{6} \div 2$$

$$= \frac{8}{12}$$

$$= \frac{2}{3}$$

The graph of $\frac{2}{3}$ is midway between the graphs of $\frac{1}{2}$ and $\frac{5}{6}$ and hence separates the line segment with the rational points corresponding to $\frac{1}{2}$ and $\frac{5}{6}$ as endpoints into two congruent parts. We now find the average of $\frac{1}{2}$ and $\frac{2}{3}$ and the average of $\frac{2}{3}$ and $\frac{5}{6}$:

$$\left(\frac{1}{2}+\frac{2}{3}\right) \div 2 = \frac{7}{6} \div 2$$

$$= \frac{7}{12}$$

$$\left(\frac{2}{3}+\frac{5}{6}\right) \div 2 = \frac{9}{6} \div 2$$

$$= \frac{9}{12}$$

$$= \frac{3}{4}$$

The graphs of $\frac{7}{12}$, $\frac{2}{3}$, and $\frac{3}{4}$ separate the given line segment into four congruent parts (Figure 10.9)

Figure 10.9

10.12. Fractions as Symbols for Division

Observe that

$$20 \div 7 = \frac{20}{1} \div \frac{7}{1}$$

$$= \frac{20}{1} \cdot \frac{1}{7}$$

$$= \frac{20}{7}$$

We see that $\frac{20}{7}$ may be viewed as a fraction or as an indicated quotient of two whole numbers. In general, if a and b are whole numbers, $b \neq 0$, then

$$a \div b = \frac{a}{1} \div \frac{b}{1}$$

$$= \frac{a}{1} \cdot \frac{1}{b}$$

$$= \frac{a}{b}$$

DEFINITION 10.12. *If a and b are whole numbers, $b \neq 0$, then $a \div b$ and $\frac{a}{b}$ are names for the same rational number.*

Definition 10.12 tells us that a fraction, for example $\frac{5}{6}$, can be viewed not only as the name for a rational number, but also as the quotient of two whole numbers, in this case $5 \div 6$. It is permissible then to say "the rational number $5 \div 6$" instead of saying "the rational number $\frac{5}{6}$."

We can extend Definition 10.12 so that the quotient of two rational numbers, for example $\frac{2}{3} \div \frac{5}{6}$, can be written as a fraction. Thus $\frac{2}{3} \div \frac{5}{6}$ may be denoted by

$$\frac{\frac{2}{3}}{\frac{5}{6}}$$

In general, the quotient of two fractional numbers, $\frac{a}{b} \div \frac{c}{d}$, $c \neq 0$, may be written as the fraction

$$\frac{\dfrac{a}{b}}{\dfrac{c}{d}}$$

which is called a **complex fraction.** A complex fraction is a fraction whose numerator or denominator or both are themselves rational numbers. Any fraction may be thought of as a complex fraction. For example

$$\frac{2}{3} = \frac{\dfrac{2}{1}}{\dfrac{3}{1}}$$

In general, the common fraction $\frac{a}{b}$ may be expressed as the complex fraction $\dfrac{\frac{a}{1}}{\frac{b}{1}}$.

Every complex fraction may be simplified and written as a common fraction, since

$$\frac{\dfrac{a}{b}}{\dfrac{c}{d}} = \frac{a}{b} \div \frac{c}{d}$$

$$= \frac{a}{b} \cdot \frac{d}{c}$$

$$= \frac{ad}{bc}$$

EXAMPLE i. Write $\dfrac{\frac{2}{3}}{\frac{16}{9}}$ as a common fraction in simplest form.

Solution:

$$\frac{\dfrac{2}{3}}{\dfrac{16}{9}} = \frac{2}{3} \div \frac{16}{9}$$

$$= \frac{2}{3} \cdot \frac{9}{16}$$

$$= \frac{18}{48}$$

$$= \frac{3}{8}$$

303

EXAMPLE ii. Write $\dfrac{\frac{5}{16}}{\frac{20}{112}}$ as a common fraction in simplest form.

Solution:

$$\frac{\frac{5}{16}}{\frac{20}{112}} = \frac{5}{16} \div \frac{20}{112}$$

$$= \frac{5}{16} \cdot \frac{112}{20}$$

$$= \frac{560}{320}$$

$$= \frac{7}{4}$$

10.13. Mixed Numerals

Since the rational numbers are ordered we may separate the set of all rational numbers into two disjoint subsets, one containing those rational numbers less than 1 and the other containing those rational numbers greater than or equal to 1. We can tell to which of these two subsets a rational number belongs by looking at the numerator of the fraction naming this rational number. A fraction with a numerator less than the denominator names a rational number less than 1. Such fractions are sometimes called **proper fractions.** A fraction whose numerator is equal to its denominator names the number 1. A fraction whose numerator is greater than its denominator names a rational number greater than 1. We recall that these fractions are sometimes called **improper fractions.**

Every rational number that is named by a fraction whose numerator is greater than its denominator is either a whole number or a number that can be expressed as the sum of a whole number and a rational number less than one. For example,

$$\frac{27}{6} = \frac{24+3}{6} = \frac{24}{6} + \frac{3}{6} = 4 + \frac{3}{6} = 4 + \frac{1}{2}$$

It is customary to write this last sum as $4\frac{1}{2}$. This symbol is read "four and one-half". The symbol $4\frac{1}{2}$ is called a **mixed numeral.**

Let us consider the fraction $\frac{23}{3}$. This fraction may be written as the indicated quotient of two whole numbers

$$\frac{23}{3} = 23 \div 3$$

$$= [(3 \times 7) + 2] \div 3$$

$$= \frac{(3 \times 7) + 2}{3}$$

$$= \frac{3 \times 7}{3} + \frac{2}{3}$$

$$= 7 + \frac{2}{3}$$

$$= 7\frac{2}{3}$$

The example above illustrates one method for changing a fraction to a mixed numeral.

EXAMPLE i. Write $\frac{29}{12}$ as a mixed numeral.

Solution:

$$\frac{29}{12} = 29 \div 12$$

$$= [(12 \times 2) + 5] \div 12$$

$$= \frac{(12 \times 2) + 5}{12}$$

$$= \frac{12 \times 2}{12} + \frac{5}{12}$$

$$= 2 + \frac{5}{12} = 2\frac{5}{12}$$

EXAMPLE ii. Write $3\frac{1}{2}$ as a common fraction.

Solution:

$$3\frac{1}{2} = 3 + \frac{1}{2}$$

$$= \frac{3}{1} + \frac{1}{2}$$

$$= \frac{3 \times 2}{1 \times 2} + \frac{1}{2}$$

$$= \frac{6}{2} + \frac{1}{2}$$

$$= \frac{7}{2}.$$

Exercise 10.4

1. Write the following as common fractions.
   - a. $7 \div 8$
   - b. $1 \div 2$
   - c. $3 \div 9$
   - d. $17 \div 16$
   - e. $242 \div 15$
   - f. $115 \div 8$
   - g. $267 \div 12$
   - h. $481 \div 25$

2. Write each of the following as the quotient of two whole numbers.
   - a. $\frac{5}{12}$
   - b. $\frac{7}{16}$
   - c. $\frac{9}{2}$
   - d. $\frac{342}{12}$
   - e. $\frac{216}{512}$
   - f. $\frac{42}{7}$

3. Write each as a complex fraction.
   - a. $\frac{2}{3}$
   - b. $\frac{5}{8}$
   - c. $\frac{3}{16}$
   - d. $\frac{27}{28}$

4. Write as common fractions in simplest form.
   - a. $\dfrac{\frac{2}{3}}{\frac{1}{2}}$
   - b. $\dfrac{\frac{5}{16}}{\frac{3}{8}}$
   - c. $\dfrac{\frac{7}{15}}{\frac{29}{30}}$
   - d. $\dfrac{\frac{9}{16}}{\frac{5}{12}}$
   - e. $\dfrac{\frac{4}{15}}{\frac{16}{45}}$
   - f. $\dfrac{\frac{7}{100}}{\frac{21}{10}}$

5. Write the following as mixed numerals. Write all fractions in simplest form.
   - a. $\frac{29}{4}$
   - b. $\frac{45}{8}$
   - c. $\frac{39}{12}$
   - d. $\frac{56}{10}$
   - e. $\frac{138}{100}$
   - f. $\frac{137}{24}$
   - g. $\frac{76}{3}$
   - h. $\frac{98}{5}$

6. Write the following as common fractions.
   - a. $2\frac{1}{2}$
   - b. $3\frac{3}{4}$
   - c. $5\frac{7}{12}$
   - d. $6\frac{7}{8}$
   - e. $9\frac{5}{16}$
   - f. $12\frac{1}{3}$
   - g. $26\frac{3}{5}$
   - h. $17\frac{5}{6}$

7. Which of the following may be named by symbols for whole numbers?
   - a. $\frac{300}{100}$
   - b. $\frac{33}{2}$
   - c. $\frac{76}{3}$
   - d. $\frac{84}{21}$
   - e. $\frac{5000}{1000}$
   - f. $\frac{192}{12}$

8. Find a rational number midway between each of the following pairs.
   - a. $\frac{1}{3}, \frac{7}{8}$
   - d. $\frac{5}{16}, \frac{11}{12}$

b. $\frac{3}{5}, \frac{9}{7}$ e. $\frac{1}{2}, \frac{13}{16}$

c. $\frac{5}{4}, \frac{17}{9}$ f. $\frac{1}{16}, \frac{2}{3}$

9. What rational number is midway between 0 and $\frac{1}{5}$?

10. What is the least rational number n such that $n \geqslant \frac{5}{6}$?

11. What is the greatest rational number n such that $n \leqslant \frac{1}{2}$?

12. How many rational numbers are greater than $\frac{1}{2}$ and less than $\frac{7}{8}$?

13. What is the multiplicative inverse of each of the following?

a. $\dfrac{1}{\frac{1}{3}}$ c. $\dfrac{1}{\frac{8}{9}}$

b. $\dfrac{1}{\frac{5}{6}}$ d. $\dfrac{1}{\frac{a}{b}}, a \neq 0$

14. What are the coordinates of the three rational points that separate the line segment with the graphs of $\frac{1}{8}$ and $\frac{7}{16}$ as endpoints into four congruent parts?

10.14. Decimal Rational Numbers

Let us now consider the subset, D, of rational numbers that may be named by fractions whose denominators are powers of ten. There will be many numbers in this set whose fraction names when written in simplest form will not have a power of ten as denominator. For example

$$\frac{1}{2} = \frac{5}{10^1}$$

$$\frac{1}{4} = \frac{25}{100} = \frac{25}{10^2}$$

$$\frac{1}{8} = \frac{125}{1000} = \frac{125}{10^3}$$

We must decide whether or not a fraction names a rational number that is a member of set D. For example, is $\frac{3}{4}$ a member of set D? Notice that

$$\frac{3}{4} = \frac{3}{2^2} = \frac{3 \cdot 5^2}{2^2 \cdot 5^2} = \frac{75}{100} = \frac{75}{10^2}$$

We see that $\frac{3}{4}$ is a member of set D.

Observe the following:

(a) $\dfrac{25}{45} = \dfrac{5}{9} = \dfrac{5}{3^2}$ Does not belong to D since there is no whole number by which we can multiply 3^2 to obtain a power of 10.

(b) $\dfrac{6}{98} = \dfrac{3}{49} = \dfrac{3}{7^2}$ Does not belong to D

(c) $\dfrac{50}{90} = \dfrac{5}{9} = \dfrac{5}{3^2}$ Does not belong to D

(d) $\dfrac{110}{125} = \dfrac{22}{25} = \dfrac{22}{5^2} = \dfrac{22 \cdot 2^2}{5^2 \cdot 2^2} = \dfrac{88}{10^2}$ Belongs to D

(e) $\dfrac{15}{24} = \dfrac{5}{8} = \dfrac{5}{2^3} = \dfrac{5 \cdot 5^3}{2^3 \cdot 5^3} = \dfrac{625}{10^3}$ Belongs to D

(f) $\dfrac{21}{35} = \dfrac{3}{5} = \dfrac{3 \cdot 2}{5 \cdot 2} = \dfrac{6}{10}$ Belongs to D

Notice (d), (e) and (f). These numbers belong to D. When each of the fractions naming these numbers was written in simplest form, the denominators had only two possible prime factors, 2 and 5. This is true in general. A rational number is a member of D if and only if the denominator of the simplest form of the fraction which names it has no prime factors other than 2 and 5.

We see that the set D contains all of the whole numbers since every whole number may be named by a fraction whose denominator is $1 = 10^0$.

DEFINITION 10.13. *The elements of the set*

$$\left\{ 0, 1, 2, \dots, \dfrac{1}{10}, \dfrac{2}{10}, \dots, \dfrac{1}{10^n}, \dfrac{2}{10^n}, \dots \right\}$$

that is, the set of rational numbers named by fractions whose denominators are powers of ten are called **decimal rational numbers.**

EXAMPLE i. Is $\frac{49}{56}$ a decimal rational number?
Solution: Reducing $\frac{49}{56}$ to simplest form we have

$$\dfrac{49}{56} = \dfrac{7}{8}.$$

Since $8 = 2^3$, $\frac{49}{56}$ is a decimal rational number.

EXAMPLE ii. Write $\frac{56}{35}$ as a common fraction whose denominator is a power of ten.

Solution:

$$\frac{56}{35} = \frac{8}{5} = \frac{8 \cdot 2}{5 \cdot 2} = \frac{16}{10}.$$

10.15. Decimal Fractions

There is another form naming decimal rational numbers that is simpler than their common fraction form. This form is called the **decimal fraction** or simply **decimal form.** It follows the same scheme as that used for naming whole numbers which are themselves decimal rational numbers. For example,

$$\frac{824}{1000} = \frac{(8 \cdot 100) + (2 \cdot 10) + (4 \cdot 1)}{1000}$$

$$= \frac{8 \cdot 100}{1000} + \frac{2 \cdot 10}{1000} + \frac{4 \cdot 1}{1000}$$

$$= \left(8 \cdot \frac{100}{1000}\right) + \left(2 \cdot \frac{10}{1000}\right) + \left(4 \cdot \frac{1}{1000}\right)$$

$$= \left(8 \cdot \frac{1}{10}\right) + \left(2 \cdot \frac{1}{100}\right) + \left(4 \cdot \frac{1}{1000}\right)$$

$$= \left(8 \cdot \frac{1}{10^1}\right) + \left(2 \cdot \frac{1}{10^2}\right) + \left(4 \cdot \frac{1}{10^3}\right)$$

If we agree that this sum should be abbreviated by the symbol 824, we will have an exact counterpart for the notation developed for the whole numbers. Here we have assigned the place-values $\frac{1}{10^1}$, $\frac{1}{10^2}$, $\frac{1}{10^3}$ to the positions occupied by the "8", "2", and "4" respectively. The only difficulty with this notation is that we now have a symbol, 824, to represent both

(1) $$(8 \cdot 10^2) + (2 \cdot 10^1) + (4 \cdot 10^0)$$

and

(2) $$\left(8 \cdot \frac{1}{10^1}\right) + \left(2 \cdot \frac{1}{10^2}\right) + \left(4 \cdot \frac{1}{10^3}\right)$$

To alleviate this duplication we use a dot, called a **decimal point.** Thus, 824 is used to represent sum (1) above and .824 (usually written 0.824) represents sum (2).

With this notation, we have extended our place-value system as a means of recording not only the whole numbers, which are decimal rational numbers, but all of the decimal rational numbers. That is, we have extended the place-value system to the right of the ones place. The place-values to the right of the ones place, that is to the right of the decimal point, are moving from left to right, $\frac{1}{10^1}$ (tenths), $\frac{1}{10^2}$ (hundredths), $\frac{1}{10^3}$ (thousandths), $\frac{1}{10^4}$ (ten thousandths), and so forth.

For example,

$$(3 \cdot 10^3) + (5 \cdot 10^2) + (1 \cdot 10^1) + (7 \cdot 10^0) + \left(5 \cdot \frac{1}{10^1}\right) + \left(2 \cdot \frac{1}{10^2}\right)$$

is written 3517.52 (read: three thousand five hundred seventeen and fifty-two hundredths). The decimal point is placed to the right of the ones place and separates the place-values of the powers of ten and the place-values of the reciprocals of the powers of ten. Notice that the ones place, that is the place-value of $10^0 = 1$, is the point of reference. We read the places to the right of the decimal point according to their place-values. Thus 0.5 is read "five tenths", and 0.03 is read "three hundredths", 0.175 is read "one hundred seventy-five thousandths".

Table 10.1 shows the place-values for the five places to the right of the

Table 10.1

Place-Values to Right of Decimal Point				
tenths	hundredths	thousandths	ten thousandths	hundred thousandths
$\frac{1}{10^1}$	$\frac{1}{10^2}$	$\frac{1}{10^3}$	$\frac{1}{10^4}$	$\frac{1}{10^5}$
$\frac{1}{10}$	$\frac{1}{100}$	$\frac{1}{1000}$	$\frac{1}{10,000}$	$\frac{1}{100,000}$

ones place in the decimal system. Each place to the right of the ones place in a numeral is called a **decimal place.**

We read the numeral 126.839 "one hundred twenty-six *and* eight hundred thirty-nine thousandths." Notice that the "and" serves to designate the decimal point. Also notice that just as we say "eight hundred thirty-nine" instead of "eight hundreds, three tens, and nine ones" we say "eight hundred thirty-nine thousandths" instead of "eight tenths, three hundredths, and nine thousandths." The place-value of the *final* digit in a decimal fraction tells us whether we say tenths, hundredths, thousandths, ten thousandths, and so forth.

EXAMPLE i. Write 26.13 in expanded notation.

Solution:

$$26.13 = (2 \cdot 10^1) + (6 \cdot 10^0) + \left(1 \cdot \frac{1}{10^1}\right) + \left(3 \cdot \frac{1}{10^2}\right)$$

EXAMPLE ii. What is the place-value of the "3" in 4.153?

Solution: The place-value of the 3 is $\dfrac{1}{10^3} = \dfrac{1}{1000}$

10.16. Changing Common Fractions to Decimal Fractions

All decimal rational numbers can be named by fractions, which, when written in simplest form, have denominators whose only prime factors are 2 and 5. All of the numbers

$$\frac{1}{2}, \frac{1}{4}, \frac{5}{8}, \frac{7}{25}, \frac{3}{40}, \frac{120}{80}, \frac{49}{98}$$

are decimal rational numbers because when the fractions naming these numbers are written in simplest form the denominators have only 2 and 5 as prime factors.

One method for expressing a common fraction as a decimal fraction is illustrated by the examples below.

$$\frac{7}{25} = \frac{7}{5^2} = \frac{7 \cdot 2^2}{5^2 \cdot 2^2} = \frac{28}{100} = \frac{28}{10^2} = 0.28$$

$$\frac{3}{40} = \frac{3}{2^3 \cdot 5} = \frac{3 \cdot 5^2}{2^3 \cdot 5 \cdot 5^2} = \frac{75}{1000} = \frac{75}{10^3} = 0.075$$

$$\frac{6}{160} = \frac{6}{2^5 \cdot 5} = \frac{6 \cdot 5^4}{2^5 \cdot 5 \cdot 5^4} = \frac{3750}{100,000} = \frac{3750}{10^5} = 0.03750$$

In the examples above we renamed each fraction by an equivalent fraction whose denominator was a power of ten. Notice that the exponent of 10 in the denominator tells the number of digits to the right of the decimal point when the common fraction is written as a decimal fraction. Thus

$$\frac{28}{10^2} = 0.28$$

two digits to the right of the decimal point

$$\frac{3750}{10^5} = 0.03750$$

five digits to the right of the decimal point.

Another method of renaming a common fraction by a decimal fraction uses division. We know that every common fraction indicates the quotient of two whole numbers. For example, $\frac{3}{4}$ indicates $3 \div 4$. Consider the division below:

$$
\begin{array}{r}
0.75 \\
4\overline{)3.00} \\
\underline{2\,8} \\
20 \\
\underline{20} \\
0
\end{array}
$$

Let us analyze and justify this algorithm.

$$\frac{3}{4} = \left(\frac{100}{100}\right)\left(\frac{3}{4}\right) = \left(\frac{1}{100}\right)\left(\frac{300}{4}\right) = \left(\frac{1}{100}\right)(75) = \frac{75}{100} = 0.75$$

When we divide the numerator of a common fraction by the denominator and the division results in a zero remainder (if carried out far enough), we know that the common fraction names a decimal number. Because of this we call decimal rational numbers **terminating decimals.**

Some rational numbers do not belong to the set of decimal rational numbers. When the numerators of the fractions naming these rational numbers are divided by the denominators the division never has a zero remainder no matter how far it is carried out. We saw earlier that the fractions naming these rational numbers have denominators which have prime factors other than 2 and 5. For example, some fractions that do not name decimal rational numbers are $\frac{1}{3}$, $\frac{5}{6}$, $\frac{3}{7}$, and $\frac{1}{12}$.

Although we cannot find decimal fractions equivalent to these fractions, it is possible to find decimals that approximate these rational numbers as closely as we desire. By "approximate" we mean that their numerals name decimal rational numbers that differ from the given rational number by an amount as small as we think desirable.

Let us now consider the computational device for dividing 1 by 3.

$$
\begin{array}{r}
0.3 \\
3\overline{)1.0} \\
9 \\
\hline
1
\end{array}
\qquad
\begin{array}{r}
0.33 \\
3\overline{)1.00} \\
9 \\
\hline
10 \\
9 \\
\hline
1
\end{array}
\qquad
\begin{array}{r}
0.333 \\
3\overline{)1.000} \\
9 \\
\hline
10 \\
9 \\
\hline
10 \\
9 \\
\hline
1
\end{array}
$$

To justify the above divisions we note the following:

$$1 = 3(0.3) + 0.1 \qquad 1 = 3(0.33) + 0.01 \qquad 1 = 3(0.333) + 0.001$$

It is intuitively clear that the above divisions will not terminate. That is, $\frac{1}{3}$ cannot be expressed by a terminating decimal. If we continue the divisions we see that $\frac{1}{3} = 0.333\ldots$. The three dots mean that when we divide 1 by 3 the division never terminates. Because of this, we call such a number a **non-terminating** decimal. Notice that $\frac{1}{3}$ is approximately 0.3, but it is closer to 0.33, and so on. Thus we see that we can approximate $\frac{1}{3}$ as closely as we like by a decimal fraction.

EXAMPLE i. Write $\frac{1}{6}$ as a decimal fraction correct to the nearest thousandth.

Solution:

$$
\begin{array}{r}
0.1666 \\
6\overline{)1.0000} \\
6 \\
\hline
40 \\
36 \\
\hline
40 \\
36 \\
\hline
40 \\
36 \\
\hline
4
\end{array}
$$

Correct to the nearest thousandth $\frac{1}{6}$ is 0.167. Notice that in order to name $\frac{1}{6}$ as a decimal fraction correct to the nearest thousandth, we carried the division out four places to the right of the decimal point, to be sure that the third place (the thousandths place) is correct.

EXAMPLE ii. Write $\frac{7}{16}$ as a terminating decimal.

Solution:

$$
\begin{array}{r}
0.4375 \\
16\overline{)7.0000} \\
6\,4 \\
\hline
60 \\
48 \\
\hline
120 \\
112 \\
\hline
80 \\
80 \\
\hline
\end{array}
$$

$$
\frac{7}{16} = 0.4376
$$

Exercise 10.5

1. Which of the following are decimal rational numbers?
   a. $\frac{7}{125}$ d. $\frac{62}{50}$
   b. $\frac{1}{3}$ e. $\frac{3}{8}$
   c. $\frac{26}{27}$ f. $\frac{9}{8}$

2. Change the following common fractions to decimal fractions.
   a. $\frac{1}{8}$ d. $\frac{37}{500}$
   b. $\frac{6}{25}$ e. $\frac{144}{240}$
   c. $\frac{23}{125}$ f. $\frac{49}{2800}$

3. Write the following decimal fractions as common fractions in simplest form.
   a. 0.042 f. 4.005
   b. 0.25 g. 1.144
   c. 1.375 h. 0.875
   d. 37.84 i. 1.64
   e. 3.625 j. 8.888

4. Write in expanded notation.

    a. 176.82　　　　　　　d. 84.987

    b. 34.567　　　　　　　e. 10.0007

    c. 0.0032　　　　　　　f. 0.000076

5. Write the following as (1) common fractions with a power of 10 as denominator and (2) as decimal fractions.

    a. $\frac{1}{4}$　　　　　　　e. $\frac{3}{5}$

    b. $\frac{1}{2}$　　　　　　　f. $\frac{7}{16}$

    c. $\frac{5}{8}$　　　　　　　g. $\frac{17}{200}$

    d. $\frac{12}{25}$　　　　　　　h. $\frac{123}{125}$

6. Which of the following are terminating decimals?

    a. $\frac{1}{4}$　　　　　　　d. $\frac{3}{25}$

    b. $\frac{2}{3}$　　　　　　　e. $\frac{7}{250}$

    c. $\frac{5}{7}$　　　　　　　f. $\frac{7}{9}$

7. Write each of the following as a decimal fraction correct to the nearest hundredth.

    a. $\frac{1}{6}$　　　　　　　e. $\frac{5}{12}$

    b. $\frac{1}{9}$　　　　　　　f. $\frac{2}{3}$

    c. $\frac{5}{6}$　　　　　　　g. $\frac{1}{7}$

    d. $\frac{3}{11}$　　　　　　　h. $\frac{8}{9}$

8. Write a decimal numeral for each of the following.

    a. Five hundred twenty-six and seven tenths.

    b. Sixty-five and forty-three hundredths.

    c. Twenty-eight and one hundred sixty-one thousandths.

    d. Two and forty-five thousandths.

    e. One hundred ninety-nine ten thousandths.

    f. Sixty-two hundred thousandths.

9. Write a word name for each of the following.

    a. 115.3　　　　　　　d. 14.465

    b. 0.134　　　　　　　e. 45.0067

    c. 1345.00145　　　　f. 70.0316

10. Write as decimal fractions correct to the nearest thousandth.

    a. $\frac{4}{11}$　　　　　　　e. $\frac{23}{17}$

    b. $\frac{9}{16}$　　　　　　　f. $\frac{15}{7}$

    c. $\frac{8}{15}$　　　　　　　g. $\frac{35}{6}$

    d. $\frac{7}{13}$　　　　　　　h. $\frac{84}{45}$

11.1. Directed Numbers

The whole numbers and the non-negative rational numbers have their origin in counting situations. So do the numbers called **integers.** All these numbers and the symbols which represent them are creations of the human mind. The integers were created to count and measure with respect to a point of reference, denoted by 0, when the direction relative to this point of reference is important. Examples of such situations are common in the physical world, as, for example, the measurement of temperature above and below zero and the measure of altitude above and below sea-level.

We may indicate a temperature of 2 degrees above zero as $^+2$ (read: positive two) degrees and a temperature of 2 degree below zero as $^-2$ (read: negative two) degrees. Instead of saying that an altitude is 100 feet above sea-level, we may say $^+100$ feet; instead of saying an altitude is 100 feet below sea-level, we may say $^-100$ feet. Similarly, a profit of 500 dollars may be indicated as $^+500$ dollars and a loss of 500 dollars may be indicated as $^-500$ dollars.

We call numbers such as $^+2$ and $^-2$ **directed numbers.** They answer "How many and in what direction?" The superscripts $+$ and $-$

The Integers

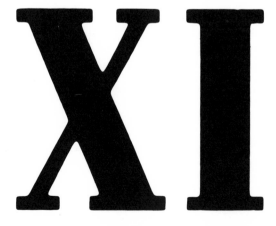

describe the relation of the number to the point of reference indicated by 0. The superscript of a directed number is called the **sign** of the number.

The elements of the set

$$\{^+1, ^+2, ^+3, \ldots\}$$

are called **positive numbers.** The elements of the set

$$\{^-1, ^-2, ^-3, \ldots\}$$

are called **negative numbers.** The set of numbers including the positive numbers, the negative numbers, and zero is called the set of **integers:**

$$\{\ldots, ^-3, ^-2, ^-1, 0, ^+1, ^+2, ^+3, \ldots\}$$

The subset of the set of integers that contains all the positive numbers is called the set of positive integers. Each element in this set is called a **positive integer.** The subset of the set of integers that contains all the negative numbers is called the set of negative integers. Each element in this set is called a **negative integer.** The integer 0 is neither positive nor negative.

The union of the set of positive integers and the set containing zero is called the set of **nonnegative integers.**

11.2. Integers and the Number Line

Just as we associate whole numbers and nonnegative rational numbers with points on a line, we also associate the integers with points on a number line. On a line we choose one point and assign it the number 0. We then mark off equal distances to the right and left of the point labeled 0. The points to the right of 0 are labeled successively $^+1, ^+2, ^+3, \ldots$; the points to the left of 0 are labeled $^-1, ^-2, ^-3,$ successively as shown in Figure 11.1. The labeling is continued indefinitely in both directions. We see then, that we have a one-to-one correspondence between the integers and points on a number line.

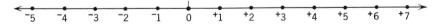

Figure 11.1

As with the whole numbers and the nonnegative rational numbers, the point associated with an integer is called its **graph,** and the integer associated with a point is called the **coordinate** of the point.

11.3. Absolute Value

Notice that the point on the number line corresponding to $^+2$ is the same distance from the point corresponding to 0 as the point corresponding to $^-2$, but in the opposite direction. Similarly, the points corresponding to $^+4$ and $^-4$ are each four units from the point corresponding to 0, but in opposite directions from it. We call pairs of integers ^+n and ^-n **opposites.** We say that each is the opposite of the other. Thus $^-9$ is the opposite of $^+9$ and $^+15$ is the opposite of $^-15$. The number 0 is its own opposite.

We observe that every integer has an opposite. *The opposite of a positive integer is a negative integer; the opposite of a negative integer is a positive integer; and the opposite of 0 is 0.*

When we are interested only in the magnitude of the distance of an integer and not in its direction from zero, we find it convenient to consider its **absolute value.**

The absolute value of $^-4$ is $^+4$; the absolute value of $^-6$ is $^+6$; the absolute value of $^+2$ is $^+2$; the absolute value of $^-8$ is $^+8$; the absolute value of 0 is 0.

The absolute value of an integer is either zero or a positive integer. The absolute value of a positive integer is the number itself. The absolute value of a negative integer is its opposite; the absolute value of zero is zero. To abbreviate "the absolute value of integer n" we write $|n|$. For example,

$$|^+3| = {}^+3 \quad \text{(read: the absolute value of positive three is positive three)}$$
$$|^-5| = {}^+5 \quad \text{(read: the absolute value of negative five is positive five)}$$
$$|0| = 0 \quad \text{(read: the absolute value of zero is zero)}$$

In general:

DEFINITION 11.1. *If x is any integer, the **absolute value** of x denoted by $|x|$ is defined as follows:*

$|x|$ *equals x if x is nonnegative*
$|x|$ *equals the opposite of x if x is negative.*

From Definition 11.1, we see that the absolute value of zero is zero, the absolute value of a positive integer is that positive integer, and the absolute value of a negative integer is its opposite, which is a positive integer. Thus

$$|^+7| = {}^+7$$
$$|^-9| = {}^-(^-9) = {}^+9$$
$$|0| = 0.$$

11.4. Ordering Integers

Just as the order of the whole numbers and the nonnegative rational numbers show up clearly by the position of their graphs on the number line, so does the order of the integers. We recall that the graph of 3 is to the left of the graph of 5 and $3 < 5$; the graph of 9 is to the right of the graph of 6 and $9 > 6$.

Similarly,

$^-3 < {}^+1$ because the graph of $^-3$ is to the left of the graph of $^+1$
$^+4 < {}^+6$ because the graph of $^+4$ is to the left of the graph of $^+6$
$^+15 > {}^-8$ because the graph of $^+15$ is to the right of the graph of $^-8$
$^-2 > {}^-4$ because the graph of $^-2$ is to the right of the graph of $^-4$.

Observe that:

(1) Every negative integer is less than 0.
(2) Every positive integer is greater than 0.
(3) Every positive integer is greater than every negative integer.

EXAMPLE i. Is $^-8$ less than, equal to, or greater than $^-5$?
Solution: Since the graph of $^-8$ is to the left of the graph of $^-5$, $^-8 < {}^-5$.

EXAMPLE ii. Order the integers below from the least to the greatest:

$$^-4, {}^+6, {}^-3, {}^+7, {}^-1.$$

Solution: $^-4, {}^-3, {}^-1, {}^+6, {}^+7.$

Exercise 11.1

1. What is the opposite of each of the following?
   a. $^+8$
   b. $^-5$
   c. $^-7$
   d. $^+11$
   e. 0
   f. $^+100$
   g. $^-16$
   h. $^-139$

2. Using symbols denote the absolute value of the following.
   a. $^+7$
   b. $^+15$
   c. 0
   d. $^-6$
   e. $^-14$
   f. $^-73$
   g. $^+46$
   h. $^-100$

3. Give the value of each of the following.
   a. $|^-56|$
   b. $|^+73|$
   c. $|0|$
   d. $|^-23|$
   e. $|^+56|$
   f. $|^-102|$
   g. $|^+14|$
   h. $|^+788|$

4. Replace * with $=$, $<$, or $>$ to make true statements.
   a. $^-5 * {}^+6$
   b. $^-8 * {}^+3$
   c. $^+11 * {}^-3$
   d. $0 * {}^-150$
   e. $^-1 * {}^+1$
   f. $^+16 * {}^+99$
   g. $^-30 * {}^-77$
   h. $^+88 * {}^+66$
   i. $^-46 * {}^-47$
   j. $^-99 * {}^-97$

5. Replace * with $=$, $<$, or $>$ to make true statements.
   a. $|^-1| * |^+1|$
   b. $|^-6| * |^+8|$
   c. $|^+9| * |^-3|$
   d. $|^-8| * |^+8|$
   e. $|^+46| * |^-32|$
   f. $|0| * |^-5|$
   g. $|^-16| * |^-67|$
   h. $|^-31| * |^-26|$

6. Find the solution sets. The domain is the set of integers.
   a. $|n| = {}^+2$
   b. $|n| = 0$
   c. $|n| = {}^+6$
   d. $|p| = {}^+5$
   e. $|x| = {}^+4$
   f. $|x| = {}^+9$
   g. $|x| = {}^+8$
   h. $|x| = {}^+15$

7. Which of the following are true statements if x and y are integers.
   a. If x is negative then $x < 0$.
   b. If x is negative and y is positive then $x < y$.
   c. If x is positive then $x < 0$.
   d. If x is negative and y is positive then $x \neq y$.

e. If x is negative then $|x|$ is positive.

f. If x is positive than the opposite of x is negative.

8. Find the solution sets. The domain is the set of integers.

a. $x > {}^+1$ e. $x \geqslant {}^-2$

b. $x < 0$ f. $x \leqslant {}^+9$

c. $x \leqslant {}^-5$ g. $x > {}^-10$

d. $x \geqslant {}^+6$ h. $x < {}^+12$

9. Which of the following statements are true if x and y are any two integers?

a. $|x| < |y|$ d. $|x| \geqslant 0$

b. $|x| = |y|$ ~   e. $|x| \leqslant 0$

c. $|x| > |y|$ f. $|x| = x$

10. What is the union of the set of positive integers and the set of negative integers?

11. What is the intersection of the set of positive integers and the set of negative integers?

12. How many units from zero is the graph of each integer given below?

a. $^+6$ d. $^+9$

b. $^-3$ e. $^-12$

c. $^+10$ f. $^-14$

11.5. Addition of Integers

We now agree on rules for adding integers. We do this by looking at the number line. We think of addition of integers as making moves along the number line. A move to the right will be described by a positive integer. A move to the left will be described by a negative integer. Thus $^+5 + {}^+3$ means to start at the point corresponding to zero, moving 5 units to the right and then 3 more units to the right. Thus

$$^+5 + {}^+3 = {}^+8.$$

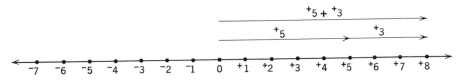

Figure 11.2

To add $^+6 + {}^-3$ means to start at the point corresponding to zero, moving 6 units to the right and then 3 units to the left. Thus

$$^+6 + {}^-3 = {}^+3.$$

Similarly,

$$^+8 + {}^-5 = {}^+3$$
$$^-6 + {}^+2 = {}^-4$$
$$^-4 + {}^-3 = {}^-7$$
$$^-2 + {}^+8 = {}^+6$$
$$^+3 + {}^-3 = 0$$

EXAMPLE i. Add $(^+7 + {}^-5) + {}^-8.$

Solution:
$$(^+7 + {}^-5) + {}^-8 = {}^+2 + {}^-8$$
$$= {}^-6$$

EXAMPLE ii. Add $(^-7 + {}^+9) + (^+2 + {}^-9)$

Solution:
$$(^-7 + {}^+9) + (^+2 + {}^-9) = {}^+2 + {}^-7$$
$$= {}^-5$$

Although the sum of any two integers can always be found using the number line, this method is not practical for finding such sums as $^+846 + {}^-327$. We now find a more practical method for adding two integers.

A few examples using the number line convince us that the sum of two positive integers is a positive integer. If a and b are two positive integers, we find their sum by adding as in the set of whole numbers. In fact, positive integers may be denoted by whole number symbols as we shall see later, and may be added, subtracted, multiplied, and divided in the same way that we add, subtract, multiply, and divide whole numbers.

In symbols we write

$$a + b = {}^+(|a| + |b|) \qquad \text{if } a > 0 \text{ and } b > 0.$$

If a and b are both negative integers, their sum is the opposite of the sum of their absolute values. In symbols

$$a + b = {}^-(|a| + |b|) \qquad \text{if } a < 0 \text{ and } b < 0.$$

The sum of a positive integer and a negative integer may be positive or negative. If a is a positive integer and b is a negative integer and $|a| > |b|$ then the sum of a and b is a positive integer and is found by subtracting the smaller absolute value, $|b|$, from the larger absolute value, $|a|$.

Symbolically we write

$$a + b = {}^+(|a| - |b|) \qquad \text{if } a > 0, b < 0, \text{ and } |a| > |b|.$$

If a is a positive integer, b a negative integer, and $|b| > |a|$ their sum is the opposite of the difference $|b| - |a|$. Symbolically,

$$a + b = {}^-(|b| - |a|) \qquad \text{if } a > 0, b < 0 \text{ and } |b| > |a|.$$

If a and b are opposites, that is, if $b = {}^-a$, then

$$a + b = a + {}^-a = 0.$$

Summarizing the above we have

Rule 1. a + b = $^+$(|a| + |b|) if a > 0 and b > 0
Rule 2. a + b = $^-$(|a| + |b|) if a < 0 and b < 0
Rule 3. a + b = $^+$(|a| − |b|) if a > 0, b < 0 and |a| > |b|
Rule 4. a + b = $^-$(|b| − |a|) if a > 0, b < 0 and |b| > |a|
Rule 5. a + b = 0 if b = $^-$a

EXAMPLE i. Add $^+186 + {}^+199$.

Solution: We add positive integers just as we add whole numbers; the sum is a positive integer. Then

$$^+186 + {}^+199 = {}^+385$$

EXAMPLE ii. Add $^+186 + {}^-342$.

Solution: Since $|^-342| > |^+186|$ we use Rule 4. Then

$$^+186 + {}^-342 = {}^-(|^-342| - |^+186|)$$
$$= {}^-156$$

EXAMPLE iii. Add $^+342 + {}^-126$.

Solution: Since $|^+342| > |^-126|$, we use Rule 3. Then

$$^+342 + {}^-126 = {}^+(|^+342| - |^-126|)$$
$$= {}^+216$$

EXAMPLE iv. Add $^-288 + {}^-156$.

Solution: Since both $^-288$ and $^-156$ are negative we add their absolute values and take the opposite of this sum (Rule 2).

$$^-288 + {}^-156 = {}^-(|^-288| + |^-156|)$$
$$= {}^-444$$

11.6. Properties of Addition of Integers

Addition of integers has all the operational properties of addition of whole numbers. If we add two integers, the sum is always an integer. Hence the set of integers is **closed** under the operation of addition.

Notice that

$$^+3 + {}^+5 = {}^+8 \quad \text{and} \quad {}^+5 + {}^+3 = {}^+8$$
$$^-7 + {}^-4 = {}^-11 \quad \text{and} \quad {}^-4 + {}^-7 = {}^-11$$
$$^+3 + {}^-7 = {}^-4 \quad \text{and} \quad {}^-7 + {}^+3 = {}^-4$$

We see that the order in which we added the integers above did not change the sum. This is true in general. **If a and b are integers, then**

$$\mathbf{a + b = b + a.}$$

We see, then, that the operation of addition of integers is **commutative.**

Observe the following:

$$(^+2 + {}^+3) + {}^+7 = {}^+5 + {}^+7 = {}^+12 \quad \text{and}$$
$$^+2 + (^+3 + {}^+7) = {}^+2 + {}^+10 = {}^+12$$
$$(^-3 + {}^+4) + {}^-5 = {}^+1 + {}^-5 = {}^-4 \quad \text{and}$$
$$^-3 + (^+4 + {}^-5) = {}^-3 + {}^-1 = {}^-4.$$

In general, **if a, b, and c are integers, then**

$$\mathbf{(a + b) + c = a + (b + c).}$$

That is, addition of integers is **associative.**

If we find the sum of 0 and any integer, the sum is that integer. For example

$$0 + {}^+2 = {}^+2 + 0 = {}^+2$$
$$0 + {}^-5 = {}^-5 + 0 = {}^-5.$$

In general, **if a is any integer**

$$\mathbf{a + 0 = 0 + a = a.}$$

We call 0 the **identity element for addition** or the **additive identity.**

The operation of addition of integers has one property that the operation of addition of whole numbers does not possess. Notice that

$$^+2 + {}^-2 = 0$$
$$^-3 + {}^+3 = 0.$$

In general, the sum of an integer and its opposite is zero. We call this the **additive inverse property** of addition of integers. The additive inverse property of integers states that every integer a has an opposite ^-a, called the **additive inverse** of a, such that

$$\mathbf{a + {^-a} = 0.}$$

We summarize by listing the properties of addition of integers.

I-1. Closure Property. If a and b are integers, then their sum $a + b$ is a unique integer.

I-2. Commutative Property. If a and b are integers, $a + b = b + a$.

I-3. Associative Property. If a, b, and c are integers, then $(a + b) + c = a + (b + c)$.

I-4. Identity Element. There is a unique integer, 0, called the **additive identity,** such that for all integers, a, $a + 0 = 0 + a = a$.

I-5. Additive Inverse Property. Every integer a has a unique opposite, ^-a, called the **additive inverse** of a, such that $a + {^-a} = {^-a} + a = 0$.

11.7. Subtraction of Integers

The operation of subtraction of integers is the inverse of the operation of addition of integers. Thus

$$^+3 - {^+7} = n$$

means

$$^+7 + n = {^+3}.$$

In general, if a, b, and n are integers

$$a - b = n$$

means

$$b + n = a.$$

Let us find $^+3 - {^+7}$. This means that we are looking for an integer n, if such an integer exists, such that

(1) $$^+7 + n = {^+3}.$$

Adding $^-7$ to both members of equation (1), we have

$$^-7 + (^+7 + n) = {^-7} + {^+3}$$

Using the associative property of addition we have

$$(^-7 + {^+7}) + n = {^-7} + {^+3}$$

But $^-7 + {^+7} = 0$ by the additive inverse property, hence we have

$$0 + n = {^-7} + {^+3}$$

Since $0 + n = n$ by the identity property of addition we have

$$n = {^-7} + {^+3} = {^+3} + {^-7}$$

Hence

$$^+3 - {^+7} = {^+3} + {^-7}$$
$$= {^-4}.$$

Notice that subtracting $^+7$ from $^+3$ gives the same result as adding the additive inverse (opposite) of $^+7$ to $^+3$. This is true in general.

DEFINITION 11.2. *If a and b are integers, their* **difference,** *denoted by a − b, is a + ^-b. The operation of finding the difference is called* **subtraction.**

We see from Definition 11.2, that

$$^+6 - {^+2} = {^+6} + {^-2} = {^+4}$$

$$^-3 - {^-2} = {^-3} + {^+2} = {^-1}$$

$$^+7 - {^-3} = {^+7} + {^+3} = {^+10}$$

Since every integer has an additive inverse (opposite) we see that the set of integers is *closed under the operation of subtraction.*

In observing Definition 11.2, we note that subtraction of an integer may be replaced by the addition of its additive inverse (opposite). This closely parallels what happened when we studied the division of nonnegative rational numbers. We found that division by any nonnegative rational number, except zero, could be replaced by multiplication of the multiplicative inverse (reciprocal) of that number. We also note at this point that the operations of addition and subtraction of integers have all the properties of the operations of addition and subtraction of whole numbers, plus the closure property for subtraction and the additive inverse property.

Exercise 11.2

1. Add

   a. $^+6 + {}^-3$ e. $^+27 + {}^-38$

   b. $^+7 + {}^-4$ f. $^-46 + {}^-12$

   c. $^+19 + {}^-26$ g. $^+46 + {}^-94$

   d. $^-14 + {}^-9$ h. $^-38 + {}^+27$

2. Subtract

   a. $^+7 - {}^-3$ e. $^+26 - {}^-34$

   b. $^+8 - {}^+4$ f. $^-15 - {}^+18$

   c. $^-9 - {}^-3$ g. $^-36 - {}^-27$

   d. $^-12 - {}^+7$ h. $^-46 - {}^+88$

3. Find the solution sets. The domain is the set of integers.

   a. $n + {}^+3 = {}^+9$ e. $^-3 - n = {}^-9$

   b. $^+5 + n = {}^-7$ f. $n - {}^+6 = {}^-5$

   c. $^-3 + n = {}^-5$ g. $^-8 - n = {}^+4$

   d. $n - {}^+7 = {}^-2$ h. $^+6 - n = {}^-5$

4. Express the following as the sum of two integers.

   a. $^+4 - {}^+3$ e. $^+6 - {}^-9$

   b. $^+5 - {}^-7$ f. $^+13 - {}^-8$

   c. $^-6 - {}^-5$ g. $^-12 - {}^+8$

   d. $^-3 - {}^+7$ h. $^-10 - {}^-10$

5. If $a > b > 0$, is $a + {}^-b$ positive or negative?

6. If $0 > a > b$, is $a + {}^-b$ positive or negative?

7. If $a > b > 0$, is $b + {}^-a$ positive or negative?

8. If $0 > a > b$ is $b + {}^-a$ positive or negative?

9. Add

   a. $^+6 + {}^-3 + {}^+4$ d. $^-26 + {}^-13 + {}^+18$

   b. $^+19 + {}^-7 + {}^-13$ e. $^+17 + {}^-19 + {}^-27$

   c. $^-16 + {}^-9 + {}^+4$ f. $^-45 + {}^-36 + {}^+77$

10. Replace the symbol $*$ with $=$, $<$, or $>$ to make true statements.

    a. $|{}^+6 + {}^+9| * |{}^+6| + |{}^+9|$

    b. $|{}^-6 + {}^-3| * |{}^-6| + |{}^-3|$

    c. $|{}^-8 + {}^+7| * |{}^-8| + |{}^+7|$

    d. $|{}^-5 + {}^+9| * |{}^-5| + |{}^+9|$

    e. $|{}^-3 - {}^+6| * |{}^-3| - |{}^+6|$

    f. $|{}^-8 - {}^-7| * |{}^-8| - |{}^-7|$

    g. $|{}^-10 - {}^+4| * |{}^-10| - |{}^+4|$

11. Perform the indicated operations.
    a. $^+8 + {}^+6 - {}^-5$ d. $^+18 + {}^-23 - {}^-91$
    b. $^-14 - {}^+57 - {}^+15$ e. $^-34 - {}^+34 - {}^-45 + {}^-66$
    c. $^+38 - {}^+19 - {}^+88$ f. $^-75 + {}^-45 - {}^-88 - {}^-123$

12. The temperature at 7:00 A.M. was $^+36°$. Between 7:00 A.M. and noon the temperature rose 20°. What was the temperature at noon?

13. In a card game a player scored 120 points. He was set 60 points each for four consecutive hands. What was his resulting score?

14. A man put $300 in his checking account when he opened it. During the month he wrote checks for $40, $25, $135, and $69. He made deposits of $20 and $34. What was his balance at the end of the month?

15. Mr. Spendthrift had a bank balance of $480 when he went on a week-end trip to Las Vegas. After he returned from his trip the bank said that he was overdrawn by $123. How much did Mr. Spendthrift spend in Las Vegas?

16. A used car dealer sold five cars with the following results: a gain of $54, a loss of $32, a loss of $86, a gain of $126, and a gain of $42. Represent his net gain or loss by means of an integer.

17. The temperature outdoors at 10:00 A.M. was 35° and 72° indoors. By noon the temperature had risen 10° outdoors and fallen 3° indoors. What was the difference at noon between the temperature outdoors and the temperature indoors?

11.8. Multiplication of Integers

By considering certain physical situations involving integers we can formulate rules for finding the product of two integers.

Suppose we have entered a turtle in a race. Since turtles are not very knowledgeable about races, our turtle may go forward toward the finish line, he may go backward in the opposite direction, and, at any moment in the race he may reverse his direction or stop entirely.

Suppose we think of the positive integers as measuring minutes in the future and the negative integers as measuring minutes in the past. According to this agreement, $^+2$ represents two minutes in the future, $^-2$ represents two minutes in the past, and 0 represents the present moment.

Let us also agree to represent the rate of a turtle traveling forward (toward the finish line) by a positive integer, the rate of a turtle traveling backward (away from the finish line) by a negative integer, and the rate of a turtle standing still by 0.

We shall represent the distance traveled forward by a positive integer and the distance traveled backward by a negative integer.

We now use the familiar formula

$$(\text{distance}) = (\text{rate}) \times (\text{time})$$

to determine the directed distance traveled by certain turtles in the race.

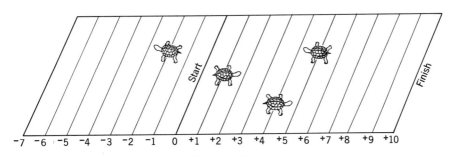

Figure 11.3

(1) A turtle traveling forward at the rate of 2 lengths per minute ($^+2$) will be 6 lengths closer ($^+6$) to the finish line 3 minutes from now ($^+3$):

$$(^+2)(^+3) = {^+6}.$$

(2) A turtle traveling backward at the rate of 2 lengths per minute ($^-2$) will be 6 lengths farther ($^-6$) from the finish line 3 minutes from now ($^+3$):

$$(^-2)(^+3) = {^-6}.$$

(3) A turtle traveling forward at the rate of 2 lengths per minute ($^+2$) was 6 lengths farther ($^-6$) from the finish line 3 minutes ago ($^-3$):

$$(^+2)(^-3) = {^-6}.$$

(4) A turtle traveling backward at the rate of 2 lengths per minute ($^-2$) was 6 lengths closer ($^+6$) to the finish line 3 minutes ago ($^-3$):

$$(^-2)(^-3) = {^+6}.$$

(5) A turtle standing still (rate: 0) will be no closer (0) to the finish line 3 minutes from now ($^+$3):

$$(0)(^+3) = 0$$

Similarly, a turtle standing still for the last 3 minutes (rate: 0) was no closer to the finish line 3 minutes ago ($^-$3) than he is now:

$$(0)(^-3) = 0.$$

The above illustrations suggest:

(a) *The product of two positive integers is a positive integer.*
(b) *The product of two negative integers is a positive integer.*
(c) *The product of a negative integer and a positive integer is a negative integer.*
(d) *The product of zero and any integer is zero.*

We summarize the above observations by writing the following formal rules using symbols:

Rule 6. ab = $^+$(|a| · |b|) if a > 0 and b > 0
Rule 7. ab = $^-$(|a| · |b|) if a > 0 and b < 0
Rule 8. ab = $^-$(|a| · |b|) if a < 0 and b > 0
Rule 9. ab = $^+$(|a| |b|) if a < 0 and b < 0
Rule 10. ab = 0 if a = 0 or b = 0.

EXAMPLE i. Find the product ($^+$6)($^-$12).
Solution: ($^+$6)($^-$12) = $^-$72

EXAMPLE ii. Find the product ($^-$7)($^-$9).
Solution: ($^-$7)($^-$9) = $^+$63

Using the rules for finding the product of two integers given above, we can show that multiplication of integers has all the properties of the operation of multiplication of whole numbers. The **closure property** certainly holds, since the product of any two integers is an integer.
Observe the following:

$$(^+2)(^+6) = {}^+12 \quad \text{and} \quad (^+6)(^+2) = {}^+12$$
$$(^-3)(^-5) = {}^+15 \quad \text{and} \quad (^-5)(^-3) = {}^+15$$
$$(^-7)(^+3) = {}^-21 \quad \text{and} \quad (^+3)(^-7) = {}^-21$$

In general, if a and b are integers, then

$$ab = ba.$$

This is the **commutative property of multiplication of integers.**

Notice the following products:

$$\left.\begin{array}{l}[(^+2)(^+3)](^-4) = (^+6)(^-4) = {}^-24 \\ (^+2)[(^+3)(^-4)] = (^+2)(^-12) = {}^-24\end{array}\right\} \text{ hence } [(^+2)(^+3)](^-4) = (^+2)[(^+3)(^-4)]$$

$$\left.\begin{array}{l}[(^-6)(^-2)](^-5) = (^+12)(^-5) = {}^-60 \\ (^-6)[(^-2)(^-5)] = (^-6)(^+10) = {}^-60\end{array}\right\} \text{ hence } [(^-6)(^-2)](^-5) = (^-6)[(^-2)(^-5)]$$

$$\left.\begin{array}{l}[(^+3)(^-4)](^+7) = (^-12)(^+7) = {}^-84 \\ (^+3)[(^-4)(^+7)] = (^+3)(^-28) = {}^-84\end{array}\right\} \text{ hence } [(^+3)(^-4)](^+7) = (^+3)[(^-4)(^+7)]$$

The above examples suggest that if a, b, and c are integers, then

$$(ab)c = a(bc).$$

This is the **associative property of multiplication of integers.**

We know that

$$(^+1)(^+5) = (^+5)(^+1) = {}^+5$$
$$(^+1)(^-6) = (^-6)(^+1) = {}^-6$$

and, in general, if a is any integer,

$$a(^+1) = (^+1)a = a;$$

hence $^+1$ is the **identity element** for multiplication of integers.

The distributive property also holds for integers. Notice that

$$(^+2)[^+3 + {}^-4] = (^+2)(^-1) = {}^-2$$

and
$$(^+2)(^+3) + (^+2)(^-4) = {}^+6 + {}^-8 = {}^-2$$

hence

$$(^+2)[^+3 + {}^-4] = (^+2)(^+3) + (^+2)(^-4).$$

In a similar fashion we can show that

$$(^-3)[^-2 + {}^-3] = (^-3)(^-2) + (^-3)(^-3)$$
$$(^-5)[^-6 + {}^+3] = (^-5)(^-6) + (^-5)(^+3).$$

In general, if a, b, and c are integers, then

$$a(b + c) = ab + ac.$$

This is the **distributive property** of multiplication of integers over addition of integers.

In summary, we state the properties of multiplication of integers:

I-6. Closure Property of Multiplication. If a and b are integers, their product ab, is a unique integer.

I-7. Commutative Property of Multiplication. If a and b are integers, $ab = ba$.

I-8. Associative Property of Multiplication. If a, b, and c are integers, then $(ab)c = a(bc)$.

I-9. Identity Element for Multiplication. There is a unique integer, $^{+}1$, called the **multiplicative identity,** such that for all integers a, $a \cdot {}^{+}1 = {}^{+}1 \cdot a = a$.

I-10. Distributive Property. If a, b, and c are integers, then $a(b + c) = ab + ac$.

11.9. Division of Integers.

We define the operation of division of integers as the inverse of the multiplication of integers. That is, if a and b are integers, $b \neq 0$,

$$a \div b = c \quad \text{means} \quad bc = a.$$

Thus

$$^{-}6 \div {}^{-}2 = n$$

means

$$^{-}2 \cdot n = {}^{-}6.$$

Since $^{-}2 \cdot {}^{+}3 = {}^{-}6$, we see that $^{-}6 \div {}^{-}2 = {}^{+}3$. However, there is no integer n such that $^{+}9 \div {}^{-}2 = n$, since there is no integer which when multiplied by $^{-}2$ gives a product of $^{+}9$. We see then, that the set of integers is *not* closed under division.

The division

$$^{+}18 \div {}^{+}2$$

requires finding the missing factor n such that

$$^{+}2 \cdot n = {}^{+}18.$$

Since $^{+}2 \cdot {}^{+}9 = {}^{+}18$, we see that $^{+}18 \div {}^{+}2 = {}^{+}9$. This example illustrates that a positive integer divided by a positive integer is a positive integer.

The division

$$^-12 \div {}^-4$$

requires finding a missing factor n such that

$$^-4 \cdot n = {}^-12.$$

Since $^-4 \cdot {}^+3 = {}^-12$, $^-12 \div {}^-4 = {}^+3$. This example illustrates that a negative integer divided by a negative integer is a positive integer.

The division

$$^-24 \div {}^+8$$

requires finding a missing factor n such that

$$^+8 \cdot n = {}^-24$$

Since $^+8 \cdot {}^-3 = {}^-24$, we see that $^-24 \div {}^+8 = {}^-3$. This example illustrates that a negative integer divided by a positive integer is a negative integer.

The division

$$^+36 \div {}^-9$$

requires finding a missing factor n such that

$$^-9 \cdot n = {}^+36.$$

Since $^-9 \cdot {}^-4 = {}^+36$, we see that $^+36 \div {}^-9 = {}^-4$. This example illustrates that a positive integer divided by a negative integer is a negative integer.

In general, *if a and b are integers, $a \div b$, $b \neq 0$ is positive if a and b are both positive or both negative and negative otherwise. If $a = 0$, than $a \div b = 0$, $b \neq 0$.*

11.10. The Integers and the Whole Numbers

The set of nonnegative rational numbers has a subset

$$\left\{ \frac{0}{1}, \frac{1}{1}, \frac{2}{1}, \frac{3}{1}, \cdots \right\}$$

which can be replaced by the whole numbers for all arithmetic purposes. We say that the system of whole numbers is a subsystem of the system of nonnegative rational numbers. We also say that the nonnegative rational number system is an extension of the whole number system.

The set of integers is also an extension of the set of whole numbers. Let us consider the set P of nonnegative integers:

$$P = \{0, {}^+1, {}^+2, {}^+3, \ldots\}$$

We can set up a one-to-one correspondence between the set of whole numbers and the elements of P:

$$
\begin{array}{ccccc}
0 & {}^+1 & {}^+2 & {}^+3 \cdots {}^+n \cdots \\
\updownarrow & \updownarrow & \updownarrow & \updownarrow \quad \updownarrow \\
0 & 1 & 2 & 3 \cdots n \cdots
\end{array}
$$

Notice that:

$$
\begin{array}{lll}
3 > 2 & \text{and} & {}^+3 > {}^+2 \\
3 + 2 = 5 & \text{and} & {}^+3 + {}^+2 = {}^+5 \\
3 - 2 = 1 & \text{and} & {}^+3 - {}^+2 = {}^+1 \\
3 \cdot 2 = 6 & \text{and} & {}^+3 \cdot {}^+2 = {}^+6 \\
6 \div 2 = 3 & \text{and} & {}^+6 \div {}^+2 = {}^+3
\end{array}
$$

So far as their order and the performance of the operations of arithmetic are concerned, the set of nonnegative integers and the set of whole numbers behave exactly alike. Because of this similarity, it is customary to use the symbols 0, 1, 2, 3, . . . , to denote both the whole numbers and the nonnegative integers.

Because a one-to-one correspondence that preserves the order and the arithmetic operations can be set up between the nonnegative integers and the whole numbers we say that the set of whole numbers is embedded in the set of integers and that the system of whole numbers is a subsystem of the set of integers.

Exercise 11.3

1. Multiply.
   a. $({}^+3)({}^+7)$
   b. $({}^-9)({}^+8)$
   c. $({}^+5)({}^-9)$
   d. $({}^-11)({}^-6)$
   e. $({}^-19)({}^+32)$

   f. $({}^-17)({}^+64)$
   g. $({}^-26)({}^-32)$
   h. $({}^+72)({}^+18)$
   i. $({}^-19)({}^-23)$
   j. $({}^-72)({}^+36)$

2. Find the products.
   a. $(^+3)(^-4)(^+6)$
   b. $(^-7)(^-3)(^-9)$
   c. $(^+7)(^-3)(^-9)$
   d. $(^+4)(^-7)(^+8)$
   e. $(^+6)(^+9)(^+3)$
   f. $(^-7)(^-8)(^-5)$
   g. $(^+12)(^-3)(^-7)$
   h. $(^-9)(^+3)(^-17)$

3. Find the quotients.
   a. $(^-36) \div (^-6)$
   b. $(^+18) \div (^-2)$
   c. $(^-69) \div (^+23)$
   d. $(^-168) \div (^-8)$
   e. $(^+39) \div (^+3)$
   f. $(^+128) \div (^-2)$
   g. $(^-945) \div (^-27)$
   h. $(^-468) \div (^+12)$

4. Fill in the blanks with numerals to make true statements.
   a. $(^-18) \div (^-2) = {}^+9$ because $(^-2) \times \underline{\hspace{2cm}} = {}^-18$
   b. $(^-24) \div (^+4) = {}^-6$ because $(^+4) \times \underline{\hspace{2cm}} = {}^-24$
   c. $(^+36) \div (^+3) = {}^+12$ because $(^+3) \times \underline{\hspace{2cm}} = {}^+36$
   d. $(^-81) \div (^-9) = {}^+9$ because $(^-9) \times \underline{\hspace{2cm}} = {}^-81$

5. Fill in the blanks to make true statements. Tell which property of multiplication of integers you used in making your choice.
   a. ${}^+6 \times \underline{\hspace{2cm}} = {}^+3 \times {}^+6$
   b. $(^+3)(^+6 + {}^-5) = (^+3)\underline{\hspace{2cm}} + (^+3)(^-5)$
   c. $(^+7)[(^-8)(^+6)] = [(^+7) \cdot \underline{\hspace{2cm}}](^+6)$
   d. $(^+3)(^+1) = \underline{\hspace{2cm}}$
   e. $[(^+7)(^+9)] + [(^+7)(^+8)] = \underline{\hspace{2cm}} [(^+9) + (^+8)]$

6. Find the solution sets. The domain is the set of integers.
   a. ${}^-6n = {}^+36$
   b. ${}^-7n = {}^+49$
   c. $n \div 9 = {}^-6$
   d. $n \div 4 = {}^-48$
   e. ${}^-56 \div n = {}^+8$
   f. ${}^-12n = {}^-48$
   g. $9n = {}^-72$
   h. $126 \div n = {}^-63$

7. Fill in the blanks with positive or negative to make true statements.
   a. The product of two negative integers is a _____ integer.
   b. The quotient of a negative integer divided by a positive integer is a _____ integer.
   c. The product of three negative integers is a _____ integer.
   d. The product of four negative integers is a _____ integer.
   e. The product of two negative integers and two positive integers is a _____ integer.

8. Perform the indicated operations.
   a. $(3 \cdot 8) \div (^-2)$
   b. $(9 \cdot {}^-6) \div (^-3)$

c. $[(8)(^-9)(7)] \div (^-4)$

d. $[(12)(^-16)(32)] \div 24$

9. Insert one of the symbols $<$, $=$, or $>$ between each pair of expressions to make true statements.

    a. $|^+3| \cdot |^-4|$; $|^+3| \cdot |^+4|$

    b. $|^-6 \cdot {}^-4|$; $|^-6| \cdot |^-4|$

    c. $|^-5 \cdot {}^-6|$; $|^-5| \cdot |^-6|$

    d. $|^-3 \cdot {}^-8|$; $|^-3| \cdot |^-8|$

    e. $|30 \cdot {}^-6|$; $|30| \cdot |^-6|$

10. A company is losing money at the rate of \$5000 per day. Compare the company's debt today with its debt (a) five days ago; (b) 10 days from now.

11. If a is a positive integer is a^2 positive or negative? Why?

12. If a is a negative integer is a^2 positive or negative? Why?

12.1. Defining the Rational Numbers

The nonnegative rational number system is an extension of the system of whole numbers. We now create the rational number system as an extension of the system of integers.

DEFINITION 12.1. *A* **rational number,** *denoted by* $\frac{a}{b}$, *is an ordered pair, (a, b), of integers, $b \neq 0$.*

Some examples of rational numbers are $\frac{-2}{3}$, $\frac{5}{-6}$, $\frac{-7}{-8}$, and $\frac{0}{3}$.

We now define equality of rational numbers.

DEFINITION 12.2. *Two rational numbers (a, b) and (c, d) are* **equal** *if and only if $ad = bc$.*

Since $(a, b) = (c, d)$ if and only if $ad = bc$, we see that

$$\frac{1}{-2} = \frac{-2}{4} \qquad \text{because } (1)(4) = (^-2)(^-2)$$

$$\frac{3}{-4} = \frac{9}{-12} \qquad \text{because } (3)(^-12) = (^-4)(9)$$

$$\frac{-3}{-6} = \frac{4}{8} \qquad \text{because } (^-3)(8) = (^-6)(4)$$

The Rational Numbers

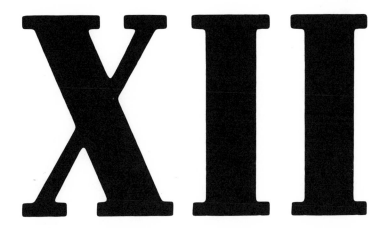

We now define the operations of addition and multiplication of rational numbers.

DEFINITION 12.3. *If (a, b) and (c, d) are two rational numbers, their* **sum,** *(a, b) + (c, d), is (ad + bc, bd).*

Using the fraction symbols $\frac{a}{b}$ and $\frac{c}{d}$ for the rational numbers (a, b) and (c, d) we see that

$$\frac{a}{b} + \frac{c}{d} = \frac{ad + bc}{bd}$$

Notice that this is the same definition given for the sum of two nonnegative rational numbers. This, of course, is as it should be, since the set of nonnegative rational numbers is a subset of the set of rational numbers.

EXAMPLE i. Find the sum $(3, {}^-2) + ({}^-4, {}^-3)$.
Solution: By Definition 12.3 we have

$$(3, {}^-2) + ({}^-4, {}^-3) = (3 \cdot {}^-3 + {}^-2 \cdot {}^-4, {}^-2 \cdot {}^-3)$$
$$= ({}^-9 + 8, 6)$$
$$= ({}^-1 \cdot 6)$$

Using fractions to denote the given rational numbers we have

$$\frac{3}{{}^-2} + \frac{{}^-4}{{}^-3} = \frac{(3 \cdot {}^-3) + ({}^-4 \cdot {}^-2)}{{}^-2 \cdot {}^-3}$$

$$= \frac{{}^-9 + 8}{6} = \frac{{}^-1}{6}$$

DEFINITION 12.4. *If (a, b) and (c, d) are rational numbers their* **product,** *(a, b) · (c, d), is (ac, bd).*

Using fractions to denote the two rational numbers we have

$$\frac{a}{b} \cdot \frac{c}{d} = \frac{ac}{bd}$$

EXAMPLE ii. Find the product $({}^-3, 2) \cdot (2, {}^-4)$.
Solution: Using Definition 12.4 we have

$$({}^-3, 2) \cdot (2, {}^-4) = ({}^-3 \cdot 2, 2 \cdot {}^-4)$$
$$= ({}^-6, {}^-8).$$

Using fractions to denote the given rational numbers we have

$$\frac{{}^-3}{2} \cdot \frac{2}{{}^-4} = \frac{{}^-3 \cdot 2}{2 \cdot {}^-4} = \frac{{}^-6}{{}^-8}$$

Exercise 12.1

1. Which of the following pairs of rational numbers are equal?
   a. $(4, ^-5), (^-3, 7)$
   b. $(2, ^-4), (^-6, 12)$
   c. $(^-1, 1), (2, ^-2)$
   d. $(3, 7), (^-21, ^-49)$
   e. $(1, ^-5), (^-1, 5)$
   f. $(^-2, ^-3), (4, 6)$
   g. $(5, ^-2), (125, 50)$
   h. $(5, ^-2), (^-125, 50)$

2. Which of the following pairs of fractions name the same rational number?
   a. $\frac{3}{8}, \frac{48}{128}$
   b. $\frac{5}{7}, \frac{-35}{-49}$
   c. $\frac{27}{36}, \frac{45}{60}$
   d. $\frac{-49}{56}, \frac{84}{-196}$
   e. $\frac{120}{-8}, \frac{15}{1}$
   f. $\frac{42}{-13}, \frac{-294}{91}$

3. Denote the following rational numbers as fractions.
   a. $(4, ^-8)$
   b. $(3, ^-5)$
   c. $(7, 11)$
   d. $(^-16, ^-17)$
   e. $(^-1, 100)$
   f. $(4, ^-7)$
   g. $(^-5, ^-9)$
   h. $(12, 16)$

4. Denote the following as ordered pairs of integers.
   a. $\frac{1}{2}$
   b. $\frac{-3}{4}$
   c. $\frac{-7}{-16}$
   d. $\frac{-3}{8}$
   e. $\frac{-4}{5}$
   f. $\frac{-18}{-3}$
   g. $\frac{1}{-4}$
   h. $\frac{5}{100}$
   i. $\frac{-50}{-25}$

5. Add.
   a. $(3, ^-2) + (^-3, ^-4)$
   b. $(^-1, ^-3) + (3, 4)$
   c. $(^-5, ^-6) + (2, ^-4)$
   d. $(3, ^-8) + (4, ^-1)$
   e. $(7, ^-2) + (3, ^-2)$
   f. $(^-1, 8) + (^-3, 5)$

6. Add.
   a. $\frac{1}{2} + \frac{-3}{4}$
   b. $\frac{2}{3} + \frac{-1}{6}$
   c. $\frac{-3}{5} + \frac{-2}{6}$
   d. $\frac{-1}{8} + \frac{-3}{4}$
   e. $\frac{-5}{7} + \frac{-1}{7}$
   f. $\frac{5}{16} + \frac{-3}{8}$
   g. $\frac{2}{3} + \frac{-1}{-3}$
   h. $\frac{-7}{2} + \frac{-5}{4}$

7. Multiply.
   a. $(3, ^-4) \cdot (2, ^-1)$
   b. $(2, 8) \cdot (^-2, ^-3)$
   c. $(1, 4) \cdot (^-2, ^-2)$
   d. $(7, 5) \cdot (^-3, 4)$
   e. $(^-5, 8) \cdot (^-3, 4)$
   f. $(^-1, ^-2) \cdot (^-4, ^-5)$

8. Multiply.
   a. $\frac{3}{4} \cdot \frac{-3}{2}$
   b. $\frac{-5}{8} \cdot \frac{-2}{-10}$
   c. $\frac{-3}{5} \cdot \frac{-7}{-10}$
   d. $\frac{-5}{8} \cdot \frac{-7}{-12}$
   e. $\frac{-7}{5} \cdot \frac{-1}{2} \cdot \frac{3}{14}$
   f. $\frac{1}{2} \cdot \frac{3}{4} \cdot \frac{-5}{9}$
   g. $\frac{5}{3} \cdot \frac{-9}{16} \cdot \frac{-8}{-9}$
   h. $\frac{-3}{4} \cdot \frac{-2}{3} \cdot \frac{-7}{8}$

9. What is $(^-3, 2) + (2, ^-3)$? What is $(2, ^-3) + (^-3, 2)$?
10. What is $(^-3, 2) \cdot [(2, 4) + (^-3, ^-2)]$? What is $(^-3, 2) \cdot (2, 4) + (^-3, 2) \cdot (^-3, ^-2)$?

12.2. The Properties of the Rational Number System

We now state the properties of the rational number system under the operations of addition and multiplication.

R*-1. Closure Properties. If (a, b) and (c, d) are rational numbers, then their sum $(a, b) + (c, d)$ and their product $(a, b) \cdot (c, d)$ are unique rational numbers.

R*-2. Commutative Properties. If (a, b) and (c, d) are rational numbers, then $(a, b) + (c, d) = (c, d) + (a, b)$ and $(a, b) \cdot (c, d) = (c, d) \cdot (a, b)$.

R*-3. Associative Properties. If (a, b), (c, d), and (e, f) are rational numbers than $[(a, b) + (c, d)] + (e, f) = (a, b) + [(c, d) + (e, f)]$ and $[(a, b) \cdot (c, d)] \cdot (e, f) = (a, b) \cdot [(c, d) \cdot (e, f)]$.

R*-4. Identity Properties. In the set of rational numbers there are unique elements $(0, 1)$ and $(1, 1)$ called the **additive identity** and the **multiplicative identity** respectively such that for every rational number (a, b), $(a, b) + (0, 1) = (0, 1) + (a, b) = (a, b)$ and $(a, b) \cdot (1, 1) = (1, 1) \cdot (a, b) = (a, b)$.

R*-5. Inverse Properties. Every rational number (a, b) has an **additive inverse** $(^-a, b)$ such that $(a, b) + (^-a, b) = (0, 1)$. Every rational number (a, b), $a \neq 0$, has a **multiplicative inverse** (b, a) such that $(a, b) \cdot (b, a) = (1, 1)$.

R*-6. Distributive Property. If (a, b), (c, d), and (e, f) are rational numbers, then $(a, b) \cdot [(c, d) + (e, f)] = (a, b) \cdot (c, d) + (a, b) \cdot (e, f)$.

We can prove that the above properties are true. We shall prove some of these properties. The reader should prove those properties which are not proved here.

Proof of R*-1: By Definition 12.3

$$(a, b) + (c, d) = (ad + bc, bd).$$

Since a, b, c, and d are integers, ad, bc, and bd are integers by the closure property of multiplication of integers. Since b and d are not zero, their product is not zero. Why? Since ad and bc are integers, their sum is an integer by the closure property of addition of integers. Hence, by Definition 12.1, $(ad + bc, bd)$ is a rational number. By Definition 12.4

$$(a, b) \cdot (c, d) = (ac, bd).$$

Since a, b, c, and d are integers, ac and bd are integers by the closure property of multiplication of integers. Since b and d are not zero, their product is not zero. Hence, by Definition 12.1, (ac, bd) is a rational number.

Proof of R*-4:

$(a, b) + (0, 1) = (a \cdot 1 + b \cdot 0, b \cdot 1)$	Definition 12.3
$= (a + b \cdot 0, b)$,	Identity property of multiplication of integers.
$= (a + 0, b)$	$b \cdot 0 = 0$
$= (a, b)$	Identity property of addition of integers

In a similar manner we can prove that

$$(0, 1) + (a, b) = (a, b).$$

We now prove the second part of property R^*-4:

$(a, b) \cdot (1, 1) = (a \cdot 1, b \cdot 1)$	Definition 12.4
$= (a, b)$	Identity property of multiplication of integers

In a similar manner we can prove that

$$(1, 1) \cdot (a, b) = (a, b).$$

Proof of R*-5:

$(a, b) + (^-a, b) = (ab + b \cdot {}^-a, b \cdot b)$	
$= (ab + {}^-ba, b^2)$	$b \cdot {}^-a = {}^-ba;$ $b \cdot b = b^2$
$= (ab + a(^-b), b^2)$	Commutative property of multiplication of integers
$= (ab + {}^-ab, b^2)$	$a \cdot {}^-b = {}^-ab$
$= (0, b^2)$	Additive inverse properties of integers
$(0, b^2) = (0, 1)$	Definition 12.2
$(a, b) \cdot (b, a) = (ab, ba)$	Definition 12.4
$(ab, ba) = (1, 1)$	Definition 12.2

By property R*-5 the additive inverse of (a, b) is $(^-a, b)$. Using fraction symbols we see that $(^-a, b) = \frac{^-a}{b}$. But

$$\frac{^-a}{b} = \frac{a}{^-b} \text{ since } (^-a)(^-b) = a \cdot b.$$

We see that we can write the additive inverse of $\frac{a}{b}$ as $\frac{a}{^-b}$ or as $\frac{^-a}{b}$. Following the notation for integers we can denote the additive inverse of (a, b) as $-(a, b)$. Using fraction symbols $-(a, b)$ is denoted by $-\frac{a}{b}$. We see then that $\frac{^-a}{b}$, $\frac{a}{^-b}$, and $-\frac{a}{b}$ are all names for the additive inverse of $\frac{a}{b}$. That is,

$$\frac{^-a}{b} = \frac{a}{^-b} = -\frac{a}{b}$$

Exercise 12.2

1. Give the additive inverse of each of the following.
   a. $(^-3, 4)$ d. $(^-5, ^-2)$
   b. $(2, ^-2)$ e. $(2, 3)$
   c. $(^-4, ^-3)$ f. $(^-2, 9)$
2. What is the additive inverse of each of the following?
   a. $\frac{2}{3}$ d. $\frac{^-3}{^-4}$
   b. $\frac{^-2}{5}$ e. $\frac{^-5}{6}$
   c. $\frac{5}{^-3}$ f. $\frac{^-5}{^-7}$
3. What is the multiplicative inverse of each of the following?
   a. $(^-3, ^-4)$ d. $(1, 2)$
   b. $(^-3, 2)$ e. $(^-7, ^-12)$
   c. $(5, ^-8)$ f. $(^-4, ^-3)$
4. What is the multiplicative inverse of each of the following?
   a. $\frac{^-2}{3}$ d. $\frac{^-3}{5}$
   b. $\frac{3}{^-4}$ e. $\frac{^-7}{^-16}$
   c. $\frac{7}{^-8}$ f. $\frac{16}{^-5}$
5. Write the additive inverse of $\frac{2}{^-7}$ in three ways.
6. Write the additive inverse of $\frac{^-3}{^-4}$ in three ways.
7. Is $\frac{^-3}{^-4}$ equal to $\frac{3}{4}$?
8. If n is a rational number and $n \cdot n = n$ what are the possible values of n?
9. Prove: If (a, b) is a rational number $(1, 1) \cdot (a \cdot b) = (a, b)$.
10. Prove: If (a, b) is a rational number, $(^-a, b) + (a, b) = (0, 1)$

11. Prove: If (a, b) and (c, d) are rational numbers, then $(a, b) + (c, d) = (c, d) + (a, b)$.

12. Prove: If (a, b) and (c, d) are rational numbers, then $(a, b) \cdot (c, d) = (c, d) \cdot (a, b)$.

13. Prove: If (a, b), (c, d), and (e, f) are rational numbers, then $[(a, b) \cdot (c, d)] \cdot (e, f) = (a, b) \cdot [(c, d) \cdot (e, f)]$.

14. Prove: If (a, b), (c, d), and (e, f) are rational numbers, then $[(a, b) + (c, d)] + (e, f) = (a, b) + [(c, d) + (e, f)]$

15. Prove: If (a, b), (c, d), and (e, f) are rational numbers, then $(a \cdot b) \cdot [(c, d) + (e, f)] = (a, b) \cdot (c, d) + (a, b) \cdot (e, f)$

12.3. Order in the Set Rational Numbers

A rational number (a, b) is defined as **positive** if and only if ab is a positive integer. We see then that $(^-3, ^-4)$ is positive because $(^-3)(^-4) = 12$, a positive integer. A rational number (a, b) is defined to be **negative** if and only if ab is a negative integer. Thus $(^-3, 5)$ is negative because $^-3 \cdot 5 = ^-15$ is a negative integer. A rational number $(0, b)$ is called **zero** and is neither positive nor negative.

DEFINITION 12.5. *A rational number (a, b) is:*

    (a) **positive** *if and only if ab is a positive integer;*
    (b) **negative** *if and only if ab is a negative integer;*
    (c) **zero** *if and only if $a = 0$.*

By Definition 12.5

$$\frac{1}{2}, \frac{^-1}{^-2}, \frac{3}{4}, \frac{^-3}{^-5}, \frac{5}{8}$$

are all positive rational numbers, and

$$\frac{1}{^-2}, \frac{^-1}{3}, \frac{3}{^-2}, \frac{^-5}{7}$$

are all negative rational numbers. The fractions

$$\frac{0}{7}, \frac{0}{^-8}, \frac{0}{2}, \frac{0}{^-9}$$

are all names for zero.

The set of rational numbers may be separated into three disjoint subsets: (1) the set of positive rational numbers; (2) the set of negative rational numbers; and (3) the set containing zero.

DEFINITION 12.6. *A rational number (a, b) is defined to be* **less than** *the rational number (c, d), denoted by $(a, b) < (c, d)$, if and only if there exists a positive rational number (e, f) such that*

$$(a, b) + (e, f) = (c, d).$$

If (a, b) is less than (c, d), then (c, d) is said to be **greater than** *(a, b), denoted by $(c, d) > (a, b)$.*

Using fractions to denote rational numbers we see that

$$\frac{a}{b} < \frac{c}{d}$$

if and only if there exists a positive rational number $\frac{e}{f}$ such that

$$\frac{a}{b} + \frac{e}{f} = \frac{c}{d}.$$

Notice that

$$\frac{3}{5} < \frac{4}{5} \quad \text{because} \quad \frac{3}{5} + \frac{1}{5} = \frac{4}{5},$$

$$\frac{-2}{-5} < \frac{3}{5} \quad \text{because} \quad \frac{-2}{-5} + \frac{1}{5} = \frac{3}{5},$$

$$\frac{2}{3} < \frac{3}{4} \quad \text{because} \quad \frac{2}{3} + \frac{1}{12} = \frac{3}{4},$$

$$\frac{-7}{8} < \frac{0}{8} \quad \text{because} \quad \frac{-7}{8} + \frac{7}{8} = \frac{0}{8},$$

$$\frac{-1}{2} < \frac{-1}{3} \quad \text{because} \quad \frac{-1}{2} + \frac{1}{6} = \frac{-1}{3},$$

Every rational number may be named by a fraction that has a positive denominator. Notice that

$$\frac{2}{-3} = \frac{-2}{3},$$

$$\frac{-2}{-3} = \frac{2}{3},$$

$$\frac{0}{-3} = \frac{0}{3},$$

Since this is true we have another method of ordering rational numbers. If $\frac{a}{b}$ and $\frac{c}{d}$ are rational numbers named by fractions with positive denominators, then

$$\frac{a}{b} > \frac{c}{d} \quad \text{if} \quad ad > bc,$$

$$\frac{a}{b} = \frac{c}{d} \quad \text{if} \quad ad = bc,$$

$$\frac{a}{b} < \frac{c}{d} \quad \text{if} \quad ad < bc.$$

DEFINITION 12.7a. *If $\frac{a}{b}$ and $\frac{c}{d}$ are rational numbers and b and d are positive integers, then*

1) $\frac{a}{b} < \frac{c}{d}$ *if and only if* $ad < bc$
2) $\frac{a}{b} = \frac{c}{d}$ *if and only if* $ad = bc$
3) $\frac{a}{b} > \frac{c}{d}$ *if and only if* $ad > bc$.

For example

$$\frac{2}{3} > \frac{3}{5} \quad \text{because} \quad 2 \cdot 5 > 3 \cdot 3$$

$$\frac{-2}{7} < \frac{9}{3} \quad \text{because} \quad (^-2)(3) < 9 \cdot 7$$

We observe that every positive rational number is greater than zero.

$$\frac{2}{3} > \frac{0}{3} \quad \text{because} \quad 2 \cdot 3 > 3 \cdot 0 = 0$$

$$\frac{9}{7} > \frac{0}{4} \quad \text{because} \quad 9 \cdot 4 > 7 \cdot 0 = 0$$

$$\frac{3}{4} > \frac{0}{5} \quad \text{because} \quad (3)(5) > (4)(0) = 0$$

Every negative rational number is less than zero.

$$\frac{-3}{4} < \frac{0}{4} \quad \text{because} \quad (^-3)(4) < 4 \cdot 0 = 0$$

$$\frac{-6}{2} < \frac{0}{3} \quad \text{because} \quad (3)(^-6) < 2 \cdot 0 = 0$$

12.4.   Subtraction and Division of Rational Numbers

We define subtraction of rational numbers as the inverse of addition of rational numbers. Thus, if (a, b) and (c, d) are rational numbers

$$(a, b) - (c, d) = (e, f) \quad \text{means} \quad (a, b) = (c, d) + (e, f)$$

We use this definition of subtraction to determine the **difference,** $(a, b) - (c, d)$, of two rational numbers. If

$$(a, b) - (c, d) = (e, f)$$

then

$$(1) \hspace{4cm} (a, b) = (c, d) + (e, f)$$

Since every rational number has an additive inverse, (c, d) has an additive inverse, $(^-c, d)$. Let us add $(^-c, d)$ to both members of equation (1):

$$
\begin{aligned}
(^-c, d) + (a, b) &= (^-c, d) + [(c, d) + (e, f)] \\
&= [(^-c, d) + (c, d)] + (e, f) \qquad &\text{associative property} \\
& &\text{of additive} \\
&= (0, 1) + (e, f) \qquad &\text{additive inverse property} \\
&= (e, f) \qquad &\text{additive inverse} \\
& &\text{property}
\end{aligned}
$$

We see that

$$
\begin{aligned}
(e, f) &= (^-c, d) + (a, b) \\
&= (a, b) + (^-c, d).
\end{aligned}
$$

Then

$$(a, b) - (c, d) = (a, b) + (^-c, d).$$

That is, subtracting a rational number gives the same result as adding its additive inverse.

DEFINITION 12.7.   *If (a, b) and (c, d) are rational numbers, then their* **difference,** $(a, b) - (c, d)$, *is* $(a, b) + (^-c, d)$.

Since every rational number has an additive inverse, we see that *the set of rational numbers is closed under subtraction.*

Division of rational numbers is defined as the inverse of multiplication of rational numbers. Thus

$$(a, b) \div (c, d) = (e, f) \qquad (c, d) \neq (0, 1)$$

means

$$(a, b) = (c, d) \cdot (e, f).$$

We use the definition of division to find the quotient of two rational numbers. Since $(c, d) \neq (0, 1)$, it has a multiplicative inverse, (d, c). Multiplying both members of

$$(a, b) = (c, d) \cdot (e, f)$$

by (d, c) we obtain

$$
\begin{aligned}
(d, c) \cdot (a, b) &= (d, c) \cdot [(c, d) \cdot (e, f)] \\
&= [(d, c) \cdot (c, d)] \cdot (e, f) \qquad \text{associative property of} \\
&\qquad\qquad\qquad\qquad\qquad\qquad \text{multiplication} \\
&= (1, 1) \cdot (e, f) \qquad\qquad\quad \text{multiplicative inverse property} \\
&= (e, f) \qquad\qquad\qquad\qquad\quad \text{multiplicative identity property}
\end{aligned}
$$

We see then that

$$
\begin{aligned}
(a, b) \div (c, d) &= (e, f) \qquad\qquad (c, d) \neq (0, 1) \\
&= (d, c) \cdot (a, b) \\
&= (a, b) \cdot (d, c)
\end{aligned}
$$

DEFINITION 12.8. *If (a, b) and (c, d) are rational numbers, $(c, d) \neq (0, 1)$, their **quotient**, $(a, b) \div (c, d)$, is $(a, b) \cdot (d, c)$.*

We see that dividing by a rational number gives the same result as multiplying by its multiplicative inverse. Since every rational number except $(0, 1)$ has a multiplicative inverse, *the set of rational numbers is closed under division except by $(0, 1)$ the additive identity.*

12.5. The Rational Number Line

In Chapter 10 we demonstrated that the nonnegative rational numbers can be graphed on a number line. Extending this idea to the left of the point labeled 0 we can graph the rational numbers on a number line. Figure 12.1 shows the rational number line on which some of the rational numbers have been graphed.

It is apparent from Figure 12.1 that every rational number has an opposite or an additive inverse. The additive inverse of $\frac{1}{2}$ is $-\frac{1}{2}$ since the graph of $\frac{1}{2}$ and $-\frac{1}{2}$ are the same distance from the graph of 0 but in the opposite directions from the graph of 0. Of course, zero is its own opposite.

It will be recalled that the set of nonnegative rational numbers was **dense.** That is, between any two nonnegative rational numbers there are

Figure 12.1

infinitely many other nonnegative rational numbers. *The set of rationl numbers is also dense.* Between any two rational numbers there are infinitely many other rational numbers.

EXAMPLE i.   Find a rational number midway between $-\frac{1}{2}$ and $\frac{3}{4}$.

Solution: We recall that the average of two rational numbers is midway between the two given numbers. Finding the average of $-\frac{1}{2}$ and $\frac{3}{4}$ we have

$$\frac{-\frac{1}{2} + \frac{3}{4}}{2} = \frac{\frac{1}{4}}{2}$$

$$= \frac{1}{4} \div 2$$

$$= \frac{1}{4} \times \frac{1}{2}$$

$$= \frac{1}{8}$$

EXAMPLE ii.   Find a rational number midway between $-\frac{3}{5}$ and $-\frac{1}{3}$.

Solution:
$$\frac{-\frac{3}{5} + \left(-\frac{1}{3}\right)}{2} = \frac{-\frac{14}{15}}{2}$$

$$= \frac{^-14}{15} \div 2$$

$$= -\frac{14}{15} \times \frac{1}{2}$$

$$= -\frac{7}{15}$$

Exercise 12.3

1.   Replace * by $<$, $=$ or $>$ to make true statements.

  a. $\frac{-1}{2}$ * $\frac{3}{4}$ e. $\frac{-2}{3}$ * $\frac{-5}{6}$

  b. $\frac{-5}{6}$ * $\frac{-7}{8}$ f. $\frac{7}{8}$ * $\frac{3}{4}$

---

c. $\frac{-4}{3} * \frac{-12}{9}$ g. $\frac{5}{16} * \frac{-7}{8}$

d. $\frac{5}{16} * \frac{-8}{9}$ h. $\frac{3}{5} * \frac{-5}{3}$

2. Compute.

  a. $\frac{-3}{4} - \frac{5}{8}$ e. $\frac{3}{5} - \frac{-7}{8}$

  b. $\frac{-2}{3} - \frac{-3}{4}$ f. $\frac{1}{4} - \frac{3}{5}$

  c. $\frac{-1}{2} - \frac{-1}{2}$ g. $\frac{2}{3} - \frac{5}{6}$

  d. $\frac{-5}{6} - \frac{-2}{3}$ h. $\frac{-1}{6} - \frac{-2}{3}$

3. Find the quotients.

  a. $\frac{-3}{7} \div 3$ d. $\frac{-2}{3} \div \frac{-3}{5}$

  b. $\frac{-1}{4} \div \frac{-2}{3}$ e. $\frac{-3}{5} \div \frac{21}{4}$

  c. $\frac{5}{2} \div \frac{10}{3}$ f. $\frac{-2}{5} \div \frac{8}{15}$

4. Find the solution sets. The domain is the set of rational numbers.

  a. $n + \frac{1}{2} = \frac{1}{3}$ e. $n - \frac{2}{5} = \frac{3}{4}$

  b. $\frac{2}{3} - n = \frac{5}{8}$ f. $\frac{-2}{3}n = \frac{7}{8}$

  c. $\frac{-7}{8}n = \frac{2}{3}$ g. $\frac{3}{4}n = \frac{-3}{10}$

  d. $n + \frac{5}{6} = \frac{3}{8}$ h. $\frac{-5}{6}n = \frac{-2}{3}$

5. Find a rational number midway between $\frac{-1}{3}$ and $\frac{3}{5}$.

6. Find a rational number midway between $\frac{-5}{6}$ and $\frac{-2}{3}$.

7. If a and b, are rational numbers and $a > 0$ and $b > 0$ is $ab > 0$ or is $ab < 0$?

8. If a, b, and c are rational numbers and $a > b$ and $c > 0$ is $ac > bc$ or is $ac < bc$?

9. If a, b, and c are rational numbers and $a > b$ and $c < 0$ is $ac > bc$ or is $ac < bc$?

10. If r is a rational number and $r < 0$ is $r^2 > 0$ or is $r^2 < 0$?

12.6. Subsystems of the Rational Number System

All the numbers of arithmetic may be studied from two different aspects. The first approaches numbers as mathematical ideas applicable to certain counting situations. We saw an illustration of this in our study of the whole numbers as viewed as the cardinal numbers associated with the elements in a set. The second examines numbers as mathematical objects whose properties are studied independently of their origin and their use in counting. This is a more abstract approach and harder to understand.

    When we view numbers as mathematical objects their properties depend upon certain relations and operations that can be performed on them. The relations which we have studied are the equality relation and the order relation. The operations which we have discussed are addition, multiplication, subtraction, and division. We viewed rational numbers in this manner.

When numbers are observed as mathematical objects, they form a **number system.** So far in our studies, we have considered three number systems: the whole number system, the system of integers, and the rational number system.

There are certain differences among these number systems, of which the most important concern the inverse operations of subtraction and division. The set of whole numbers is closed under the operations of addition and multiplication, but not under the operations of subtraction and division. The set of integers is closed under the operations of addition, multiplication, and subtraction, but not under division. Finally, the set of rational numbers is closed under addition, multiplication, subtraction, and, except for zero, division.

Both the system of rational numbers and the system of integers are extensions of the system of whole numbers. Hence they may be thought of as including the whole number system. In the case of integers, the whole numbers may be identified with the nonnegative integers. In the case of rational numbers, the whole numbers may be identified with the nonnegative rational numbers named by fractions with denominators of one. The relationship of these three number systems is shown in Figure 12.2.

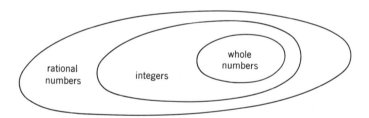

Figure 12.2

12.7. Number Fields

The set, R^*, of rational numbers

$$R^* = \{(a, b) \mid a \text{ and } b \text{ are integers and } b \neq 0\}$$

together with the operations of addition and multiplication and the

---

properties listed in Section 12.2 is called by mathematicians a **number field** or, simply, **a field.**

Any set of elements on which an addition and a multiplication is defined (they may be different from ordinary addition and multiplication as we shall see in Chapter 14) and having all the properties listed in Section 12.2 is a field. In Chapter 14 we shall study other examples of fields. Of the many fields used in mathematics, the field of rational numbers, called the **rational number field,** is the most important in elementary school mathematics.

EXAMPLE. Is the set of whole numbers with the operations of addition and multiplication a field? Why or why not?

Solution: The set of whole numbers with the operations of addition and multiplication is not a field because R*-5, the inverse property, is not satisfied.

Exercise 12.4

1. Consider the set of natural numbers

$$N = \{1, 2, 3, 4, \ldots\}$$

   a. Check the properties of a field and state whether or not they also hold for the natural numbers. State the properties, if there are any, which do not hold for N.
   b. Is N a field?
2. Consider the set of integers

$$I = \{\ldots, {}^-2, {}^-1, 0, 1, 2, \ldots\}$$

   a. Check the properties of a field and state whether or not they also hold for the integers. State the properties, if there are any, which do not hold for I.
   b. Is I a field?
3. Consider the set of nonnegative rational numbers.

$$R = \left\{ \frac{a}{b} \,\middle|\, a \text{ and } b \text{ are whole numbers and } b \neq 0 \right\}$$

   a. Check the properties of a field and state whether or not they also hold for the nonnegative rational numbers. State the properties, if there are any which do not hold for R.

b. Is R a field?

4. Consider the set of even integers.

$$E = \{\ldots, {}^{-}4, {}^{-}2, 0, 2, 4, \ldots\}$$

a. Check the properties of a field and state whether or not they also hold for the even integers. State the properties, if there are any which do not hold for E.

b. Is E a field?

5. Consider the set of odd integers.

$$O = \{\ldots, {}^{-}3, {}^{-}1, 1, 3, \ldots\}$$

a. Check the properties of a field and state whether or not they also hold for the odd integers. State the properties, if there are any, which do not hold for O.

b. Is O a field?

6. Consider the finite number system which contains only two elements 0 and 1. The addition and multiplication tables for this system are given below.

+	0	1
0	0	1
1	1	0

×	0	1
0	0	0
1	0	1

a. Check the properties of a field and state whether or not they also hold for this finite number system. State the properties, if there are any, which do not hold for this system.

b. Is this finite system a field?

13.1. An Extension of the Rational Number System

Although the rational number system is sufficient for most purposes of elementary arithmetic, in geometry, algebra, and other branches of mathematics, there is a need for another kind of number. We meet this need by an extension of the rational number system to one called the **real number system.**

We need numbers that are not rational numbers to find square roots of positive rational numbers. The **square root** of a positive rational number a, denoted by $\sqrt{a}$ is that positive number c such that $c^2 = a$. Thus

$$\sqrt{4} = 2 \quad \text{since} \quad 2^2 = 4$$
$$\sqrt{25} = 5 \quad \text{since} \quad 5^2 = 25$$
$$\sqrt{\frac{4}{9}} = \frac{2}{3} \quad \text{since} \quad \left(\frac{2}{3}\right)^2 = \frac{4}{9}$$

Although the square root of each positive rational number above was a positive rational number, there are some positive rational numbers, for example 2, whose square roots are not positive rational numbers.

Real Numbers

XIII

Other examples are 3, 5, and 10; $\sqrt{3}$, $\sqrt{5}$, and $\sqrt{10}$ are not positive rational numbers.

We can show that $\sqrt{2}$ is not a rational number by proving the following theorem.

THEOREM 13.1.   There exists no rational number whose square is 2.
Proof: We are trying to prove that $\sqrt{2}$ is not a rational number. Let $\frac{a}{b}$ represent a rational number in simplest form; that is, $(a, b) = 1$. Let us assume that

$$2 = \frac{a^2}{b^2}$$

Then

$$2b^2 = a^2$$

by the multiplication property of equality. Since $2b^2 = a^2$, a^2 must be an even number, and therefore a must be an even number. For, if a were odd, a^2 would also be odd because the product of two odd numbers is an odd number. Since all even numbers may be written in the form $2n$, let $a = 2n$. Then

$$a^2 = 2n \cdot 2n = 2(2n^2)$$

Now

$$2b^2 = a^2 = 2(2n^2)$$

and

$$b^2 = 2n^2$$

an even number. Since b^2 is even, b is also even. Why? We conclude from the above that both a and b are even numbers and hence have a factor 2. This contradicts our assumption that a and b were relatively prime, and hence our assumption that $\sqrt{2}$ is a rational number.

13.2.   The Real Number Line

We have showed that the nonnegative rational numbers are dense; that is, between any two of them there are infinitely many others. This is also true for all the rational numbers.

From this it follows that there are infinitely many rational numbers, and corresponding to them on the number line, infinitely many rational points. These points are spread throughout the number line. Any segment of the number line, no matter how small, contains infinitely many rational points. This might lead us to the conclusion that every point on the number line is a rational point. This is not so, for there are many points on the line that are not rational points. These points are called **irrational points** and correspond to numbers called **irrational numbers.**

We can show that there is at least one point on the number line that is not a rational point.

Figure 13.1 shows a square, $ABCD$, whose side is two units long. The area of this square region is four square units. The area of the shaded square region is one-half the area of the square region bounded by $ABCD$. Hence the area of this region is two square units. Each of its sides is therefore $\sqrt{2}$ units in length.

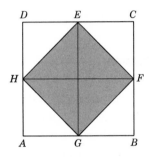

Figure 13.1

With the point labeled 0 as one endpoint, we lay off on the number line a line segment congruent to $\overline{EF}$, the side of the shaded square. The other endpoint of this line segment is the point on the number line labeled P (Figure 13.2). The point P is the graph of the irrational number $\sqrt{2}$. Since $\sqrt{2}$ is not a rational number, we see that there is *at least* one point on the number that is not a graph of a rational number. This is, there is at least one irrational point on the number line.

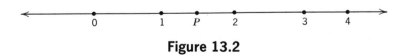

Figure 13.2

13.3.   Irrational Numbers

The explanation of these so-called "holes" in the number line, that is, points that are not rational points, is a very sophisticated mathematical idea and is not easy to understand. We have shown in Section 13.2 that at least one of the points on the number line does not correspond to a rational number and is the graph of an irrational number, $\sqrt{2}$. We shall not attempt to show that there are infinitely many points on the number line that are graphs of other irrational numbers. We shall accept without proof that there are infinitely many points on the number line that are irrational points, that is are graphs of **irrational numbers.**

The irrational numbers have a misleading name; "irrational" in this case does not mean unreasonable, but merely "not rational." Some irrational numbers are $\sqrt{2}$, $\sqrt{3}$, $5 + \sqrt{7}$, and π.

All the points on the number line fall into two non-empty sets: the set of points that are rational points and the set of points that are irrational points.

13.4.   The Real Numbers

The union of the set of rational numbers and the set of irrational numbers is called the set of **real numbers.** Every point on the number line is associated with either a rational number or an irrational number. Hence every point on the number line is associated with a real number and is called the **real number line.**

Having now filled all the gaps in the line, we say that the real number line is not only dense, but continuous. A geometric line may now be interpreted in terms of the real numbers. There is a one-to-one correspondence between the points on the line and the real numbers.

The real number system is an extension of the rational number system. It is used extensively in algebra, trigonometry, and the calculus and its applications.

Exercise 13.1

1.   Which of the following are rational numbers?
   a. $\sqrt{25}$ e. $\sqrt{\frac{9}{25}}$

---

b. $\sqrt{\frac{1}{16}}$ f. $\sqrt{169}$

c. $\sqrt{3}$ g. $\sqrt{7}$

d. $\sqrt{\frac{1}{5}}$ h. $\sqrt{\frac{5}{4}}$

2. Which of the following are irrational numbers?

   a. $\sqrt{2}$ e. $5 + \sqrt{2}$

   b. $\sqrt{27}$ f. $\sqrt{\frac{2}{3}}$

   c. π g. $3 + \sqrt{16}$

   d. $\sqrt{\frac{1}{9}}$ h. $\sqrt{25} + \sqrt{9}$

3. If R is the set of real numbers, Q the set of rational numbers, and S the set of irrational numbers, what is

   a. $Q \cup S$?   e. $R \cap Q$?

   b. $Q \cap S$?   f. $R \cap S$?

   c. $R \cup S$?   g. S'?

   d. $R \cup Q$?   h. Q'?

4. If n is a rational number is n^2 always a rational number? Why?

5. If n is an irrational number is n^2 always an irrational number? Give a numerical example to substantiate your answer.

6. Prove that there is no rational number whose square is 3. That is, show that $\sqrt{3}$ is irrational.

7. Which of the following are true statements?

   a. The set of rational numbers and the set of irrational numbers are disjoint sets.

   b. Every real number is an irrational number.

   c. Every rational number is a real number.

   d. If n is a whole number then $\sqrt{n}$ is always an irrational number.

   e. The set of irrational numbers is a subset of the set of real numbers.

8. If r is a rational number, then it is a real number.

   a. Is the given statement true or false?

   b. What is the converse of the given statement?

   c. Is the converse true or false?

9. If r is an irrational number, then it is a real number.

   a. Is the given statement true or false?

   b. What is the converse of the given statement?

   c. Is the converse true or false?

10. If a is a rational number, then a^2 is a rational number.

   a. Is the given statement true or false?

   b. What is the converse of the given statement?

   c. Is the converse true or false?

14.1. Finite Systems

A **mathematical system** consists of:

(1) A non-empty set of elements.
(2) One or more operations which associate with each ordered pair, (a, b), of elements of the system an element c of the system.
(3) A set of properties, called **postulates** or **axioms,** which the elements and the operations satisfy.

We have studied several mathematical systems: (1) the whole number system; (2) the system of integers; and (3) the rational number system. All of these systems were infinite systems. We shall now study some mathematical systems that are **finite systems,** that is, the number of elements in the set of the system is finite.

Suppose we have a system that consists of only five elements, the numbers 0, 1, 2, 3, and 4. We call this system the **modulo-five system** because it contains five elements. We shall define two operations of this system called **addition modulo 5,** denoted by $+$, and **multiplication modulo 5,** denoted by $\times$. We define the operation of addition

Mathematical Systems

XIV

modulo 5 by using the five-hour clock shown in Figure 14.1.

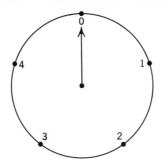

Figure 14.1

The addition $2 + 1$ is defined thus: the hand of the clock starts at 0; it travels two spaces in a clockwise direction and then it travels one more space in a clockwise direction. Thus $2 + 1 = 3$. Using this definition of addition modulo 5 we see that

$$2 + 2 = 4$$
$$2 + 3 = 0$$
$$4 + 3 = 2$$
$$3 + 3 = 1$$

An addition table for this system is shown in Table 14.1.

Table 14.1

+	0	1	2	3	4
0	0	1	2	3	4
1	1	2	3	4	0
2	2	3	4	0	1
3	3	4	0	1	2
4	4	0	1	2	3

We call Table 14.1 the **addition table modulo 5** (abbreviated **mod 5**). What are some of the properties of addition mod 5?

Only the numbers 0, 1, 2, 3, and 4 appear in the table. In other words, the sum of any two elements of the modulo-five system is an element of the system. We see that the system is **closed** under the operation of addition mod 5. We call this the **closure property of addition.**

In examining the addition table mod 5, we see that

$$3 + 4 = 4 + 3$$
$$2 + 1 = 1 + 2$$
$$3 + 2 = 2 + 3$$

If we examine all cases we find that for any two numbers a and b of the modulo-five system $a + b = b + a$. This is the **commutative property of addition.**

Again examining the table we find

$$0 + 0 = 0$$
$$1 + 0 = 1$$
$$2 + 0 = 2$$
$$3 + 0 = 3$$
$$4 + 0 = 4$$

If 0 is added to any element a in the modulo-five system, the sum is a. That is, for all a, $a + 0 = 0 + a = a$. We call zero the **identity element for addition mod 5** or the **additive identity.**

Notice

$$0 + 0 = 0$$
$$1 + 4 = 0$$
$$2 + 3 = 0$$
$$3 + 2 = 0$$
$$4 + 1 = 0$$

In each sentence above the sum of two addends is the additive identity. We see then that each element in the modulo-five system has an **additive inverse.** The additive inverse of 0 is 0; the additive inverse of 1 is 4; the additive inverse of 2 is 3; the additive inverse of 3 is 2; and the additive inverse of 4 is 1.

It is not obvious from the addition table, but it can be proved that addition mod 5 is **associative.** Notice that

$$\left.\begin{array}{l}(2 + 3) + 4 = 0 + 4 = 4 \\ 2 + (3 + 4) = 2 + 2 = 4\end{array}\right\} \text{ hence } (2 + 3) + 4 = 2 + (3 + 4)$$

$$\left.\begin{array}{l}(3 + 4) + 4 = 2 + 4 = 1 \\ 3 + (4 + 4) = 3 + 3 = 1\end{array}\right\} \text{ hence } (3 + 4) + 4 = 3 + (4 + 4)$$

In general, if a, b, and c are elements of the modulo-five system, then $(a + b) + c = a + (b + c)$. This is called the **associative property of addition.**

We now define multiplication modulo 5 (abbreviated mod 5). We define 2×3 thus: the hand of the clock starts at 0 and moves three spaces in a clockwise direction, then it moves three more spaces in the same direction. Thus $2 \times 3 = 1$. We see that 2×3 means two clockwise turns of three spaces each. Similarly

$2 \times 4 = 3$ (two clockwise turns of four spaces each)
$3 \times 3 = 4$ (three clockwise turns of three spaces each)
$4 \times 4 = 1$ (four clockwise turns of four spaces each).

The multiplication table mod 5 is shown in Table 14.2.

Table 14.2

$\times$	0	1	2	3	4
0	0	0	0	0	0
1	0	1	2	3	4
2	0	2	4	1	3
3	0	3	1	4	2
4	0	4	3	2	1

We now ask: What are the properties of multiplication mod 5? We see from the table that the product of any two numbers in the modulo-five system is a number in the system. The modulo-five system is **closed** with respect to multiplication mod 5. This is the **closure property of multiplication.**

We also see that multiplication mod 5 is commutative since for every a and b in the system $a \times b = b \times a$. This is the **commutative property of multiplication.**

Observe that

$$1 \times 0 = 0$$
$$1 \times 1 = 1$$
$$1 \times 2 = 2$$
$$1 \times 3 = 3$$
$$1 \times 4 = 4$$

If *a* is any element in the modulo-five system then $a \times 1 = 1 \times a = a$. The number 1 is called the **identity element** for multiplication mod 5 or the **multiplicative identity.**

From the sentences

$$1 \times 1 = 1$$
$$2 \times 3 = 1$$
$$3 \times 2 = 1$$
$$4 \times 4 = 1$$

we see that every element, except 0, has a **multiplicative inverse.** The multiplicative inverse of 1 is 1; the multiplicative inverse of 2 is 3; the multiplicative inverse of 3 is 2; and the multiplicative inverse of 4 is 4.

Again it is not obvious, but it can be proved, that multiplication mod 5 is **associative.** If *a*, *b*, and *c* are elements of the modulo-five system, then $(ab)c = a(bc)$. This is the **associative property of multiplication.**

That multiplication mod 5 distributes over addition mod 5 is demonstrated by the following examples:

$$2 \times (3 + 4) = (2 \times 3) + (2 \times 4)$$
$$2 \times 2 = 1 + 3$$
$$4 = 4$$
$$3 \times (4 + 1) = (3 \times 4) + (3 \times 1)$$
$$3 \times 0 = 2 + 3$$
$$0 = 0$$

If *a*, *b*, and *c* are elements of the modulo-five system then $a(b + c) = ab + ac$. This is the **distributive property of multiplication over addition.**

The system described above is an example of a **modular system.** Since there are five elements in the system it is called the modulo-five system, and five is called the **modulus** of the system.

We now list the properties of the modulo-five system.

M-1. Closure Property of Addition. If *a* and *b* are elements of the system then their sum $a + b$ is a unique element of the system.

M-2. Commutative Property of Addition. If *a* and *b* are elements of the system then $a + b = b + a$.

M-3. Associative Property of Addition. If *a*, *b*, and *c*, are elements of the system, then $(a + b) + c = a + (b + c)$.

M-4. Additive Identity Property. There is a unique element, 0, in the system, called the **additive identity,** such that for all elements *a*, $a + 0 = 0 + a = a$.

M-5. Additive Inverse Property. For every element a there is a unique element in the system, denoted by ^-a, called the **additive inverse** of a, such that $a + {}^-a = {}^-a + a = 0$.

M-6. Closure Property of Multiplication. If a and b are elements of the system, then their product ab is a unique element of the system.

M-7. Commutative Property of Multiplication. If a and b are elements of the system, then $ab = ba$.

M-8. Associative Property of Multiplication. If a, b, and c are elements of the system, then $(ab)c = a(bc)$.

M-9. Multiplicative Identity Property. There is a unique element, 1, in the system called the **multiplicative identity,** such that for all a, $a \times 1 = 1 \times a = a$.

M-10. Multiplicative Inverse Property. For every element $a \neq 0$ there is a unique element in the system, denoted by a^{-1}, called the **multiplicative inverse** of a, such that $a \times a^{-1} = a^{-1} \times a = 1$.

M-11. Distributive Property. If a, b, and c are elements of the system then $a(b + c) = ab + ac$.

Notice that the modulo-five system is a field according to the definition given in Chapter 12, Section 12.7.

14.2. Subtraction and Division Mod 5

We define subtraction mod 5 as the inverse of addition mod 5. Thus $a - b = c$ means $b + c = a$. Notice

$$
\begin{array}{lll}
4 - 3 = 1 & \text{because} & 3 + 1 = 4 \\
2 - 4 = 3 & \text{because} & 4 + 3 = 2 \\
1 - 3 = 3 & \text{because} & 3 + 3 = 1
\end{array}
$$

By the definition of subtraction mod 5, $4 - 3 = 1$. The additive inverse of 3 is 2. Observe that $4 + 2 = 1$. Similarly, $2 - 4 = 3$. The additive inverse of 4 is 1 and $2 + 1 = 3$. In general, subtracting a modulo-five number is the same as adding its additive inverse, that is $a - b = a + ({}^-b)$.

We define division mod 5 as the inverse of multiplication. Thus $a \div b = c$, $b \neq 0$, means $bc = a$. Notice that as in the other number systems we studied we exclude division by zero. Then

$$
\begin{array}{lll}
3 \div 2 = 4 & \text{because} & 2 \times 4 = 3 \\
4 \div 3 = 3 & \text{because} & 3 \times 3 = 4
\end{array}
$$

The multiplicative inverse of 2 is 3. Notice that $3 \div 2 = 4$ and $3 \times 3 = 4$. Similarly, the multiplicative inverse of 3 is 2 and $4 \div 3 = 3$ and $4 \times 2 = 3$. This is true in general; $a \div b = a \times b^{-1}$, $b \neq 0$.

Since every element in the modulo-five system has an additive inverse, we see that *the modulo-five system is closed under the operation of subtraction.* Since every element except zero has a multiplicative inverse, we see that *the modulo-five system is closed under division except by zero.*

EXAMPLE i. Find the solution set of $3n = 4$ in the modulo-five system. That is find n so that

$$3n = 4 \quad (\mathrm{mod}\ 5).$$

Solution: Using the multiplication table mod 5 we see that $3 \times 3 = 4$, hence the solution set is $\{3\}$.

EXAMPLE ii. Find the solution set of

$$n - 3 = 4 \quad (\mathrm{mod}\ 5).$$

Solution: $n - 3 = 4$ means $n = 3 + 4 = 2$, hence the solution set is $\{2\}$.

Exercise 14.1

1. Perform the indicated operations mod 5.
   a. $(3 + 4) + 3$ d. $(3 + 3) \times (2 + 4)$
   b. $4 \times 4 \times 2$ e. $(4 + 2) \times (4 + 3)$
   c. $(4 \times 4) + (3 \times 2)$ f. $(3 \times 4) + (4 + 4)$
2. Find the solution sets mod 5.
   a. $n + 4 = 3$ e. $0 + n = 3$
   b. $3 - n = 4$ f. $4n + 2 = 3$
   c. $2n = 4$ g. $2n - 3 = 4$
   d. $4 + n = 2$ h. $2n + 3 = 2$
3. Using a seven-hour clock construct an addition table for the modulo-seven system.
4. Using a seven-hour clock construct a multiplication table for the modulo-seven system.
5. Check the properties of the modulo-five system and state whether or not they hold for the modulo-seven system. State the properties, if there are any, which do not hold for the modulo-seven system.

6. Find the indicated sums in addition mod 7.
   a. 3 + 4
   b. 5 + 6
   c. 3 + 4 + 2
   d. 3 + 5 + 4
   e. 5 + 5 + 1
   f. 3 + 6 + 4 + 5
7. Find the solution sets mod 7.
   a. $4 + n = 2$
   b. $3 + n = 1$
   c. $2 - n = 5$
   d. $3n = 5$
   e. $2n + 1 = 4$
   f. $5n = 6$
   g. $4n = 1$
   h. $3n - 4 = 5$
8. Find the sums in the indicated modular systems.
   a. 3 + 4 + 7   (mod 9)
   b. 6 + 7 + 3   (mod 12)
   c. 1 + 1 + 1   (mod 3)
   d. 3 + 3 + 2 + 1   (mod 4)
   e. 1 + 1 + 1 + 1   (mod 2)
   f. 4 + 6 + 3   (mod 8)
   g. 2 + 2 + 2 + 2   (mod 4)
   h. 6 + 9 + 10 + 8   (mod 11)
9. Find the products in the indicated modular systems.
   a. 3 × 4 × 4   (mod 7)
   b. 8 × 9 × 3   (mod 11)
   c. 2 × 2 × 2   (mod 3)
   d. 9 × 9 × 8   (mod 10)
   e. 2 × 2 × 3   (mod 4)
   f. 8 × 10 × 11   (mod 12)
   g. 6 × 8 × 3   (mod 9)
   h. 11 × 11 × 11   (mod 12)
10. Using a four-hour clock construct addition and multiplication tables mod 4.
11. Check the properties of the modulo-five system and state whether or not they hold for the modulo-four system. State the properties, if there are any, which do not hold for the modulo-four system.
12. With what phase of everyday life would the modulo-seven system be associated? Describe some other modular systems in everyday use.

14.3. Prime and Non-Prime Modular Systems

We found in Section 14.1 that the modulo-five system is a field. In Exercise 14.1, Problem 5, we discovered that the modulo-seven system is also a field. The modulo-five and the modulo-seven systems are called **prime modular systems** because their moduli (plural of modulus) are prime numbers. *Every prime modular system is a field.*

Let us consider the modulo-six system and inspect the addition mod 6 and the multiplication mod 6 tables (Table 14.3). The modulo-six system is not a prime modular system because its modulus, 6, is not a prime number.

Table 14.3

+	0	1	2	3	4	5
0	0	1	2	3	4	5
1	1	2	3	4	5	0
2	2	3	4	5	0	1
3	3	4	5	0	1	2
4	4	5	0	1	2	3
5	5	0	1	2	3	4

×	0	1	2	3	4	5
0	0	0	0	0	0	0
1	0	1	2	3	4	5
2	0	2	4	0	2	4
3	0	3	0	3	0	3
4	0	4	2	0	4	2
5	0	5	4	3	2	1

Now let us see whether or not this modular system is a field. We must check the following field properties.

F-1. Closure Properties. If a and b are elements of a field, F, then their sum $a + b$ and their product ab are unique elements of the field.

F-2. Commutative Properties. If a and b are elements of a field, F, then $a + b = b + a$ and $ab = ba$.

F-3. Associative Properties. If a, b, and c are elements of a field, F, then $(a + b) + c = a + (b + c)$ and $(ab)c = a(bc)$.

F-4. Identity Properties. There exist elements 0 and 1 in the field F called respectively the **additive identity** and the **multiplicative identity** such that for all a in F, $a + 0 = 0 + a = a$ and $a \times 1 = 1 \times a = a$.

F-5. Inverse Properties. For every a in a field F there exists in F a unique element ^-a, called the **additive inverse** of a, such that $a + {}^-a = {}^-a + a = 0$. For every $a \neq 0$ in F there exists a unique element a^{-1} of F called the **multiplicative inverse** of a such that $a \times a^{-1} = a^{-1} \times a = 1$.

F-6. Distributive Property. If a, b, and c are elements of a field F, then $a(b + c) = ab + ac$.

Since both the tables for addition mod 6 and multiplication mod 6 contain only elements of the modulo-six system, certainly F-1 is satisfied, that is, the modulo-six system is closed under both addition mod 6 and multiplication mod 6.

Notice that both tables are symmetric with respect to the diagonal from the upper left hand corner to the lower right hand corner, that is, the entry in the ith row and the jth column is the same as that in the jth row and the ith column. This means, for example, in the addition mod 6 table, that $i + j = j + i$, for all numbers i and j in the table. Similarly, in the multiplication mod 6 table, this means that $i \times j = j \times i$ for all numbers i and j in the table. We conclude that addition mod 6 and multiplication mod 6 are commutative operations and F-2 is satisfied.

To check F-3, the associative properties, we must replace each of a, b, and c in turn by 0, 1, 2, 3, 4, and 5 in

$$(a + b) + c = a + (b + c) \quad \text{and} \quad (ab)c = a(bc),$$

and see whether or not in each case we obtain a true statement.

We see that in the modulo-six system there are $6 \times 6 \times 6 = 6^3 = 216$ cases to test. We can replace a with any one of the six elements; we can replace b with any one of the six elements; and we can replace c with any one of the six elements. Since we have six choices for a and six choices for b, we have $6 \times 6 = 36$ choices for a and b. We now have six choices for c; so we have $6 \times 6 \times 6 = 6^3 = 216$ choices in all to test.

If we test each of these 6^3 cases we find that $(a + b) + c = a + (b + c)$ and $(ab)c = a(bc)$ are true statements. Hence addition mod 6 and multiplication mod 6 are associative.

The additive identity is 0, since, for all numbers a in the modulo-six system, $a + 0 = 0 + a = a$. The multiplicative identity is 1 since, for all numbers a in the modulo-six system, $a \times 1 = 1 \times a = a$. We see then that F-4 is satisfied.

Again, replacing a, b, and c in turn by 0, 1, 2, 3, 4, and 5 in

$$a(b + c) = ab + ac$$

we always obtain a true statement, and hence F-6 is satisfied.

Now let us see whether or not F-5 is satisfied. If every element in the system has an additive inverse, for every element a there is a unique element ^-a such that $a + {}^-a = {}^-a + a = 0$. We see that

$$0 + 0 = 0$$
$$1 + 5 = 0$$
$$2 + 4 = 0$$
$$3 + 3 = 0$$
$$4 + 2 = 0$$
$$5 + 1 = 0$$

Hence every element has an additive inverse and the first part of F-5 is satisfied.

Now we must check to see whether or not every element, except 0, has a multiplicative inverse. If every element, except 0, has a unique multiplicative inverse, for all $a \neq 0$, the equation

$$an = 1$$

has one and only one solution. We see that

$$1 \times 1 = 1 \quad \text{and} \quad 5 \times 5 = 1$$

hence 1 and 5 are their own multiplicative inverses. Now let us see whether or not

$$2n = 1$$

has a solution. Looking at the multiplication mod 6 table we see

$$2 \times 0 = 0$$
$$2 \times 1 = 2$$
$$2 \times 2 = 4$$
$$2 \times 3 = 0$$
$$2 \times 4 = 2$$
$$2 \times 5 = 4$$

Since no replacement for n makes $2n = 1$ a true statement, 2 has no multiplicative inverse in the modulo-six system. The modulo-six system is not a field since F-5 is not satisfied.

Checking we see that $3n = 1$ and $4n = 1$ have no solutions and therefore 3 and 4 do not have multiplicative inverses in the modulo-six system.

In general, if m is a prime number the modulo-m system is a field; if m is a composite number, then the modulo-m system is not a field.

EXAMPLE i. Find the solution set of $2n = 3$ (mod 6)
Solution: Inspecting the multiplication mod 6 table (Table 14.3) we see that there is no number n in the modulo-six system such that $2n = 3$. Hence the solution set of the given equation is the empty set, ϕ.

EXAMPLE ii. Find the solution set of $2n = 4$ (mod 6)
Solution: Inspecting the multiplication mod 6 table we see that

$$2 \times 2 = 4 \quad \text{and} \quad 2 \times 5 = 4$$

hence the solution set is $\{2, 5\}$.

Exercise 14.2

1. Check the field properties for the following modular systems. State the properties, if there are any, that do not hold for each system.
   a. modulo-two system   c. modulo-eight system
   b. modulo-three system   d. modulo-twelve system
2. Which elements other than zero do not have multiplicative inverses in the modulo-eight system?
3. Which elements other than zero do not have multiplicative inverses in the modulo-twelve system?
4. Find the solution sets over the indicated modular systems.
   a. $3n = 5 \pmod 8$ e. $5n = 7 \pmod{12}$
   b. $4n = 0 \pmod{12}$ f. $6n = 5 \pmod{12}$
   c. $2n = 1 \pmod 3$ g. $4n = 3 \pmod 8$
   d. $5n = 3 \pmod 7$ h. $6n = 3 \pmod 7$
5. Judging from your experiences in finding the solution sets in Problem 4, how many solutions does $ax = b \pmod m$ have if m is a prime number?

14.4. Groups

A mathematical group is a mathematical system defined as follows.

DEFINITION 14.1. *A* **group** *is a non-empty set* $G = \{a, b, c, \ldots\}$ *and one binary operation denoted by* * *defined on the elements of G and satisfying the following properties:*

(1) *The set G is closed under the operation* *
(2) *There is an element e in G called the* **identity element,** *such that for all a in G*

$$a * e = e * a = a$$

(3) *The operation* * *is associative*
(4) *Every element a in G has an inverse in G, denoted by* a^{-1}, *such that* $a * a^{-1} = a^{-1} * a = e$.

When a set G and a binary operation * defined on the elements of G satisfy the properties listed in Definition 14.1 we say that G is a group under the operation * or that G is a group with respect to the operation *.

If, in addition to the properties listed in Definition 14.1, the group

---

Mathematical Systems   369

operation is commutative, we say that G is a **commutative group** or an **Abelian group.**

EXAMPLE i.   Is the modulo-five system a group with respect to addition mod 5?

Solution: The modulo-five system is closed under addition mod 5; hence property (1) is satisfied. The identity element for addition mod 5 is 0; hence property (2) is satisfied. Since addition mod 5 is associative, property (3) is satisfied. Since every element has an additive inverse, property (4) is satisfied. Hence the modulo-five system is a group under addition mod 5. Since addition mod 5 is commutative, the modulo-five system is a commutative group under addition mod 5.

EXAMPLE ii.   Is the set of integers a group with respect to multiplication?

Solution: Since the set of integers is closed under multiplication, property (1) is satisfied. The identity element for multiplication is 1, so property (2) is satisfied. Multiplication of integers is associative; hence property (3) is satisfied. Only two integers, 1 and ⁻1, have multiplicative inverses; hence property (4) is not satisfied and the set of integers is not a group under multiplication.

Some very interesting groups can be developed through movements of geometric figures. Suppose a rectangle is cut from a piece of plywood as shown in Figure 14.2. The rectangle is lifted from the plywood, turned into

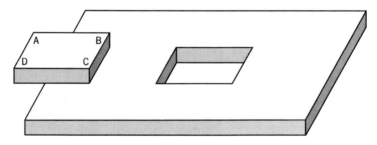

Figure 14.2

a new position, called a **symmetry** of the rectangle, and replaced. There are four possible ways of turning the rectangle before replacing it as described as follows:

(1) Leave unturned. We shall call this movement I.
(2) Turn around the horizontal axis. After this movement the rectangle looks like Figure 14.3. Call this movement H.

Figure 14.3

(3) Turn through 180° in a clockwise direction. After this movement the rectangle looks like Figure 14.4. Call this movement *R*.

Figure 14.4

(4) Turn around the vertical axis. After this movement the rectangle looks like Figure 14.5. Call this movement *V*.

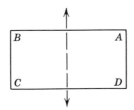

Figure 14.5

Let us consider the set whose elements are these movements: $\{I, H, R, V\}$. We now define an operation, denoted by *, on these elements. We define $H * V$ as meaning to perform movement H and follow it by movement V. (A good way to perform this operation is to use a cardboard rectangular region and do the actual movements with it. Be sure that you put the letters in the corners on both sides of the card.)

We are now ready to construct the operation table (Table 14.4) for this set and the operation *.

The elements in the left column of the operation table represent the first term in the operation; those in the top row represent the second term. Thus $H * R$ is found by finding H in the left column and R in the top row. Looking across the row beginning with H until we are directly under the heading R we find V. Thus $H * R = V$.

Table 14.4

Second Term

	I	R	V	H
I	I	R	V	H
R	R	I	H	V
V	V	H	I	R
H	H	V	R	I

First Term

We now ask ourselves is this set a group under the operation *? We see from examination of the table that the set is closed under the operation, since every result of performing the operation on two elements of the set is an element of the set.

The identity is I since

$$I * R = R * I = R$$
$$I * V = V * I = V$$
$$I * H = H * I = H$$
$$I * I = I.$$

To check whether or not the operation * is associative we must replace a, b, and c in $(a * b) * c = a * (b * c)$ with I, H, V, and R in turn. Checking a few cases we find:

$$(R * V) * H = R * (V * H)$$
$$H * H = R * R$$
$$I = I$$
$$(V * V) * H = V * (V * H)$$
$$I * H = V * R$$
$$H = H$$

If we check each of these possibilities we find that the operation * is associative.

Does every element have an inverse? Since

$$I * I = I$$
$$V * V = I$$
$$H * H = I$$
$$R * R = I$$

we conclude that every element in the set is its own inverse.

Since the four properties of Definition 14.1 are satisfied, this set is a group under the operation of * defined in Table 14.4. This group is called the **transformation group** of the symmetries of a rectangle.

In a similar fashion we can construct transformation groups of the symmetries of other geometric figures.

14.5.   A Non-Commutative Group

All of the groups that we have discussed have been commutative groups. Now let us study a group that is non-commutative. Consider the following: A professor has three graduate assistants. He is anxious that each assistant should have experience teaching algebra, trigonometry, and calculus. The three students, Ross, Smith, and Taggart are each assigned a particular class which is considered to be his own. Each teaches his own class for one week. The next week Ross's class is assigned to Taggart, Smith's class is assigned to Ross, and Taggart's class is assigned to Smith. These changes are made by the professor by placing the following diagram on the bulletin board outside his office door.

$$\begin{bmatrix} R & S & T \\ T & R & S \end{bmatrix}$$

The top row of letters in the diagram represent Ross (R), Smith (S), and Taggart (T). The letter under R is T; this means that Taggart is to teach Ross's class. Similarly the letter under the S is R; this means that Ross is to teach Smith's class. Finally the S under the T means that Smith is to teach Taggart's class.

Each week the professor puts a similar diagram on the bulletin board. Since this rotation of classes is to continue all semester, the three students soon begin to guess which class each will be teaching the following week. It was not long before they discovered that there were only six possible changes:

$$I = \begin{bmatrix} R & S & T \\ R & S & T \end{bmatrix} \quad X = \begin{bmatrix} R & S & T \\ S & T & R \end{bmatrix} \quad Y = \begin{bmatrix} R & S & T \\ S & R & T \end{bmatrix}$$

$$Z = \begin{bmatrix} R & S & T \\ T & S & R \end{bmatrix} \quad W = \begin{bmatrix} R & S & T \\ R & T & S \end{bmatrix} \quad V = \begin{bmatrix} R & S & T \\ T & R & S \end{bmatrix}$$

As the weeks passed the professor decided to have a little fun with the students and indicated their teaching assignments as follows: *ZX*. This meant that the students were to find the final result (called a product for convenience) by first performing arrangement *Z* and then arrangement *X*. Similarly, *WY* means teaching that class assignment which results from doing arrangement *W* followed by arrangement *Y*. Thus to figure the week's teaching assignment *ZX* means the result of performing *Z* and then following this by performing *X*. Thus *ZX* would be found as follows.

$$ZX = \begin{bmatrix} R & S & T \\ T & S & R \end{bmatrix}\begin{bmatrix} R & S & T \\ S & T & R \end{bmatrix}$$

In assignment *Z*, *R* is replaced by *T* but assignment *X* replaces *T* by *R*. Hence the product *ZX* replaces *R* by *R*. Similarly assignment *Z* replaces assignment *S* by *S* but assignment *X* replaces *S* by *T*. Hence *ZX* replaces *S* by *T*. Assignment *Z* replaces *T* by *R* but assignment *X* replaces *R* by *S*, hence *ZX* replaces *T* by *S*. Performing *ZX* could be charted as follows:

(a) $R \rightarrow T \rightarrow R$
(b) $S \rightarrow S \rightarrow T$
(c) $T \rightarrow R \rightarrow S$

We see then that

$$ZX = \begin{bmatrix} R & S & T \\ T & S & R \end{bmatrix}\begin{bmatrix} R & S & T \\ S & T & R \end{bmatrix} = \begin{bmatrix} R & S & T \\ R & T & S \end{bmatrix} = W$$

Similarly

$$YZ = \begin{bmatrix} R & S & T \\ S & R & T \end{bmatrix}\begin{bmatrix} R & S & T \\ T & S & R \end{bmatrix} = \begin{bmatrix} R & S & T \\ S & T & R \end{bmatrix} = X$$

The students decided that they could save a lot of time, which they could certainly use for study, by constructing an operation table for this system. This operation table is shown in Table 14.5.

The set of elements $\{I, X, Y, Z, W, V\}$ and the operation defined by Table 14.5 give us a mathematical system. Is this system a group, that is, does it satisfy the group properties? Let us check.

If this set is a group under the defined operation, then it is closed under the operation. Since only elements of the set appear in the operation

Table 14.5

Second Term

	I	X	Y	Z	W	V
I	I	X	Y	Z	W	V
X	X	V	W	Y	Z	I
Y	Y	Z	I	X	V	W
Z	Z	W	V	I	X	Y
W	W	Y	X	V	I	Z
V	V	I	Z	W	Y	X

First Term (label on left side)

table, the set is certainly closed under the operation. Is there an identity element? We see that the identity is I since

$$II = I$$
$$IX = XI = X$$
$$IY = YI = Y$$
$$IZ = ZI = Z$$
$$IW = WI = W$$
$$IV = VI = V$$

To assure ourselves that the defined operation is associative we must check all cases; that is, we must replace a, b, and c with each element I, X, Y, Z, W, and V in turn in $(ab)c = a(bc)$ and see whether or not in each case the resulting equation is a true statement. Let us check a few cases:

$$(WY)Z = W(YZ)$$
$$XZ = WX$$
$$Y = Y$$
$$(VW)X = V(WX)$$
$$YX = VY$$
$$Z = Z$$
$$(VY)W = V(YW)$$
$$ZW = VV$$
$$X = X$$

When each case is checked (there are $6^3 = 216$ cases to check) we find that the operation is associative.

Since

$$VX = XV = I$$
$$YY = I$$
$$ZZ = I$$
$$WW = I$$
$$II = I$$

we see that every element in the set has an inverse.

Since all four of the properties of Definition 14.1 are satisfied, the set is a group under the given operation. This group is called a **permutation group.**

Is this group a commutative group? Since, in particular, $XZ = Y$ and $ZX = W$, that is, $XZ \neq ZX$, we conclude that this is not a commutative group.

Exercise 14.3

1. Which of the following sets are groups over the indicated operations? State the group properties, if there are any, that do not hold for each.
   a. $\{0, 1, 2, 3, 4\}$; addition mod 5
   b. $\{\ldots, {}^-3, {}^-1, 1, 3, \ldots\}$; addition
   c. $\{0, 1, 2, 3, \ldots\}$; multiplication
   d. The set of rational numbers; multiplication
   e. The set of all rational numbers except zero; multiplication
   f. $\{0, 1, 2, 3, \ldots, 11\}$; multiplication mod 12
   g. The set of integers; addition
2. Make an operation table for the transformation group of an isosceles triangle. There will be two elements in this group: (1) I—leave unturned; (2) V—turn around the vertical axis. Verify that this set is a group under the operation defined by the table you constructed.
3. Make an operation table for the transformation group of an equilateral triangle. There will be six elements in the group: (1) R_0—leave unturned; (2) R_1—turn through $120°$ in a clockwise direction; (3) R_2—turn through $240°$ in a clockwise direction; (4) V_1—turn around axis labeled v_1 in the figure; (5) V_2—turn around axis labeled v_2 in the figure; (6) V_3—turn around the axis labeled v_3 in the figure. Verify that this set is a group under the operation defined by the table you constructed.

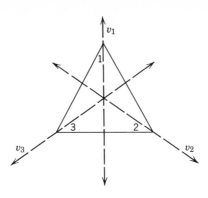

4. Make an operation table for the transformation group of a square. There will be eight elements in this group: (1) I—leave unturned; (2) R_1—turn through $90°$ in a clockwise direction; (3) R_2—turn through $180°$ in a clockwise direction; (4) R_3—turn through $270°$ in a clockwise direction; (5) H—turn around the horizontal axis; (6) V—turn around the vertical axis; (7) D—turn around upper right-lower left hand axis; (8) D'—turn around upper left-lower right hand axis. Verify that this set is a group under the operation defined in the operation table you constructed.

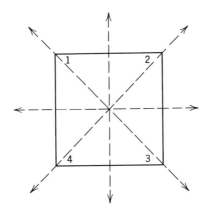

5. Use the table you constructed in problem 4 and find

   a. $H * D$ e. $(V * R_1) * D'$
   b. $R_1 * R_3$ f. $(R_1 * R_2) * D$
   c. $(V * D) * D'$ g. $(H * D') * D$
   d. $(D * R_3) * D'$ h. $(H * R_3) * (R_2 * V)$

6. Is the transformation group of the symmetries of a square a commutative group? Give a reason for your answer.

7. A finite mathematical system can be created by considering the rotation of the four tires on an automobile. Let the operation be denoted by *. The set of elements in the system is $\{I, R, S, T\}$ as explained in the figure below.

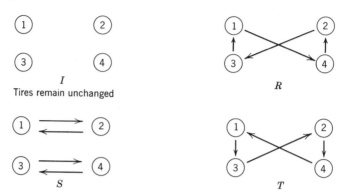

a. Make an operation table for this system.
b. Check the group properties for this system. State whether each of the properties is satisfied. Is the system a group under the operation?
c. If this system is a group, is it a commutative group?

8. Consider the miniature checkerboard in the figure below. Call a move of the checker across or back "H"; a move up or down "V"; a diagonal move "D"; and no move at all "N". Complete the table

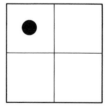

Second Term

*	N	H	V	D
N				D
H				V
V		D		
D		H		

First Term

on page 378. $H * D$ means a move across the board followed by a diagonal move.

a. Is this operation commutative?

b. What is the identity element of this operation?

c. Is the operation associative?

d. Is this set a group with respect to the operation *?

9. Prove: If x and y are elements of a group G then $(x * y) * y^{-1} = x$.

10. Prove: If x and y are elements of a group G and $x * y = x$, then $y = e$.

11. Prove: If x and y are elements of a group G then $(x * y)^{-1} = y^{-1} * x^{-1}$.

15.1. Measurement of Line Segments

To **measure** a line segment is to assign a number to it. Measurement is one of the links between the physical world around us and mathematics. So is counting, but in a different way. We *count* the number of steaks we need for a dinner party, but we *measure* the amount of coffee or wine.

Suppose we wish to measure the line segment $\overline{AB}$ pictured in Figure 15.1. A line segment is measured by comparing it to another

A ———————————————————— B

Figure 15.1

line segment called the **unit of measurement.** Suppose we choose PQ (Figure 15.2) to serve as our unit of measurement. This means that we agree that the measure of $\overline{PQ}$ is exactly the number 1. The selection of a unit is an arbitrary choice we make. Different people may

Measurement of Geometric Figures

XV

Figure 15.2

choose different lengths as their unit, and before the adoption of standard units of measurement they did. For example, in 1324, Edward II of England decreed that three barley corns taken from the end of the ear and placed end to end equaled an inch; Henry I decreed the distance from the end of his nose to the end of his thumb as the lawful yard.

Now that our unit of measurement is chosen, we lay it off along $\overline{AB}$ as shown in Figure 15.3. We see that $\overline{PQ}$ is contained in $\overline{AB}$ exactly

Figure 15.3

two times. We therefore say the **measure** of $\overline{AB}$ is 2. We denote this by

$$m(AB) = 2$$

We say the measurement of $\overline{AB}$ is 2 units. We also say that the **distance** from point A to point B is 2 units and that the **length** of $\overline{AB}$ is 2 units.

Now suppose we have $\overline{ST}$ (Figure 15.4) such that our unit of measurement $\overline{PQ}$ does not "fit into" it a whole number of times. We see that starting at S our unit of measurement can be laid off three times along $\overline{ST}$ reaching point R which is between S and T; if it were laid off four times it

Figure 15.4

would arrive at point W which is beyond T. We see that $m(ST)$ is greater than 3 but less than 4. In this case we usually estimate and say that $m(ST)$ is nearer 3 than 4; so that to the nearest unit $m(ST) = 3$. This is the best we can do without considering fractional parts of our unit of measurement or shifting to a smaller unit.

A statement such as "$m(AB) = 2$" is meaningless unless the unit of measurement is known. For this reason, certain units of measurement, called **standard units,** have been adopted. Some standard units for measuring lengths are the inch, the foot, the yard, the mile, the centimeter, the meter, and the kilometer.

The measurement of $\overline{PQ}$ is 1 inch, and hence $m(AB) = 2$ and the measurement of $\overline{AB}$ is 2 inches. We see that 2 is the **measure** of $\overline{AB}$, inch is the **unit of measure,** and 2 inches is the **measurement.**

It must be observed that although we wrote $m(AB) = 2$, so far as the practical process of measurement is concerned it is impossible to draw a physical model of a line segment that is exactly 2 inches long.

Let us consider the line segment $\overline{XY}$ pictured in Figure 15.5. In Figure 15.5a the ruler used to measure $\overline{XY}$ is marked only in inches. We see that

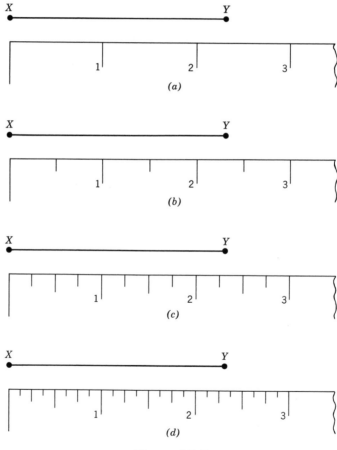

Figure 15.5

the measurement of $\overline{XY}$ is 2 inches correct to the nearest inch. In Figure 15.5b, the ruler has been made more precise by marking every half-inch on the scale. We see, using this ruler, the measurement of $\overline{XY}$ is $2\frac{1}{2}$ inches correct to the nearest half-inch. Similarly, Figure 15.5c and Figure 15.5d show rulers marked every fourth-inch and every eighth-inch, respectively. We see that the measurement of $\overline{XY}$ is $2\frac{1}{4}$ inches correct to the nearest fourth-inch and $2\frac{3}{8}$ inches correct to the nearest eighth-inch.

We see from the above that all measurement is approximate in nature. The exactness of the measurement depends upon the precision of the instrument used and the accuracy of the person doing the measuring.

15.2. Measurement of Angles

We measure an angle, such as $\angle ABC$ in Figure 15.6, by choosing a basic angle called the **unit of measurement** with which to compare the given

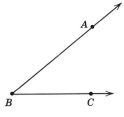

Figure 15.6

angle. Figure 15.7 suggests how we determine the measure of $\angle ABC$ when $\angle XYZ$ is the unit of measurement.

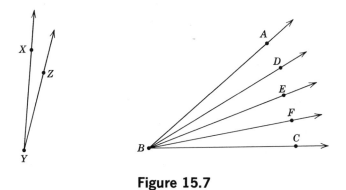

Figure 15.7

Notice that rays $\overrightarrow{BD}$, $\overrightarrow{BE}$, and $\overrightarrow{BF}$ have been chosen so that

$$\angle ABD \cong \angle DBE \cong \angle EBF \cong \angle FBC$$

and so that the union of the interiors of these four angles and the half lines BD, BE, and BF is the interior of $\angle ABC$. If the measure of $\angle XYZ$ is 1, the **measure** of $\angle ABC$ is 4. We write

$$m(\angle ABC) = 4.$$

A common standard unit of measurement of angles is the **degree** (denoted by °). Suppose point P is a point on $\overleftrightarrow{AB}$ (Figure 15.8) and $\overrightarrow{PQ}$ is drawn so that $\angle QPA \cong \angle QPB$. In such a case, the two angles are said to be **right angles** and the lines $\overleftrightarrow{AB}$ and $\overrightarrow{PQ}$ are said to be **perpendicular.**

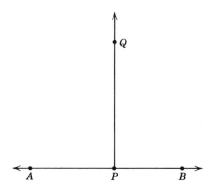

Figure 15.8

A right angle has a measure of 90 and a measurement of 90 degrees (written: 90°). Thus it would take 90 angles of measurement 1° to cover either $\angle QPA$ or $\angle QPB$ as explained above.

The instrument most commonly used to measure angles is a **protractor** (Figure 15.9). The upper and lower scales of a protractor are marked in degrees. Figure 15.10 shows a protractor used to measure $\angle ABC$ and $\angle ABD$. We see, using the lower scale of the protractor, that

$$m(\angle ABC) = 30$$
$$m(\angle ABD) = 120$$

Hence the measurement of $\angle ABC$ is 30° and the measurement of $\angle ABD$ is 120°.

An angle whose measurement is between 0° and 90° is called an **acute angle**. In Figure 15.11, $\angle PQR$ is an acute angle. An angle whose

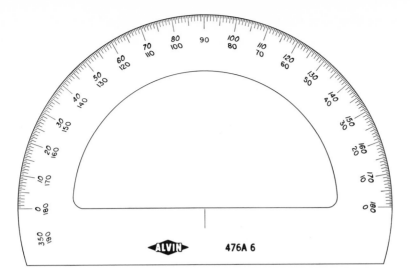

Figure 15.9

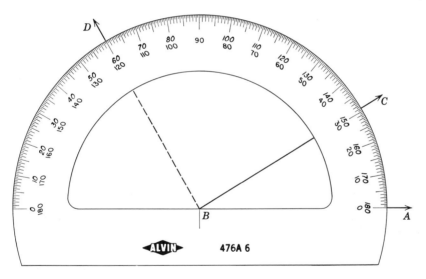

Figure 15.10

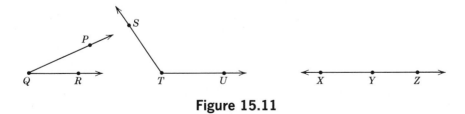

Figure 15.11

measurement is between 90° and 180° is called an **obtuse angle.** In Figure 15.11, ∠*STU* is an obtuse angle. An angle whose measurement is 180° is called a **straight angle.** ∠*XYZ* is a straight angle. A straight angle is an angle such that the two rays forming it lie on the same line. An angle of measurement 0° is an angle such that the two rays forming it actually coincide.

If the sum of the measures of two angles is 90, the angles are called **complementary angles.** Each is called the **complement** of the other. If the sum of the measures of two angles is 180, the angles are called **supplementary.** Each angle is called the **supplement** of the other.

15.3. Classification of Triangles

If three sides of a triangle are congruent, the triangle is called an **equilateral** triangle. Triangle *ABC* in Figure 15.12 is an equilateral triangle. If two sides of a triangle are congruent, the triangle is called **isosceles.** Triangle *PQR* in Figure 15.12 is isosceles. A triangle is called **scalene** if no two of its sides are congruent.

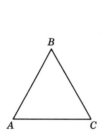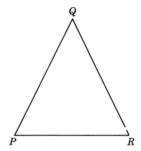

Figure 15.12

If three angles of a triangle have the same measure, the triangle is called **equiangular.** If three sides of a triangle are congruent, then its three angles have the same measure. Conversely, if three angles of a triangle have the same measure, then its three sides are congruent. Thus we see that every equilateral triangle is an equiangular triangle and conversely.

A **right triangle** is a triangle one of whose angles is a right angle. An **obtuse triangle** is a triangle one of whose angles is an obtuse angle. An **acute triangle** is a triangle all of whose angles are acute angles.

15.4. Classification of Quadrilaterals

A quadrilateral is a polygon with four sides. If two sides of a quadrilateral are parallel, the quadrilateral is called a **trapezoid.** All of the figures shown in Figure 15.13 are trapezoids.

If a trapezoid has two pairs of parallel sides, it is called a **parallelogram.** Figures 15.13b, 15.13c, and 15.13d are parallelograms.

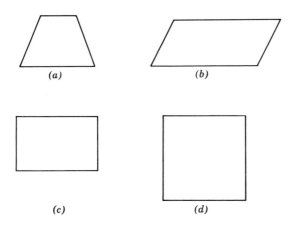

(a)　　　　　　　　　(b)

(c)　　　　　　　　(d)

Figure 15.13

If all the angles of a parallelogram are right angles, the parallelogram is a **rectangle.** Figures 15.13c and 15.13d are rectangles. A **square** is a rectangle with four congruent sides. Figure 15.13d is a square.

Exercise 15.1

1. Determine the measure (to the nearest unit) of $\overline{AB}$ if $\overline{XY}$ is the unit of measure.

X　　　　Y　A　　　　　　　　　　　　　　　　B

2. Determine the length of $\overline{AB}$ to the nearest inch.
3. Determine the length of $\overline{AB}$ to the nearest half-inch.
4. Determine the length of $\overline{AB}$ to the nearest fourth-inch.
5. If the measurement of $\overline{PQ}$ is 36 inches, determine its measurement in feet; in yards.

6. If the measurement of *KL* is 42 inches, determine its measurement in feet; in yards.

7. Which of the angles pictured below are acute angles? right angles? obtuse angles? straight angles?

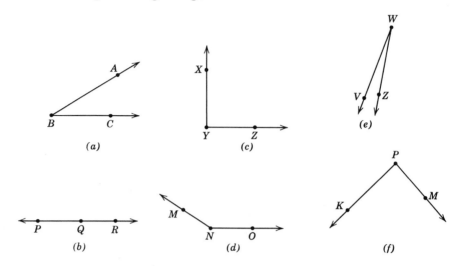

(a)   (c)   (e)

(b)   (d)   (f)

8. Use a protractor to draw angles with the following measurements.

a. 30°	c. ·45°	e. 150°	g. 160°
b. 60°	d. 90°	f. 180°	h. 100°

9. Find the measurement of each angle in Problem 7 correct to the nearest degree.

10. Classify the triangles in the drawings as acute, right, or obtuse.

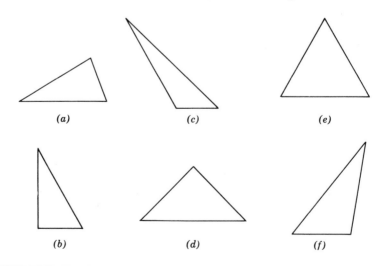

(a)   (c)   (e)

(b)   (d)   (f)

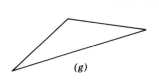
(g)

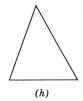
(h)

11. Classify the quadrilaterals below as trapezoids, parallelograms, rectangles, or squares.

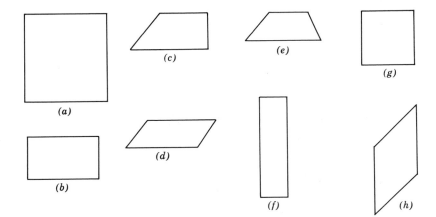

12. Draw a triangle *ABC* and cut it out carefully. Then cut off the vertices as shown in the diagram below. Put the corners together on a line so that they fit without overlapping and *A*, *B*, and *C* touch the same point on the line. What does this experiment suggest about the sum of the measures of the angles of a triangle?

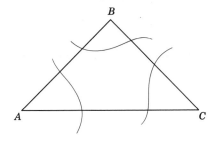

13. What is the measurement of the third angle of a triangle if two of the angles have the measurements given? (The sum of the measures of the angles of a triangle is 180.)

a. 30° and 40°   d. 100° and 60°
b. 45° and 45°   e. 80° and 70°
c. 70° and 35°   f. 110° and 45°

14. Give the measurement of the complement of angles having the following measurements.

a. 30°   c. 50°   e. 80°
b. 45°   d. 75°   f. 90°

15. Give the measurement of the supplement of angles having the following measurements.

a. 100°   d. 140°   g. 130°
b. 45°   e. 75°   h. 150°
c. 90°   f. 120°   i. 88°

16. Fill in the blanks to make true statements.

a. The measure of a right angle is _____ .
b. The measurement of a straight angle is _____ .
c. The measurement of an acute angle is _____ 90°.
d. The measurement of an obtuse angle is _____ 90° and _____ 180°.
e. A rectangle with four sides of equal lengths is a _____ .
f. A parallelogram with four right angles is a _____ .
g. A triangle that has three congruent sides is called _____ .

17. Which of the following are true statements?

a. All parallelograms are trapezoids.
b. All rectangles are squares.
c. Two angles are complementary if the sum of their measures is 90.
d. Every equilateral triangle is an isosceles triangle.
e. Every right triangle is isosceles.
f. Opposite sides of a parallelogram are congruent to each other.

18. The sum of the measures of the angles of a triangle is 180. Find the sum of the measures of the polygons pictured below.

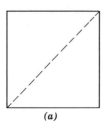

(a)

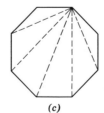

(c)

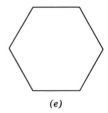

(e)

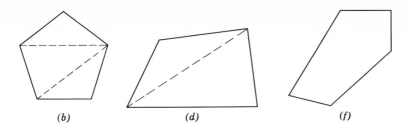

(b)   (d)   (f)

19. If two lines intersect in such a way as to form a right angle, the lines are said to be perpendicular. If you are working in a plane,
    a. How many lines are perpendicular to a given line at a given point on the line?
    b. How many lines are perpendicular to a given line and contain a given point on the line?

20. In the figure below lines $\overleftrightarrow{AB}$ and $\overleftrightarrow{CD}$ are parallel. Line $\overleftrightarrow{KE}$, called a **transversal,** intersects $\overleftrightarrow{AB}$ at K and $\overleftrightarrow{CD}$ at E.

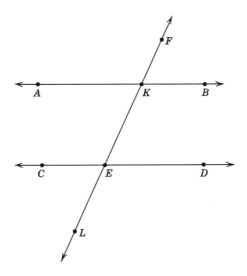

    a. $\angle FKB$ and $\angle AKE$ are called **vertical angles.** Name two other pairs of vertical angles.
    b. $\angle BKE$ and $\angle KEC$ are called **alternate interior** angles. Name another pair of alternate interior angles.

15.5. Perimeters of Polygons

The **perimeter** of a polygon is the sum of the lengths of its sides. Suppose the measurements of $\overline{AB}$, $\overline{BC}$, $\overline{CD}$, and $\overline{DA}$ of polygon $ABCD$ (Figure 15.14)

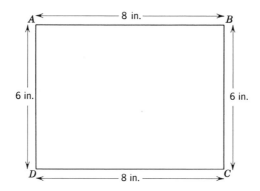

Figure 15.14

are 8 inches, 6 inches, 8 inches, and 6 inches respectively. Then the perimeter of $ABCD$ is found by adding the measures of its sides:

$$m(AB) + m(BC) + m(CD) + m(DA) = 8 + 6 + 8 + 6 = 28$$

The perimeter of $ABCD$ is 28 inches.

The perimeter of any polygon is the sum of the measures of its sides. In general, if a, b, c, . . . , are the measures of the sides of the polygon, the formula for its perimeter, P, is

$$P = a + b + c + \cdots$$

EXAMPLE i. Find the perimeter of the polygon shown in Figure 15.15.

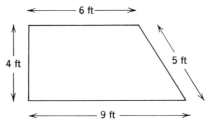

Figure 15.15

Solution: We have, using the formula,

$$P = 4 + 6 + 5 + 9 = 24$$

The perimeter is 24 ft.

EXAMPLE ii.   Find the perimeter of a square whose sides are 10 in. long.
Solution: Since a square is a polygon with four congruent sides, its perimeter is four times the measure of each side:

$$P = 4 \times 10 = 40$$

The perimeter is 40 in.

15.6.   Circumference of a Circle

The perimeter of a circle is called its **circumference.** Measuring the circumference of a circle is more difficult than measuring a line segment. We cannot lay off our unit of measurement along the circumference. Using a physical object, for example, a tire, we could find the approximate circumference by wrapping a string around it and then measuring the length of the string. Using this technique, we get a good approximation of the circumference of a circular object.

In mathematics we want to determine the circumferences of circles with absolute accuracy. Consider the circle in Figure 15.16a. We can find the perimeters of square $ABCD$ and square $WXYZ$. We say that square $ABCD$ is **inscribed** in the circle and square $WXYZ$ is **circumscribed** about

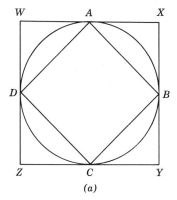
(a)

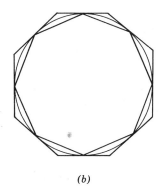
(b)

Figure 15.16

the circle. The perimeter of $ABCD$ is less than and the perimeter of $WXYZ$ is greater than the circumference of the circle.

Now let us circumscribe and inscribe octagons about the circle. (Figure 15.16b) We again intuitively see that the perimeter of the circumscribed octagon is greater than the circumference of the circle and the perimeter of the inscribed octagon is less than the circumference of the circle.

Now let us replace the circumscribed and inscribed octagons by sixteen-sided polygons. Again we notice that the perimeter of the circumscribed polygon is greater than and the perimeter of the inscribed polygon is less than the circumference of the circle.

Continuing in this fashion replacing the circumscribed and inscribed sixteen-sided polygons with polygons of thirty-two sides, the thirty-two sided polygons by sixty-four sided polygons, and so forth, we see that the perimeters of the polygons may be considered to be upper and lower bounds of the circumference of the circle. The difference between these two perimeters can be made as small as we wish by increasing sufficiently the number of sides of the polygons.

It can be proved, although we shall not do it here, that the circumference, C, of a circle is the product of the diameter, d, of the circle and an irrational number, $3.1415926535\ldots$, called π (pi):

$$\mathbf{C = \pi d}$$

Since the diameter of a circle is twice as long as a radius of the circle, the formula above may be written

$$\mathbf{C = 2\pi r}$$

In computation, we usually approximate the value of π and use 3.14 or $\frac{22}{7}$ for its value.

EXAMPLE i. Find the circumference of a circle whose diameter is 14 inches. Use $\frac{22}{7}$ for π.

Solution: Using the formula

$$C = \pi d$$

we have

$$C = \frac{22}{7} \times 14 = 44$$

The circumference is 44 inches.

EXAMPLE ii. Find the circumference of a circle whose radius is 2.15 feet. Use 3.14 for π.

Solution: Using the formula

$$C = 2\pi r$$

we have

$$C = 2 \times 3.14 \times 2.15$$
$$= 13.5020$$

The circumference is 13.5020 feet.

Exercise 15.2

1. Find the perimeters of the polygons pictured.

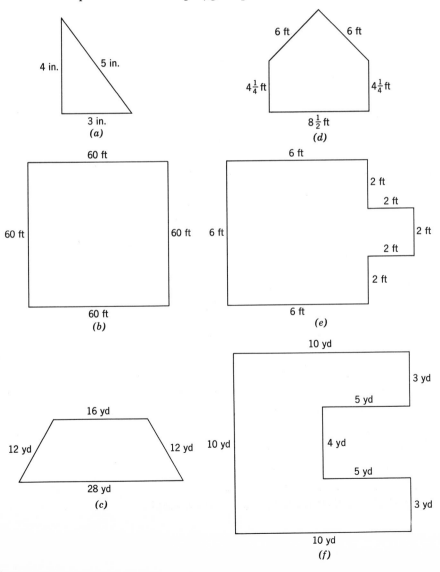

2. Match each figure with the general statement for its perimeter. The same statement may be matched with more than one figure.

  i. $4s$ iv. $a + b + a + b$
  ii. $5s$ v. $a + b + 4s$
  iii. $2a + 2b$ vi. $10s$

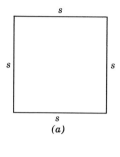
(a)

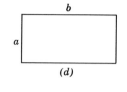
(d)

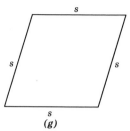
(g)

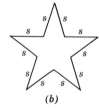
(b)

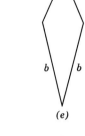
(e)

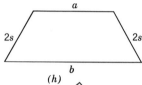
(h)

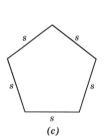
(c)

(f)

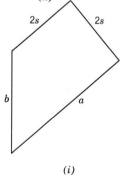
(i)

3. Find the perimeters of the following polygons.
   a. A rectangle with sides 6 ft and $3\frac{1}{2}$ ft.
   b. A square with side $4\frac{1}{2}$ ft long.
   c. A pentagon with five congruent sides; each side $1\frac{1}{2}$ ft long.
   d. An octagon with all sides of equal length of 12 in.

4. Find the circumferences of circles with the following diameters. Use $\frac{22}{7}$ for π.
   a. 7 in.   c. 56 yd
   b. 42 ft   d. 35 mi

5. The distance between the bases of a softball diamond is 60 ft. How many feet does a batter run when he hits a home run? How many yards does he run?

6. The Fisher's garden is in the shape of a polygon. The sides have lengths 12 ft, 20 ft, 10 ft, and 9 ft.
   a. How much fencing is needed to enclose the garden?
   b. If fencing costs 34 cents per running foot, how much will the fencing cost?

7. Find the perimeter of the figure pictured below. The curve above the rectangle is a **semicircle,** that is, half of a circle. Use $\frac{22}{7}$ for π.

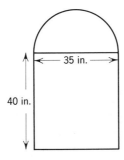

8. What is the difference between the measure of the perimeter of the square and the measure of the circumference of the circle in the figure below? Use $\frac{22}{7}$ for π.

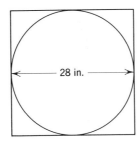

9. How long is each side of a square if its perimeter is 21 ft?

10. The perimeter of a rectangle is 252 ft. Its length is twice as long as its width. What are the dimensions of the rectangle?

11. An isosceles triangle has perimeter 44 inches. Its two congruent sides are each 15 in. long. How long is the third side of the triangle?

12. The perimeter of an equilateral triangle is $10\frac{1}{2}$ ft. What is the length of each of its sides?

13. The perimeter of an isosceles triangle is 50 in. The third side is 5 in. longer than the two congruent sides. What are the lengths of the three sides of the triangle?

14. The diameter of a circle is 13.2 inches. What is its circumference? Use 3.14 for π.

15. The circumference of a circle is 8.792 in. What is the length of a radius of the circle? Use 3.14 for π.

15.7. Area and Its Measure

We recall that a plane region is the union of a simple closed curve and its interior. To measure a plane region is to assign a number to it. We call this number the **area** of the region.

The first step in measuring a line segment is to choose a unit of measurement. That is, we select a certain arbitrary line segment and assign it the measure 1. We now proceed in the same fashion to find areas of plane regions. First we choose a unit of area, that is, a region whose area is assigned the measure 1. We must choose a unit region such that enough of them—placed so that they touch, but do not overlap—will together cover either exactly or with some excess any given plane region. Some shapes will not do this; for example, a circular region does not have this property. Thus in Figure 15.17, if we try to cover a triangular region with small non-

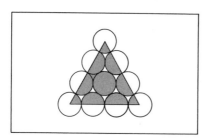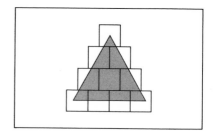

Figure 15.17

overlapping congruent circular regions, there are always parts of the triangular region left uncovered. On the other hand, we can always completely cover a triangular region, or any plane region, by using enough non-overlapping square regions.

While a square region is not the only kind of plane region with this

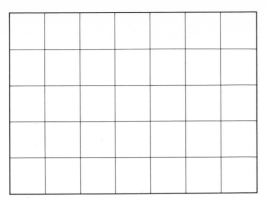

Figure 15.18

covering property, it has the advantage of being simply shaped. Some common standard units of measurement for measuring areas are the square inch (a region bounded by a square each side of which is 1 inch long), the square foot (a region bounded by a square each side of which is 1 foot long); the square yard (a region bounded by a square each side of which is 1 yard long); the square centimeter (a region bounded by a square each side of which is 1 centimeter long); the square meter (a region bounded by a square each side of which is 1 meter long).

We have instruments, for example rulers and tapes, for helping to measure line segments. Corresponding measuring instruments are not available for measuring areas. We make a grid as shown in Figure 15.18 for this purpose. A **grid** is a regular arrangement of non-overlapping square unit regions. Each unit region of the grid is taken as our unit of measurement. Each of these unit regions has an area of 1 square unit.

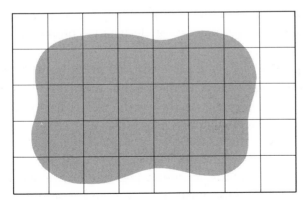

Figure 15.19

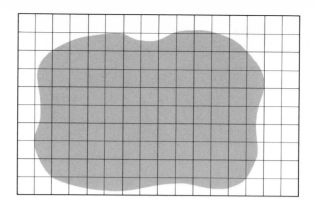

Figure 15.20

To use such a grid in estimating the area of a given region, we think of it as superimposed on the region as shown in Figure 15.19. We can verify by counting that the area of the region in the figure is at least 15 square units. A better approximation of the area of this region can be obtained by dividing each square unit in our grid into four congruent square regions as shown in Figure 15.20.

From the above discussion, we may make the basic assumption that to any plane region a unique number, depending on the unit of measurement, can be assigned. This number, together with the unit of measure, is called the **area** of the region.

15.8. Area of a Rectangle

For some of the more common plane regions such as those bounded by polygons and circles, formulas can be derived to compute the areas of the regions.

If the sides of a rectangle are measured in terms of the same unit, it can be proved that the area, A, of the region bounded by a rectangle whose sides have measures b and h is the product bh. The formula, then, for finding the area of a rectangle is

$$A = bh$$

We can easily verify this formula when b and h are whole numbers. If a rectangle has sides whose measures are b and h, b and h being whole numbers, we have a b by h array of unit squares. We know by the definition of multiplication of whole numbers that the number of squares in this array is bh. We can check this by actual count if we wish.

It can be proved that the formula $A = bh$ holds for all real numbers b and h. We shall accept this without proof, that is, we shall accept without proof the formula $A = bh$ for finding the area of a rectangle.

EXAMPLE i.   Find the area of a rectangle whose sides are 6 inches and 8 inches.

Solution: Using the formula $A = bh$ with $b = 6$ and $h = 8$ we have

$$A = 6 \times 8 = 48$$

The area of the rectangle is 48 sq in.

EXAMPLE ii.   Find the area of a rectangle $2\frac{1}{2}$ ft by $3\frac{1}{2}$ ft.

Solution: Using the formula $A = bh$ with $b = 2\frac{1}{2}$ and $h = 3\frac{1}{2}$ we obtain

$$A = 2\frac{1}{2} \times 3\frac{1}{2} = 8\frac{3}{4}.$$

The area of the rectangle is $8\frac{3}{4}$ sq ft.

EXAMPLE iii.   The area of a rectangle is 3.22 sq yd. One of its sides is 1.4 yd. What is the length of the other side?

Solution: Using the formula $A = bh$ with $A = 3.22$ and $b = 1.4$ we obtain

$$3.22 = 1.4h$$

and $h = 2.3$. The length of the other side is 2.3 yd.

15.9.   The Area of a Parallelogram

The diagram in Figure 15.21 shows how to cut the region bounded by a parallelogram into two parts and fit them together to form a rectangle whose sides are b units and h units.

We call any pair of parallel sides of a parallelogram the **bases** of the parallelogram. Any segment congruent to one which extends from one base to the other and is perpendicular to both bases is called a corresponding **altitude** of the parallelogram. In parallelogram $ABCD$ in Figure 15.22, any one of the congruent line segments shown by dotted lines is an altitude corresponding to the bases $\overline{CD}$ and $\overline{AB}$. We now state a formula for finding

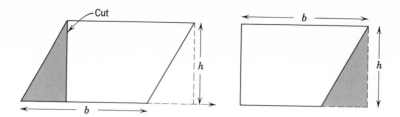

Figure 15.21

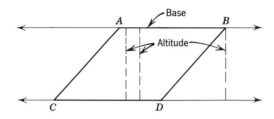

Figure 15.22

the area of a parallelogram in which the measure of a base is b and the measure of the corresponding altitude is h. The diagram in Figure 15.21 suggests that

$$A = bh.$$

EXAMPLE i. Find the area of a parallelogram whose base is 8 in. long and the corresponding altitude is 5 in. long.

Solution: Using the formula $A = bh$ we have

$$A = 8 \times 5 = 40$$

The area is 40 sq in.

EXAMPLE ii. The area of a parallelogram is 36 sq ft and a base is 9 ft long. What is the distance between the bases?

Solution: We have $A = 36$; and $b = 9$. Using the formula $A = bh$ we have

$$36 = 9h$$

and $h = 4$. The distance between the bases is 4 ft.

15.10. Area of a Triangle

The diagram in Figure 15.23 shows how to fit two congruent triangles together to form a parallelogram with base b units long and altitude h units long.

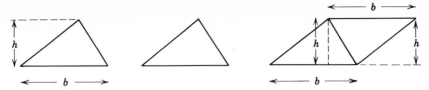

Figure 15.23

When one of the sides of a triangle is called the **base,** then the segment that extends from the vertex, not on the base, perpendicular to the line containing the base is called the corresponding **altitude** of the triangle. (Figure 15.24)

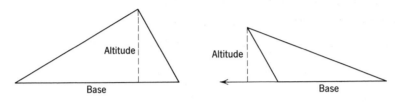

Figure 15.24

Notice that the area of the parallelogram in Figure 15.23 is twice the area of each of the congruent triangles in the figure. The area of the parallelogram is given by $A = bh$, hence the area, A, of a triangle may be found using the formula

$$A = \frac{1}{2}bh$$

where b is the length of the base and h is the length of the corresponding altitude.

EXAMPLE i.   Find the area of a triangle whose base is 5 in. long and whose corresponding altitude is $2\frac{1}{2}$ in. long.
Solution: Using the formula $A = \frac{1}{2}bh$ we have

$$A = \frac{1}{2} \times 2\frac{1}{2} \times 5$$

$$= 6\frac{1}{4}$$

The area is $6\frac{1}{4}$ sq in.

EXAMPLE ii.   The area of a triangle is 32 sq ft. Its base is 8 ft. Find the length of the corresponding altitude.

Solution: Using the formula $A = \frac{1}{2}$ bh with $b = 8$ and $A = 32$ we have

$$32 = \frac{1}{2} \times 8 \times h$$

and $h = 8$. The altitude is 8 in.

15.11. Area of a Trapezoid

The two parallel sides of a trapezoid are called its **bases.** A line segment that has one of its endpoints on the line of one of the bases and its other endpoint on the line of the opposite base and is perpendicular to these bases is called an **altitude** of the trapezoid. In Figure 15.25a, $\overline{AD}$ and $\overline{BC}$ are bases of the trapezoid and $\overline{BE}$ and $\overline{FD}$ are congruent altitudes.

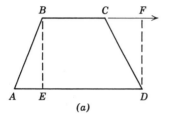 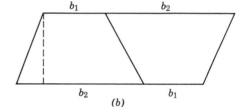

Figure 15.25

Fitting two congruent trapezoids together as shown in Figure 15.25b, we see that a parallelogram is formed whose base has length $b_1 + b_2$, where b_1 and b_2 denote the lengths of the bases of the trapezoid and whose altitude has length h, the altitude of the trapezoid. The area of this parallelogram is $h(b_1 + b_2)$. But the area of this parallelogram is twice the area of the trapezoid, hence the area of the trapezoid is

$$A = \tfrac{1}{2}h(b_1 + b_2)$$

Figure 15.26

where b_1 and b_2 are the lengths of the bases and h denotes the length of the altitude.

EXAMPLE. Find the area of the trapezoid shown in Figure 15.26.
Solution: Using the formula $A = \frac{1}{2}h(b_1 + b_2)$ with $b_1 = 4$, $b_2 = 7$ and $h = 2$ we have

$$A = \frac{1}{2} \times 2 \times (4 + 7)$$

$$= \frac{1}{2} \times 2 \times 11$$

$$= 11$$

The area is 11 sq ft.

15.12. Area of a Circle

The formula for the area of a circle is more difficult to derive than the formulas for the areas of the polygonal regions discussed.

Suppose we cut a circle into sixteen congruent parts and rearrange them as shown in Figure 15.27. If we divide the circle into more and more parts the resulting figure (Figure 15.27b) looks more and more like a parallelogram. The base of this figure is approximately equal in length to one-half of the circumference of the circle as it is made up of the arcs of half of the pieces. The length of the altitude of the figure is approximately equal to the length of the radius of the circle. The area of this figure can be approximated as the product of one-half the measure of the circum-

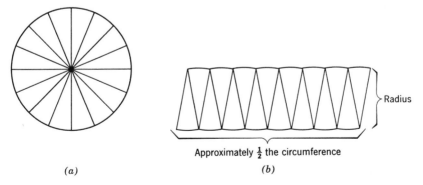

Approximately $\frac{1}{2}$ the circumference

(a) (b)

Figure 15.27

ference and the measure of the radius of the circle. Since the area of this figure is the same as the area of the circle, we obtain the formula

$$A = \frac{1}{2}Cr.$$

But the measure of the circumference of a circle is $\pi d = 2\pi r$, where d is the measure of a diameter and r is the measure of a radius. Hence

$$A = \frac{1}{2}(2\pi r)r$$

$$\mathbf{A = \pi r^2}$$

EXAMPLE. Find the area of a circle whose radius is 7 in. Use $\frac{22}{7}$ for π.
Solution: Using the formula $A = \pi r^2$ we have

$$A = \pi r^2$$

$$= \frac{22}{7} \times 7^2$$

$$= \frac{22}{7} \times 49$$

$$= 154$$

The area is 154 sq in.

Exercise 15.3

1. Name a base and the altitude to that base for each triangle shown.

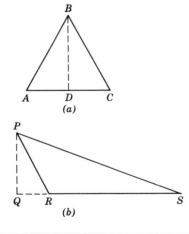
(a)

(b)

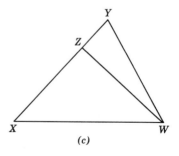
(c)

---

2. Find the area of each figure shown below.

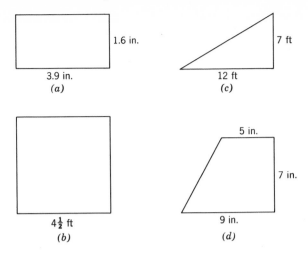

1.6 in.

3.9 in.

(a)

7 ft

12 ft

(c)

$4\frac{1}{2}$ ft

(b)

5 in.

7 in.

9 in.

(d)

3. The area of a rectangle is 56 sq ft. The base is 8 ft long. What is the height of the rectangle?

4. (a) Two triangles are congruent. Do they have the same area?
   (b) Two triangles have the same area. Are they congruent?
   (c) Two triangles do not have the same area. Are they congruent?
   (d) Two triangles are not congruent. Do they have the same area?

5. Two triangles have the same height. The base of one is twice as long as the base of the other. The area of the smaller triangle is 36 sq in. What is the area of the other triangle?

6. Show that if a line segment is drawn from any vertex of a triangle to the midpoint of the opposite side, it divides the triangle into two triangles whose areas are equal.

7. How many rectangles $\frac{1}{2}$ in. by $\frac{1}{3}$ in. are needed to cover a rectangle whose length is $3\frac{1}{2}$ in. and whose width is $2\frac{2}{3}$ in.?

8. Find the area of a circular flower bed 12.25 ft in diameter. Use 3.14 for π.

9. A circular pond is surrounded by a walk 6 ft wide. The diameter of the pond is 30 ft. What is the surface area of the walk? Use 3.14 for π.

10. In the figure below, 6 circular disks each of whose diameter is 2 in. are placed on a rectangular region 4 in. by 6 in. What part of the area is not covered? Use 3.14 for π.

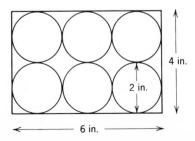

4 in.

2 in.

6 in.

15.13. Volume

By the **volume** of a solid we mean the measurement of the amount of space occupied by the solid. To find the volume of a solid we choose a basic solid with which to compare the given solid. We assign the measure 1 to this basic solid. The usual choice for this basic solid, called the unit of measurement, is a cube one unit on each side. The volume of this basic cube is 1 cubic unit.

Some of the standard units of measurement for measuring volume are the cubic inch, the cubic foot, the cubic yard, the cubic centimeter, and the cubic meter.

Consider a large number of basic cubes. If we arrange three such basic cubes in a row and make two rows of them we have a layer of six basic cubes (Figure 15.28).

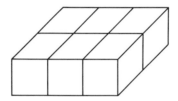

Figure 15.28

If four such layers are piled on top of each other, the result is a rectangular prism made up of 24 basic cubes. Its volume is 24 cubic units. (Figure 15.29).

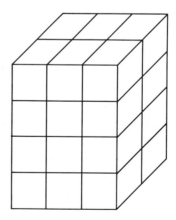

Figure 15.29

Thus for a rectangular prism constructed in this way, the volume appears to be

$$(2 \times 3) \times 4$$

In general, for a rectangular prism the volume, V, is

$$V = lwh$$

where l and w are the length and width of the base of the prism respectively and h is the height or measure of the altitude of the prism.

Notice that lw gives the area of the rectangular base of the prism. Letting B represent the area of the base of the rectangular prism we may write

$$V = Bh$$

for the formula for finding the volume of a rectangular prism.

The derivation of formulas for finding the volumes of other solids can be found by using calculus. We shall simply state the following formulas.

The formula for finding volume of any **prism** is

$$V = Bh$$

where B is the area of the base and h is the height or the measure of the altitude of the prism.

The formula for the volume of a **cylinder** is

$$V = Bh$$

where B is the area of the base and h is the height or altitude. (Figure 15.30)

The volume of a **pyramid** is given by the formula

$$V = \frac{1}{3} Bh$$

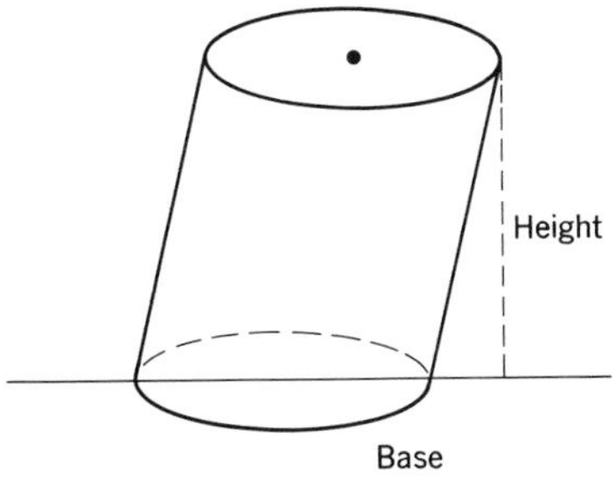

Height

Base

Figure 15.30

where B is the area of the base and h is the height or altitude. (Figure 15.31)

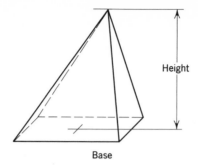

Height

Base

Figure 15.31

The volume of a **cone** is given by the formula

$$V = \frac{1}{3} Bh$$

where B is the area of the base and h is the height or altitude. (Figure 15.32)

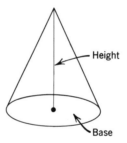

Height

Base

Figure 15.32

The formula for finding the volume of a **sphere** is

$$V = \frac{4}{3} \pi r^3$$

where r is the radius of the sphere.

EXAMPLE i.   Find the volume of a cone whose altitude is 10 ft and whose base is a circle whose radius is 7 ft. Use $\frac{22}{7}$ for π.

Solution: The volume of a cone is given by

$$V = \frac{1}{3} Bh$$

where h is the measure of the altitude and B is the area of the base. Since the base of the given cone is a circle of radius 7 ft, the area of the base is

$$B = \pi r^2$$

$$= \frac{22}{7} \times 7 \times 7$$

$$= 154$$

Then

$$V = \frac{1}{3} \times 154 \times 10$$

$$= 513\frac{1}{3}$$

The volume is $513\frac{1}{3}$ cu ft.

EXAMPLE ii.   A pyramid has a square base with sides each 15 ft long and a height of 7 ft. What is its volume?

Solution: Since the base of the pyramid is a square, the area of the base is

$$B = 15 \times 15 = 225$$

Then

$$V = \frac{1}{3} Bh$$

$$= \frac{1}{3} \times 225 \times 7$$

$$= 525$$

The volume is 525 cu ft.

Exercise 15.4

1.   A tin can has height 3 in. and a base of 15 sq in. What is its volume?
2.   A truck is called a 5-ton truck if its capacity is 5 cu yd. How big is a dump truck which has a body 6 ft wide, 8 ft long, and 5 ft high? (There are 3 ft in one yard.)
3.   A triangular prism has as its base a triangle whose base is 12 ft and whose altitude is 6 ft. The height of the prism is 16 ft. What is the volume of the prism?

4. A pyramid has a rectangular base with dimensions 10 in. by 18 in. The height of the prism is 7 in. What is its volume?
5. A cone has altitude 12 in. and base a circle of radius 4 in. What is its volume? Use 3.14 for π.
6. Find the volume of a sphere whose radius is 1.3 in. Use 3.14 for π.
7. Find the volume of a sphere whose radius is 14 ft. Use $\frac{22}{7}$ for π.
8. If each of l, w, and h is doubled for a rectangular prism, what is the effect on the volume?
9. If the radius of a sphere is tripled, what is the effect on the volume?
10. If the sides of the square base of a pyramid are doubled and the altitude is unchanged, what is the effect on the volume?

16.1. The Fundamental Counting Principle

Let us examine the following problem. There are two roads, A and B, which connect Circle Ranch with the town of Sleepyville and three highways, 1, 2, and 3, connecting Sleepyville and Big City (Figure 16.1). How many different routes can be chosen to get from Circle Ranch to Big City by way of Sleepyville? From Figure 16.1 we see

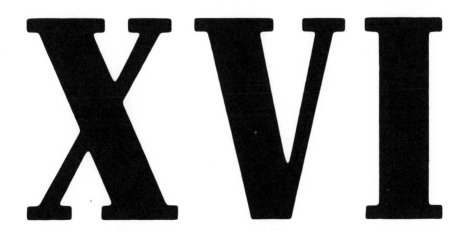

Figure 16.1

that we can travel from Circle Ranch to Sleepyville by either of two routes, and then for each of these routes we have three choices for traveling on to Big City. Hence we have $3 \times 2 = 6$ possible ways in all.

A generalization of the above problem is the fundamental counting principle.

Probability

XVI

DEFINITION 16.1. FUNDAMENTAL COUNTING PRINCIPLE.

If a first choice can be made in any one of n_1 ways, and following this, a second choice can be made in n_2 ways, and so on, until the kth choice can be made in n_k ways, then the r choices in the order indicated can be made in $n_1 n_2 \ldots n_k$ ways.

EXAMPLE i.   A committee of two children is to be appointed. One of the members is to be chosen from a group of 6 boys and the other member from a group of 5 girls. How many possible committees can be chosen?

Solution: Here we have $n_1 = 6$ and $n_2 = 5$. Hence by the fundamental counting principle we have

$$n_1 n_2 = 6 \times 5 = 30$$

There are 30 possible committees.

EXAMPLE ii.   A penny, a nickel and a dime are tossed. How many different outcomes are possible?

Solution: Here n_1, n_2, and n_3 are all 2 since each coin can fall in two possible ways, heads or tails. Using the fundamental counting principle we have

$$n_1 n_2 n_3 = 2 \times 2 \times 2 = 8$$

There are eight possible outcomes.

16.2.   Permutations

Suppose we have a set with four elements:

$$A = \{1, 2, 3, 4\}$$

How many different arrangements of the elements of set A can we make?

Let us suppose that we have four boxes into which we can put one of the elements of Set A (Figure 16.2). In Box I we can put any one of the four elements of the set. Hence the choice of the first element in the arrangement of a set of four elements can be made in four ways. To show this we write "4" in Box I (Figure 16.3).

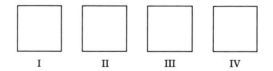

I   II   III   IV

Figure 16.2

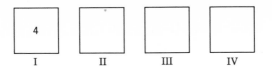

Figure 16.3

Having put one element in Box I we have three elements left. We can put any one of these three elements in Box II; that is, we can fill Box II in three ways. To show this we write "3" in Box II (Figure 16.4). This means that altogether there are 4 × 3 or 12 ways of filling Boxes I and II.

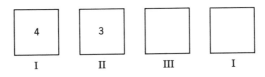

Figure 16.4

Having put elements in Boxes I and II we have two elements left. We can put either of these in Box III, that is, we can fill Box III in two ways. To show this we write "2" in Box III (Figure 16.5). This means that altogether there are 4 × 3 × 2 or 24 ways of filling Box I, II, and III.

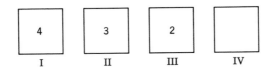

Figure 16.5

Since there is now only one element left, there is only one way to fill Box IV. To show this we write "1" in Box IV (Figure 16.6). We now have 4 × 3 × 2 × 1 or 24 arrangements that we can make from the elements of set *A*.

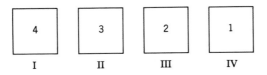

Figure 16.6

We call each of the twenty-four arrangements of the elements of set A a **permutation.** Since the elements themselves had no bearing on the number of arrangements we can make from the elements, we see that twenty-four permutations that can be made from a set of four elements, regardless of what the elements are.

Now let us generalize the above example and find how many permutations we can make with the elements of a set of n distinct elements. Suppose A is a set and $n(A) = n$. Let us imagine that we have n boxes into each of which we can put one element of set A. In Box I we can put any one of n elements. Having put one element in Box I we have $(n - 1)$ elements left; so we can put any one of $(n - 1)$ elements in Box II. This means that together there are $n(n - 1)$ ways to fill Boxes I and II. There are now $(n - 2)$ elements left. We can fill Box III with any one of these; so we can fill Box III in $(n - 2)$ ways. This means that we can fill Boxes I, II, and III in $n(n - 1)(n - 2)$ ways. Continuing in this fashion we find that the number of permutations of the elements of set A if $n(A) = n$ is

$$n(n - 1)(n - 2) \ldots (3)(2)(1).$$

We use the symbol $n!$ to denote the product $n(n - 1)(n - 2) \ldots$ $(3)(2)(1)$. We read this symbol "n factorial." Thus we see that

$$4! = 4 \times 3 \times 2 \times 1 = 24$$
$$5! = 5 \times 4 \times 3 \times 2 \times 1 = 120$$
$$2! = 2 \times 1 = 2$$

We define 1! to be 1. From the above we see that the number of permutations of n elements taken n at a time is

$$\mathbf{n! = n(n - 1)(n - 2) \ldots (3)(2)(1)}$$

EXAMPLE iii. Evaluate $\dfrac{4!}{3!}$.

Solution:
$$\frac{4!}{3!} = \frac{4 \times 3 \times 2 \times 1}{3 \times 2 \times 1} = 4$$

EXAMPLE iv. In how many ways can we serve six steaks on six plates? *Solution:* We are looking for the number of permutations of 6 elements taken 6 at a time. Hence

$$6! = 6 \times 5 \times 4 \times 3 \times 2 \times 1$$
$$= 720$$

There are 720 ways to serve the steaks.

Exercise 16.1

1. A salesman plans a trip to San Francisco from St. Louis by way of Denver. From St. Louis there are four possible routes that he can take, and from Denver there are three possible routes to San Francisco. In how many different ways can the trip be made?

2. In a library there are 14 science fiction books, 20 novels, and 12 history books. In how many ways can someone choose three books, one of each kind?

3. A penny, a dime, a nickel, and a quarter are tossed. In how many possible ways can they fall?

4. We roll a red die and a green die. In how many possible ways can this pair of dice fall?

5. A girl has six skirts and eight blouses. Each skirt can be worn with any of the blouses. How many possible outfits can she select?

6. How many ways can a group photograph of 8 people be taken if the people are to stand in a line?

7. In how many ways can 10 tin soldiers be arranged in a line?

8. In how many ways can 12 students be seated in a row of 12 seats?

9. How many different signals can be made by displaying three signal flags in a vertical line?

10. Mrs. Tuttle and her six children are to be lined up in a row for their photograph to be taken. In how many ways can they be lined up if Mrs. Tuttle is to stand in the center?

11. Evaluate.

   a. $6!$

   b. $\dfrac{8!}{3!}$

   c. $\dfrac{4!\,3!}{2!}$

   d. $\dfrac{(10-6)!}{6!}$

   e. $\dfrac{30!}{29!}$

   f. $(2!)(3!)$

12. List the 24 possible four-letter arrangements of the letters X, Y, Z, and W.

13. How many different batting orders are possible for a nine-man baseball team if the pitcher must bat last?

14. How many two-digit numerals can be formed from the digits 2, 3, 4 if no repetition of digits is allowed? What are they?

15. An ice cream store has 31 different flavors of ice cream and 12 different sauces. How many different sundaes (ice cream covered with sauce) can be made?

16.3.   Sample Spaces and Events

Probability questions arise when we think of situations where we cannot be certain of the outcomes. For example, let us perform an experiment of tossing a coin. There are two possible outcomes of this experiment; either the coin lands "heads" or it lands "tails."

In actual practice, we occasionally find other outcomes to this experiment. The coin may roll off of the table and get caught in the side of someone's shoes, or perhaps the coin stands on edge in a crack in the floor. In the experiment described here, we shall rule out such outcomes. The coin lands either heads or tails and it has an equal chance of falling either way.

When a coin is tossed there are two possible outcomes, either heads, which we denote by H, or tails, which we denote by T. Then each outcome of this experiment corresponds to exactly one of the elements in the set

$$S = \{H, T\}.$$

This set is called a **sample space** of the experiment of tossing a coin. We say that each outcome is **equally likely** to occur. We do not define equally likely, but expect the reader to have an intuitive idea of this notation.

Now let us perform an experiment of rolling a die. A die is a small cube, presumably symmetrical in shape and uniform in material (not loaded) and whose faces are marked with 1, 2, 3, 4, 5, and 6 dots as shown in Figure 6.7.

Figure 16.7

When a die is rolled the possible outcomes are 1, 2, 3, 4, 5, and 6. A sample space for this experiment is

$$S_2 = \{1, 2, 3, 4, 5, 6\}$$

DEFINITION 6.2.   *A **sample space** of an experiment is a set of elements that contains all possible outcomes of the experiment.*

Suppose we perform the experiment of tossing two coins, a penny and a nickel. There are four possible outcomes: both coins show heads (HH), the penny shows heads and the nickel shows tails (HT), the penny shows tails and the nickel shows heads (TH), and both coins show tails (TT). A sample space for this experiment is

$$S_3 = \{HH, HT, TH, TT\}$$

We recognize S_3 as the Cartesian product $A \times A$ where $A = \{H, T\}$ and where we have introduced a simplified notation for ordered pairs with HH for (H, H), HT for (H, T), and so forth.

We call any element of a sample space an **outcome.** A subset of a sample space is called an **event** of the sample space.

DEFINITION 16.3. *An event of an experiment is a subset of the sample space of the experiment.*

EXAMPLE i. A die is rolled. List the elements of the event, E, "an even number showing."
Solution: A sample space of the experiment is

$$S = \{1, 2, 3, 4, 5, 6\}$$

The event, E, "an even number showing" is

$$E = \{2, 4, 6\}$$

A **simple event** is an event containing only one element. Any event that is not a simple event is called a **compound event.** Every compound event may be expressed as the union of simple events. For example, the event E in Example i above may be written

$$E = \{2, 4, 6\} = \{2\} \cup \{4\} \cup \{6\}$$

EXAMPLE ii. A coach wishes to choose two volleyball players from a set of seven students, five men and two women. What is the subset defining the event, "a man and a woman"?
Solution: Let a, b, c, d, and e represent the five men and A and B represent the two women. The subset of the sample space defining a man and woman is

$$\{aA, bA, cA, dA, eA, aB, bB, cB, dB, eB\}$$

Exercise 16.2

1. A die is tossed.
   a. What is a sample space?
   b. What is the event, E, "a number less than 3 showing"?
   c. What is the event, F, "an odd number showing"?
   d. What is the event, G, "a prime number showing"?
   e. What is the event, H, "a multiple of 3 showing"?

2. A nickel and a dime are tossed.
   a. What is a sample space?
   b. What is the event, E, "two heads showing"?
   c. What is the event, F, "a head and a tail showing"?
   d. What is the event, G, "a head on the dime"?

3. A pair of dice are tossed. One die is red and one is green. An outcome is denoted by $(1, 2)$, when 1 is showing on the red die and 2 is showing on the green die.
   a. What is a sample space?
   b. What is the event, E, "the sum of the two numbers showing is 7"?
   c. What is the event, F, "the sum of the two numbers showing is 11"?
   d. What is the event, H, "a number greater than 4 showing on one die"?

4. A card is selected from a standard deck of 52 playing cards. What is the subset defining each of the following events?
   a. A heart
   b. A deuce
   c. A jack, queen, or king
   d. The ace of hearts
   e. A black card

5. A committee of three men is picked from a group of six men denoted by A, B, C, D, E, and F.
   a. List the elements of a sample space S.
   b. List the elements of the event, "A is a member of the committee."
   c. List the elements of the event, "A and B are members of the committee."

6. A box contains 3 green beads and 2 red beads. We select two beads at random. (An item is selected at **random** from a group of items if the selection procedure is such that each item in the group is equally likely to be selected.) How many elements are in the sample space?

7. Three coins, a penny, a dime, and a nickel, are tossed. An outcome is denoted by *HTH* when heads shows on the penny, tails shows on the dime, and heads shows on the nickel. Thus *HHT* means that the penny and the dime show heads and the nickel shows tails.
   a. How many elements are in a sample space?
   b. List the elements in a sample space.
   c. List the elements in the event "two heads and a tail."
   d. List the elements in the event "the penny shows heads."

8. The names California, Oregon, Washington, Arizona, Colorado, and Alaska are written on six separate cards. The cards are put into a hat and one is drawn at random. List the elements in the following events.
   a. The name of the state begins with A.
   b. The state borders the Pacific Ocean.

9. Write the following events as the union of simple events.
   a. $\{2, 3, 4\}$
   b. $\{HT, TH\}$
   c. $\{(6, 4), (4, 6), (5, 5)\}$

10. An employer wishes to hire two persons from a group of five equally qualified applicants, three women, denoted by A, B, and C, and two men denoted by X and Y.
    a. How many elements are in a sample space?
    b. List the elements in a sample space.
    c. List the elements in the event "two men."
    d. List the elements in the event "a man and a woman."
    e. List the elements in the event "two women".

16.4. Probability of an Event

When an experiment is performed, there are many events in whose occurrence we may be interested. If we are given a sample space

$$S = \{0_1, 0_2, 0_3, \ldots 0_n\}$$

containing n outcomes, $0_1, 0_2, \ldots, 0_n$, then there are 2^n different subsets of S and since every subset of S is an event, there are 2^n different events.

For each simple event $\{0_i\}$ of S, we define a number denoted by $P(\{0_1\})$. This number is called the **probability** of the event $\{0_i\}$. These numbers, or probabilities may be assigned arbitrarily, but they must satisfy two conditions:

(1) The probability of each simple event of a sample space S is a nonnegative number. That is,

$$P(\{0_i\}) \geqslant 0 \qquad i = 1, 2, 3, \ldots, n$$

(2) The sum of the probabilities of all simple events of S is 1. That is,

$$P(\{0_1\}) + P(\{0_2\}) + \cdots + P(\{0_n\}) = 1$$

We see from these two conditions that the probability of each simple event is at least 0 but at most 1.

If each of n outcomes of an experiment is **equally likely** to occur we shall agree to assign probability $\frac{1}{n}$ to each of the n simple events of a sample space S. In this case, if E is a subset of S, then it can be proved that

$$P(E) = \frac{\text{number of elements in } E}{\text{number of elements in } S}$$

But the number of elements in E is $n(E)$ and the number of elements in S is $n(S)$, hence

$$P(E) = \frac{n(E)}{n(S)}$$

In this book we shall use the above equality as the definition of the probability of an event E. The symbol $P(E)$ is read "the probability of event E."

In the following examples and throughout the rest of this chapter, when there is no reason to doubt the "equally likely" property for the probability of each simple event, we shall use the value $\frac{1}{n}$ as the probability of each simple event.

EXAMPLE i. Consider the rolling of a die with sample space

$$S = \{1, 2, 3, 4, 5, 6\}.$$

What is the probability of the event, E, "a prime number showing?"
Solution: The event "a prime number showing" is

$$E = \{2, 3, 5\}$$

Since $n(E) = 3$ and $n(S) = 6$, we have

$$P(E) = \frac{3}{6} = \frac{1}{2}$$

EXAMPLE ii. An urn contains 2 green balls and 3 red balls. We select a ball at random. What is the probability of drawing a red ball?

Solution: We set up a sample space by numbering the green balls g_1, g_2 and the red balls r_1, r_2, r_3. Then a sample space is

$$S = \{g_1, g_2, r_1, r_2, r_3\}$$

The event, E, "drawing a red ball" is

$$E = \{r_1, r_2, r_3\}$$

Since $n(E) = 3$ and $n(S) = 5$ we have

$$P(E) = \frac{3}{5}$$

In an experiment we designate the set of outcomes in which we are interested as an event of the experiment. Those elements of the sample space that are not elements of the event in question are elements of another subset of the sample space called the **complementary event** of the first event. If E is an event, we denote its complementary event by E'. Thus if

$$S = \{1, 2, 3, 4, 5, 6\}$$

and
$$E = \{2, 4, 6\}$$

then
$$E' = \{1, 3, 5\}.$$

We observe that

$$E \cup E' = S \quad \text{and} \quad E \cap E' = \phi$$

We now prove

THEOREM 16.1. **If E is an event in a sample space S and E' is the complementary event of E, then**

$$\mathbf{P(E) + P(E') = 1}$$

Proof: Since E' is the complementary event of E,

$$E \cup E' = S \quad \text{and} \quad E \cap E' = \phi$$

Since E and E' are disjoint sets,

$$n(E) + n(E') = n(E \cup E') = n(S)$$

But

$$P(E) = \frac{n(E)}{n(S)} \quad \text{and} \quad P(E') = \frac{n(E')}{n(S)}$$

Then

$$P(E) + P(E') = \frac{n(E)}{n(S)} + \frac{n(E')}{n(S)}$$
$$= \frac{n(E) + n(E')}{n(S)}$$
$$= \frac{n(S)}{n(S)} = 1$$

EXAMPLE iii. The probability that a certain horse will win a race is $\frac{1}{4}$. What is the probability that the horse will not win the race?

Solution: The event "the horse will not win the race" is the complementary event of the event "the horse will win the race." Since the sum of the probabilities of two complementary events is 1, we have

$$P(E) + P(E') = 1$$
$$P(E') = 1 - P(E)$$
$$= 1 - \frac{1}{4} = \frac{3}{4}$$

The probability that the horse will not win the race is $\frac{3}{4}$.

Exercise 16.3

1. A die is tossed. Find the probability of the following events.
   a. A 6 showing
   b. An even number showing
   c. A prime number showing
   d. A number greater than 3 showing
2. Two coins are tossed. Find the probability of each of the following events.
   a. Two heads
   b. One head and one tail
3. A pair of dice is rolled. Find the probability of each of the following events.
   a. A sum of 7   d. A sum of 10
   b. A sum of 11   e. A sum of 8
   c. A 6 on one die   f. A number greater than 4 on both dice
4. A bag contains 10 balls, 2 red, 3 green, and 5 white. A ball is drawn

at random from the bag. Find the probability of each of the following events.

    a. Drawing a green ball

    b. Drawing a red ball

    c. Drawing a white ball

5. A card is drawn at random from a standard deck of 52 playing cards. Find the probability of each of the following events.

    a. A queen   d. A red jack

    b. A black card   e. A heart

    c. An ace   f. The king of spades

6. A bag contains 10 marbles, 5 white, 3 black, and 2 green. On two successive draws without replacing the marbles in the bag each time, a white and a black marble respectively are drawn. What is the probability that a green marble will be drawn on the third draw?

7. A letter is chosen from the word Mississippi. What is the probability of choosing a vowel?

8. A committee of three is selected from six men designated by A, B, C, D, E, and F.

    a. List the elements of a sample space

    b. What is the probability that C is on the committee?

    c. What is the probability that D is on the committee?

    d. What is the probability that C is not on the committee?

    e. What is the probability that D is not on the committee?

9. The probability that it will snow on a certain day in a certain city is $\frac{2}{5}$. What is the probability that it will not snow?

10. Of 1,000,000 income tax returns that come into a branch of the state income tax bureau, 800,000 are checked for arithmetic errors and 100,000 are checked thoroughly. If Fred Clark's return is sent to this branch, find the probability that

    a. It will be checked for arithmetic errors.

    b. It will be checked thoroughly.

    c. It will not be checked for arithmetic errors.

11. A bag contains 25 gum drops of which 10 are pink, 6 are red, 2 are black, and the rest are white. A gum drop is drawn at random from the bag. Find the probabilities of the following events.

    a. A red one is chosen   c. A white one is chosen

    b. A black one is chosen   d. A pink one is chosen

12. A hat contains twenty slips of paper numbered consecutively 1 through

20. A slip is drawn at random from the hat. What is the probability of each of the following events?
   a. Drawing 7
   b. Drawing an even number
   c. Drawing a number greater than 9
   d. Drawing a number less than 16
13. There are one hundred ping pong balls in a large bowl. Fifty are numbered with one of the numerals 1 through 50 inclusive and the rest are blank. One ball is drawn at random from the bowl. What is the probability of
   a. Drawing the ball numbered 50?
   b. Drawing a blank ball?
   c. Drawing a even numbered ball?
14. Three coins are tossed. Find the probability of each of the following events.
   a. Three heads
   c. Tails on at least one coin
   b. Heads on exactly one coin
   d. Heads on exactly two coins
15. Two burned-out bulbs are accidentally put in a box containing four good bulbs. It cannot be determined by looking which bulbs are good. When two bulbs are selected at random, what is the probability that
   a. One good bulb and one bad bulb are drawn?
   b. Both bulbs drawn are good?
16. A set of cards consists of 3 green and 4 red cards. Two cards are drawn at random. What is the probability of each of the following events?
   a. Two red cards are drawn
   b. Two green cards are drawn
   c. A red and a green card are selected
17. A poll indicates that the probability that Ramon Weltch will win the election for student body president is 0.7, while the probability that he will lose the election is 0.5. Is this possible? Why?
18. A student estimates that his probability of passing a mathematics course is 0.3. What is his probability of failing the course?
19. Two dice are rolled. What is the probability of not getting a double?
20. A box contains 5 bags of potatoe chips, 2 bags of popcorn, 3 bags of corn chips, and 2 bags of pretzels. All of the bags are identical. One bag is selected at random and put into a lunch box. What is the probability that it is not a bag of potatoe chips?

16.5. Probability of E or F

In computing probabilities, the relation of events to each other must be taken into consideration. If S is a sample space and E and F are two events of S, we have one of the following two cases: (1) E and F have no elements in common; or (2) E and F have some common elements.

If E and F are two events of a sample space and they have no elements in common, that is, they are disjoint sets, they are called **mutually exclusive events.**

If E and F are two events of a sample space S and we ask the probability of the event E or F, we are using "or" in the inclusive sense, that is we are asking that event E or event F or both events E and F occur.

Suppose we toss a pair of dice. A sample space for the experiment has thirty-six ordered pairs as outcomes:

$$S = \{(1, 1), (1, 2), (1, 3), (1, 4), (1, 5), (1, 6),$$
$$(2, 1), (2, 3), (3, 3), (2, 4)\ (2, 5), (2, 6),$$
$$(3, 1), (3, 2), (3, 3), (3, 4), (3, 5), (3, 6),$$
$$(4, 1), (4, 2), (4, 3), (4, 4), (4, 5), (4, 6),$$
$$(5, 1), (5, 2), (5, 3), (5, 4), (5, 5), (5, 6),$$
$$(6, 1), (6, 2), (6, 3), (6, 4), (6, 5), (6, 6)\ \}$$

Suppose we ask: "What is the probability of the event 'obtaining a sum of 7 or a sum of 11'?" We are interested in the event, E, "a sum of 7" or the event, F, "a sum of 11" or both of these events occurring. In set language this means that we are interested in the union of the two events E and F.

The event, E, "a sum of 7" is

$$E = \{(1, 6), (2, 5), (3, 4), (4, 3), (5, 2), (6, 1)\}$$

The event, F, "a sum of 11" is

$$F = \{(5, 6), (6, 5)\}$$

We see that $E \cap F = \phi$, that is, E and F are mutually exclusive events. The union of E and F is

$$E \cup F = \{(1, 6), (2, 5), (3, 4), (4, 3), (5, 2), (6, 1), (5, 6), (6, 5)\}$$

The probability of the event "a sum of 7 or a sum of 11", that is the probability of the event $E \cup F$ is

$$P(E \cup F) = \frac{n(E \cup F)}{n(S)}$$

$$= \frac{8}{36} = \frac{2}{9}$$

Since E and F are disjoint sets,

$$n(E \cup F) = n(E) + n(F)$$

Now

$$P(E \cup F) = \frac{n(E \cup F)}{n(S)}$$

$$= \frac{n(E)}{n(S)} + \frac{n(F)}{n(S)}$$

$$= P(E) + P(F)$$

In general, we have

THEOREM 3.2. If E and F are two events of a sample space S and E ∩ F = φ, then

$$\mathbf{P(E \cup F) = P(E) + P(F)}$$

Now for the same experiment let us find the probability of the event "a sum of 4 or a 3 on one of the dice." Again we are interested in one event or the other event or both events occurring at the same time. If E is the event "a sum of 4" and F is the event "a 3 on one of the dice", we are interested in $E \cup F$. Now

$$E = \{(1, 3), (2, 2), (3, 1)\}$$
$$F = \{(3, 1), (3, 2), (3, 3), (3, 4), (3, 5), (3, 6), (1, 3),$$
$$(2, 3), (4, 3), (5, 3), (6, 3)\}$$

In this case E and F are not mutually exclusive events, since they have some elements in common. Now

$$E \cup F = \{(1, 3), (2, 2), (3, 1), (3, 2), (3, 3), (3, 4), (3, 5), (3, 6),$$
$$(2, 3), (4, 3), (5, 3), (6, 3)\}$$

The probability of $E \cup F$ is

$$P(E \cup F) = \frac{n(E \cup F)}{n(S)}$$

$$= \frac{12}{36} = \frac{1}{3}$$

Notice that in this case

$$P(E \cup F) \neq P(E) + P(F)$$

since $P(E) = \frac{3}{36}$ and $P(F) = \frac{11}{36}$ and $\frac{3}{36} + \frac{11}{36} \neq \frac{12}{36}$.

This is true in general. If E and F are not mutually exclusive events, Theorem 3.2 does not hold.

Suppose we throw a die. A sample space is

$$S = \{1, 2, 3, 4, 5, 6\}$$

What is the probability of the event "an even number or an odd number showing"? We are interested in the probability of the event $E \cup F$, where E is the event "an even number showing" and F is the event "an odd number showing." Now

$$E = \{2, 4, 6\} \quad \text{and} \quad F = \{1, 3, 5\}$$

Since E and F are mutually exclusive events, that is, $E \cap F = \phi$, using Theorem 3.2 we have

$$P(E \cup F) = P(E) + P(F)$$

$$= \frac{3}{6} + \frac{3}{6} = 1$$

When rolling a die, the only numbers that can show are 1, 2, 3, 4, 5, and 6. All of these numbers are either odd or even. We see then that the occurrence of an even or an odd number showing when a die is rolled is a certainty. *If an event is certain to occur, its probability is 1.*

16.6. Probability of E and F

Suppose we again throw a pair of dice. What is the probability of the event "obtaining a sum of 10 and a 4 on one of the dice"? Now we are looking for the probability of E and F, where E is the event "a sum of 10" and F is the event "a 4 on one of the dice." In set language this means that we are looking for the probability of $E \cap F$.

The event E "a sum of 10" is

$$E = \{(4, 6), (5, 5), (6, 4)\}.$$

The event, F, "a 4 on one of the dice" is

$$F = \{(4, 1), (4, 2), (4, 3), (4, 4), (4, 5), (4, 6), (1, 4), (2, 4)$$
$$(3, 4), (5, 4), (6, 4)\}$$

Then

$$E \cap F = \{(6, 4), (4, 6)\}$$

We see that the probability of the event E and F, that is the probability of $E \cap F$ is

$$P(E \cap F) = \frac{n(E \cap F)}{n(S)}$$

$$= \frac{2}{36} = \frac{1}{18}$$

Now let us find the probability of the event "a sum of 10 and a sum of 12". We are again concerned with two events happening at the same time. Let us call A the event "a sum of 10" and B the event "a sum of 12."

Now

$$A = \{(4, 6), (5, 5), (6, 4)\} \quad \text{and} \quad B = \{(6, 6)\}$$

We see that A and B are mutually exclusive sets. That is $A \cap B = \phi$. Since $A \cap B = \phi$, $n(A \cap B) = 0$ and

$$P(A \cap B) = \frac{n(A \cap B)}{n(S)}$$

$$= \frac{0}{36} = 0$$

If two events have no elements in common, that is, if the two events are mutually exclusive, then the probability of both of these events happening at the same time is 0. We see then that *the probability of an event that cannot happen* is 0.

Exercise 16.4

1. A pair of dice is rolled. What is the probability of the following events?
   a. A sum of 7 or a 3 on one of the dice.
   b. A sum of 7 and a 3 on one of the dice.
   c. A sum of 12 or a sum of 8.
   d. A 4 one die and a sum that is an even number.
   e. A sum of 6 or a sum of 8.
2. A card is selected at random from a standard deck of 52 playing cards. What is the probability of each of the following events?
   a. A red card or a black card.
   b. An ace or a king.
   c. A deuce or a four.
   d. A jack, queen, or king.
3. Two coins are tossed. What is the probability of each of the following events?
   a. Two heads or two tails.
   b. Two heads or a head on the first coin.
4. Three coins are tossed. What is the probability of each of the following events?
   a. Three heads or heads on the second coin.
   b. Three tails or tails on the first coin.
   c. Exactly two heads or tails on the third coin.
   d. Exactly two tails or heads on the third coin.
5. A bag contains 25 jelly beans of which 10 are red, 6 are white, 8 are pink, and 1 is green. A jelly bean is drawn at random from the bag. What is the probability of each of the following events?
   a. A red or a white is drawn.
   b. A red or a pink is drawn.
   c. A green or a pink is drawn.
   d. A black is drawn.
6. Each card in a pack of twenty-six is labeled with a different letter of the English alphabet. One card is drawn at random. What is the probability of each of the following events?
   a. Drawing a card marked with the letters from the names Alice or Albert.
   b. Drawing a card marked with a letter from the names Joyce or Janice.

c. Drawing a card marked with a letter from the words "horse" or "house."

7. A set of cards consists of 3 green cards and 2 red cards. Two cards are drawn at random. What is the probability of drawing:
   a. One red card and one green card?
   b. Two green cards or two red cards?

8. The names of 2 Yorkshire terriers, 8 poodles, 4 scotties, and 6 boxers are written on identical cards and placed in a hat. A prize of a sweater is awarded by drawing at random a name from the hat. What is the probability that the winner will be:
   a. A poodle?
   b. A scottie?
   c. A Yorkshire terrier?
   d. A poodle or a Yorkshire terrier?
   e. A boxer or a scottie?

9. An employer has three positions to fill. He decides to fill them by selecting at random from five equally qualified applicants, X, Y, Z, T, and W. What is the probability that:
   a. Both X and Z will be chosen?
   b. That X or Y will be chosen?
   c. That both X and Y or both Z and T will be chosen?

10. A poll is taken among 50 students at a college on the question of whether grades should be eliminated. The results of the poll are shown in the table below. What is the probability that a student picked at random will be a
    a. A senior who opposes eliminating grades?
    b. A freshman or a sophomore?
    c. A junior who has no opinion on the subject?
    d. A junior or a senior who favors eliminating grades?

Response	Freshman	Sophomores	Juniors	Seniors	Total
Favor	4	2	5	2	13
Oppose	1	6	15	7	29
No Opinion	1	2	3	2	8
Total	6	10	23	11	50

Answers to Odd-Numbered Problems

CHAPTER 1

Exercise 1.1

1. {April, August}
3. {Nebraska, Nevada, New Hampshire, New Jersey, New Mexico, New York, North Carolina, North Dakota}
5. {California, Arizona, New Mexico, Texas}
7. {Sunday, Monday, Tuesday, Wednesday, Thursday, Friday, Saturday}
9. {Tuesday, Thursday}
11. {Sunday, Monday, Tuesday, Wednesday, Thursday}
13. $\{x \mid x$ is a state of the United States and x borders the Pacific Ocean$\}$
15. $\{x \mid x$ is one of the first six months of the year$\}$
17. $\{x \mid x$ is the 49th or 50th states of the United States$\}$
19. a; c; e; f; h; i
21. ϕ; { }
23. There are six possible ways. Two of them are $x \leftrightarrow a, y \leftrightarrow b, z \leftrightarrow c$; $x \leftrightarrow b$, $y \leftrightarrow c, z \leftrightarrow a$.
25. There are 24 possible ways. Two of them are: Mary $\leftrightarrow$ Joe, Jane $\leftrightarrow$ Charles, Bee $\leftrightarrow$ Walter, Ann $\leftrightarrow$ Oscar; Mary $\leftrightarrow$ Oscar, Jane $\leftrightarrow$ Joe, Bee $\leftrightarrow$ Charles, Ann $\leftrightarrow$ Walter.
27. $A = D$; $E = F$
29. a; b
31. a. Equal and equivalent; b. neither equal nor equivalent; c. equivalent; d. neither equal nor equivalent; e. neither equal nor equivalent.
33. A is equivalent to C
35. Two sets are equivalent if the elements of each set can be put into one-to-one correspondence with the elements of the other.

Exercise 1.2

1. $\{11, 12, 13\}$; $\{11\}$; $\{12\}$; $\{13\}$; $\{11, 12\}$; $\{11, 13\}$; $\{12, 13\}$; ϕ
3. {Marilyn, Sharon}; {Marilyn, Joan}; {Marilyn, Ruth}; {Sharon, Joan}; {Sharon, Ruth}; {Joan, Ruth}
5. a. B; b. ϕ; c. $\{0, 6\}$; d. $\{8, 10\}$
7. a. $\{1, 2, 3, 4\}$; $\{1\}$; $\{2\}$; $\{3\}$; $\{4\}$; $\{1, 2\}$; $\{1, 3\}$; $\{1, 4\}$; $\{2, 3\}$; $\{2, 4\}$; $\{3, 4\}$; $\{1, 2, 3\}$; $\{1, 2, 4\}$; $\{1, 3, 4\}$; $\{2, 3, 4,\}$; ϕ; b. $\{a, b, c\}$; ϕ; $\{a\}$;

$\{b\}; \{c\}; \{a, b\}; \{a, c\}; \{b, c\};$ c. $\{5, 7, 9, 11, 13\}; \phi; \{5\}; \{7\}; \{9\};$
$\{11\}; \{13\}; \{5, 7\}; \{5, 9\}; \{5, 11\}; \{5, 13\}; \{7, 9\}; \{7, 11\}; \{7, 13\};$
$\{9, 11\}; \{9, 13\}; \{11, 13\}; \{5, 7, 9\}; \{5, 7, 11\}; \{5, 7, 13\}; \{5, 9, 11\};$
$\{5, 9, 13\}; \{7, 9, 13\}; \{5, 11, 13\}; \{7, 9, 11\}; \{7, 11, 13\}; \{9, 11, 13\};$
$\{5, 7, 9, 11\}; \{5, 9, 11, 13\}; \{7, 9, 11, 13\}; \{5, 7, 9, 13\}; \{5, 7, 11, 13\}$

9. b; c; d
11. a. 2^{100};   b. 2^{26};   c. 2^4;   d. 2^7 e. 2^2
13. one
15. 56
17. a; c

Exercise 1.3

1. a.   Set of animals;   b.   Set of plants;   c.   Set of games

3. a. 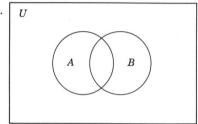   b.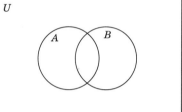

  c.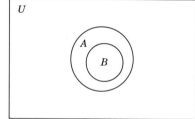

5.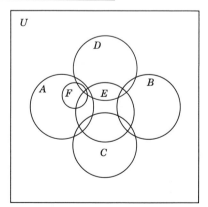

7.

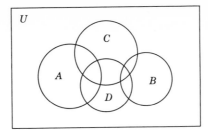

9.

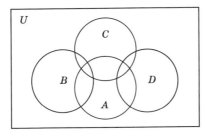

Exercise 1.4

1. a. 4; b. 5; c. 3
3. 2
5. Infinite b; i; Finite: a; c; d; e; f; g; h
7. a. 7; b. 5; c. 465; d. 50; e. 100; f. 25; g. 23
9. 10
11. The rule for correspondence is $n \leftrightarrow 3n$, n a natural number.

Exercise 1.5

1. a. 4; b. 7; c. 1; d. 6; e. 2; f. 0
3. a. ϕ; b. P; c. ϕ; d. $P \cap R$ or $P \cap Q$
5. a. A; b. B; c. B; d. A or B
7. a. The set of elementary school teachers who have visited Canada or Mexico;
   b. The set of elementary school teachers who have visited Canada and Mexico;
   c. The set of elementary school teachers who have not visited Canada;
   d. The set of elementary school teachers who have not visited Mexico.
9. a. U; b. U; c. A; d. A; e. ϕ; f. A; g. ϕ;
   h. B
11. a. $\{2, 4, 5, 6, 7, 9\}$; b. ϕ; c. $\{1, 2, 3, 4, 6, 8, 10\}$;
    d. $\{1, 3, 5, 7, 8, 9, 10\}$; e. $\{2, 3, 6, 5, 8, 9\}$; f. $\{1, 3, 8, 10\}$;
    g. U; h. $\{7\}$; i. $\{1, 2, 3, 4, 5, 6, 8, 9, 10\}$; j. $\{2, 4, 5, 6, 7, 9\}$
13. a. Introvert and antisocial; b. emotional and introvert;
    c. introvert and antisocial and emotional;

d. disliked and aggressive and emotional;
e. extrovert and disliked and aggressive.

Exercise 1.6

1. a. $\{(c, 9), (c, 7), (b, 9), (b, 7), (a, 9), (a, 7)\}$;
    b. $\{(9, c), (9, a), (9, b), (7, c), (7, b), (7, a)\}$;
    c. $\{(c, c), (c, b), (c, a), (b, c), (b, b), (b, a), (a, c), (a, a), (a, b)\}$;
    d. $\{(9, 9), (9, 7), (7, 9), (7, 7)\}$
3. b; d; f
5. a. $\{(a, 1)\}$;   b. $\{(1, p), (2, p), (3, p), (4, p), (1, y), (2, y), (3, y), (4, y)\}$;
    c. $\{(\text{red}, 3), (\text{red}, 4), (\text{green}, 3), (\text{green}, 4), (\text{blue}, 3), (\text{blue}, 4)\}$;
    d. $\{(w, 1), (w, 2), (m, 1), (m, 2), (f, 1), (f, 2)\}$

7.

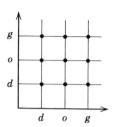

9. $5k$
11. 180
13. a. $\{(1, 1), (1, 2), (1, 3), (1, 4), (2, 1), (2, 2), (2, 3), (2, 4), (3, 1), (3, 2),$
    $(3, 3), (3, 4), (4, 1), (4, 2), (4, 3), (4, 4)\}$;

b. 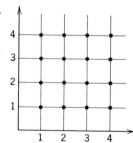   c. $\{(1, 2), (2, 2)\}$

15. a. $\{(1, 1), (1, 2), (1, 3), (1, 4), (1, 5), (2, 1), (2, 2), (2, 3), (2, 4), (2, 5),$
    $(3, 1), (3, 2), (3, 3), (3, 4), (3, 5), (4, 1), (4, 2), (4, 3), (4, 4), (4, 5), (5, 1),$
    $(5, 2), (5, 3), (5, 4), (5, 5)\}$;

b. 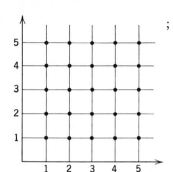   ;   c. $\{(2, 1), (4, 2)\}$

Exercise 1.7

3.   a. $\{0, 1, 2, 3, 4, 5, 6, 7, 8, 9\}$ or A;   b. $\{0, 2, 3, 4, 6, 9\}$;
     c. $\{0, 2, 3, 4, 6, 9\}$;   d. $\{6\}$;   e. ϕ;   f. $\{4, 8\}$

5.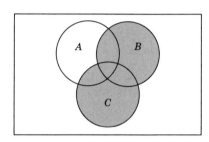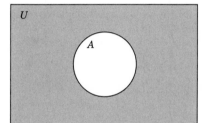

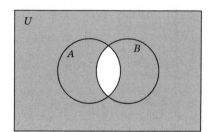

CHAPTER 2

Exercise 2.1

1.   a.   True;   b.   true;   d.   true;   e.   false;   f.   false
     i.   false;   j.   false
3.   a.   If the sky is blue, then the grass is green;   b.   If today is Monday, then yesterday was Sunday;   c.   If the train is late, then we shall be late to work;   d.   If all birds have wings, then all dogs are intelligent.

5. The following are listed with antecedent first and conclusion second.
   a. I go skiing; I have fun; b. He cooperates with his boss; He will probably get a promotion; c. A number is divisible by two; A number is even; d. You were born in 1952; You were born in leap year; e. Charles likes Elaine and Elaine likes Charles; Miracles do happen.

7. a. $\sim p$; b. $p \wedge q$; c. $q \to (\sim p)$; d. $p \vee q$

9. a. True; b. true; c. true; d. true; e. true; f. false; g. true; h. false; i. true; j. true; k. true

11. a. $p \vee (\sim q)$; b. $q \to p$; c. $p \wedge (\sim q)$; d. $(\sim p) \to q$

Exercise 2.2

1.
p	r	$p \vee r$	$p \to (p \vee r)$
T	T	T	T
T	F	T	T
F	T	T	T
F	F	F	T

3.
p	q	r	$q \vee r$	$p \to (q \vee r)$
T	T	T	T	T
T	T	F	T	T
T	F	T	T	T
T	F	F	F	F
F	T	T	T	T
F	T	F	T	T
F	F	T	T	T
F	F	F	F	T

5.
p	q	r	$\sim q$	$p \vee (\sim q)$	$[p \vee (\sim q)] \wedge r$
T	T	T	F	T	T
T	T	F	F	T	F
T	F	T	T	T	T
T	F	F	T	T	F
F	T	T	F	F	F
F	T	F	F	F	F
F	F	T	T	T	T
F	F	F	T	T	F

7.
p	q	$\sim p$	$\sim q$	$p \to q$	$(\sim q) \to (\sim p)$	$(p \to q) \to [(\sim q) \to (\sim p)]$
T	T	F	F	T	T	T
T	F	F	T	F	F	T
F	T	T	F.	T	T	T
F	F	T	T	T	T	T

9.

p	q	r	$p \wedge q$	$(p \wedge q) \vee r$
T	T	T	T	T
T	T	F	T	T
T	F	T	F	T
T	F	F	F	F
F	T	T	F	T
F	T	F	F	F
F	F	T	F	T
F	F	F	F	F

11. Tautology
13. Tautology
15. Not a tautology
17. Tautology
19. Not a tautology
21. a. $\sim[p \wedge (\sim p)]$;   b. $p \vee (\sim p)$;
    c. $[(p \rightarrow q) \wedge (q \rightarrow r)] \rightarrow (p \rightarrow r)$.
23. Symmetric
25. Symmetric; transitive
27. Reflexive; symmetric; transitive
29. Reflexive; symmetric
31. Reflexive; symmetric; transitive
33. 22; 27; 28; 30; 31

Exercise 2.3

1. a. If Tim is Alice's brother, then Alice is Tim's sister;   b. If Alice is not Tim's sister, then Tim is not Alice's brother;   c. If Tim is not Alice's brother then Alice is not Tim's sister.
3. a. If Jack oversleeps, then the alarm does not go off;   b. If the alarm goes off, then Jack does not oversleep;   c. If Jack does not oversleep, then the alarm goes off.
5. a. If Jay was born in leap year, then he was born on February 29;
   b. If Jay was not born on February 29, then he was not born in leap year;
   c. If Jay was not born in leap year, then he was not born on February 29.
7. a. If set B is equal to set A, then set A is equal to set B;   b. If set A is not equal to set B, then set B is not equal to set A;   c. If set B is not equal to set A, then set A is not equal to set B.
9. a. If a polygon is a regular pentagon, then it has five congruent sides;
   b. If a polygon does not have five congruent sides, then it is not a regular pentagon;   c. If a polygon is not a regular pentagon, then it does not have five congruent sides.

11.

p	q	$\sim(p \wedge q)$	$(\sim p) \vee (\sim q)$
T	T	F	F
T	F	T	T
F	T	T	T
F	F	T	T

13.

p	q	$p \to q$	$\sim p \vee q$
T	T	T	T
T	F	F	F
F	T	T	T
F	F	T	T

15.

p	q	$\sim[(\sim p) \vee q]$	$p \wedge (\sim q)$
T	T	F	F
T	F	T	T
F	T	F	F
F	F	F	F

Exercise 2.4

1. $q \to (\sim s)$ and q are true by hypothesis. Hence $(\sim s)$ is true by the law of detachment.

3. Since $(\sim p)$ is true, p is false. Since p is false and $p \vee q$ is true, q is true by the definition of a disjunction.

5. Since $p \to q$ and $q \to t$ and $t \to r$ are true, $p \to r$ is true by the law of syllogisms. Since $p \to r$ and p are true, r is true by the law of detachment.

7. Let p: $ABCD$ is a square; q: $ABCD$ is a rectangle; r: $ABCD$ is a rhombus. We have given $p \to q$, $p \vee r$ and $(\sim r)$. We are asked to prove q. Proof: Since $(\sim r)$ is true r is false. Since $p \vee r$ is true and r is false p is true by the definition of disjunction. Since $p \to q$ and p are true, q is true by the law of detachment.

9. Let p: $ABCD$ is a rectangle; q: $ABCD$ is a parallelogram; q: $ABCD$ is a quadrilateral. Given: $p \vee q$, $p \to q$ and $q \to r$. Prove $q \vee r$. Proof: Suppose $q \vee r$ is false. Then both q and r are false by definition of a disjunction. Since q is false and $p \to q$ is true, p must be false by definition of an implication. If p and q are both false, $p \vee q$ is false. But $p \vee q$ is true by hypothesis, hence we have a contradiction, $p \vee q$ both true and false. Hence our assumption is false and $q \vee r$ is true.

11. Let p: Mary is telling the truth; q: John is a true friend; r: Helen is not reliable. We have given $p \to q$, $r \to (\sim q)$. We are asked to prove $p \to (\sim r)$. Proof: Since $r \to (\sim q)$ is true, its contrapositive $q \to (\sim r)$ is true. Since $p \to q$ and $q \to (\sim r)$ are true, $p \to (\sim r)$ is true by the law of syllogisms.

13. Let p: The White Rabbit wears white gloves; q: The Queen of Hearts is giving a party; r: The Duchess is angry. We have given $p \to q$, $q \to r$ and

$(\sim r)$. We are asked to prove $(\sim p)$. Proof: Since $p \to q$ and $q \to r$ are true, $p \to r$ is true by the law of syllogisms. Since $p \to r$ is true, its contrapositive $(\sim r) \to (\sim p)$ is true. Since $(\sim r) \to (\sim p)$ and $(\sim r)$ are true, $(\sim p)$ is true by the law of detachment.

15. Let p: Gene advocates peace; q: Dick advocates war; r: Carrie marries Dick. We are given $p \to q$ and $q \to (\sim r)$. We are asked to prove $r \to (\sim p)$. Proof: Since $p \to q$ and $q \to (\sim r)$ are true, $p \to (\sim r)$ is true by the law of syllogisms. Since $p \to (\sim r)$ is true, its contrapositive $r \to (\sim p)$ is true.

CHAPTER 3

Exercise 3.1

1. a. $\{1, 3, 5, 7, 9\}$;   b. $\{11, 12, 16, 34, 56, 95\}$;
   c. $\{11, 22, 44, 66, 88, 101\}$;   d. $\{0, 4, 6, 7, 18, 45, 93\}$;
   e. $\{0, 7, 12, 16, 101, 404, 666\}$
3. X: 3; Y: 7; Z: 12
5. a. 10;   b. 54;   c. 17;   d. none;   e. 300;   f. 600
7. a. 10;   b. 15;   c. 14;   d. 20;   e. 14;   f. 21
9. a. 22;   b. 70;   c. 32;   d. 0;   e. 96;   f. 56
11. See Definition 3.2
13. Subtraction and division
15. The set of natural numbers
17. Addition; subtraction; multiplication; and division
19. a. 0;   b. 0;   c. 2;   d. 0
21. a. Symmetric;   b. addition;   c. substitution;   d. division;
    e. transitive;   f. subtraction;   g. multiplication
23. $\{0, 1, 2, 3, 4, 5, 6, 7\}$
25. $\{10, 11\}$
27. 0, 1, 2, 3, 4, 5
29. 0, 1, 2, 3, 4, 5, 6, 7

Exercise 3.2

1. a. Commutative property of addition;   b. identity property of addition;   c. associative property of addition;   d. identity property of addition;   e. commutative property of addition;   f. commutative property of addition.
3. a and e; c and d; because of the commutative property of addition
5. a; b; e; f
7. a; c

Exercise 3.3

1. a. 32;   b. 63;   c. 12;   d. 32;   e. 12;   f. 192;
    g. 552;   h. 63;   i. 18,544;   j. 5976;   k. 18,544;
    l. 552

3. a. · · ·   b. · · · ·   c. · · · · · ·
        · · ·   · · · ·
                      · · · ·

    d. · ·   e. · ·   f. · · ·
        · ·   · ·   · · ·
        · ·   · ·   · · ·
        · ·   · ·   · · ·
        · ·   · ·   · · ·
        · ·   · · ·
        · ·

5. a. $(8 \times 12) + (8 \times 26)$;   b. $(15 \times 98) + (15 \times 36)$;
    c. $(24 \times 38) + (45 \times 38)$;   d. $(67 \times 107) + (93 \times 107)$
7. a. 2;   b. 6;   c. 0;   d. 1;   e. 0, 1, 2, 3, 4, 5, 6;
    f. 0, 1, 2, 3;   g. 0, 1, 2, 3, 4, 5, 6, 7;   h. any whole number
    greater than 40
9. Because of the identity property of multiplication.
11. a. 2×12;   b. 4×6;   c. 6×4;   d. 8×3;
    e. 12×2;   f. 24×1
13. 20
15. a. 26;   b. 60;   c. 130;   d. 567;   e. 522;   f. 564;
    g. 312;   h. 228

Exercise 3.4

1. a. Walking 5 blocks north;   b. closing the window;
    c. taking off a sweater;   d. picking up the pencil;
    e. lowering the left hand;   f. erasing "4";
    g. uncorking the bottle
3. Division
5. $a(bc) = a(cb)$ commutative property of multiplication
          $= (cb)a$ commutative property of multiplication
7. $(a + b)(c + d) = (a + b)c + (a + b)d$ distributive property
                   $= ac + bc + ad + bd$ distributive property
9. $a = b$ hypothesis
    $ac = bc$ multiplication property of equality
    $ac = bd$ substitution property of equality

Exercise 3.5

1. a; b; d; e; f; h; k
3. a; b; e; h
5. a. $3 + 6 = 9$;   b. $7 + 5 = 12$;   c. $6 + 8 = 14$;
   d. $9 + 10 = 19$;   e. $12 + 14 = 26$;   f. $29 + 58 = 87$;
   g. $38 + 58 = 96$;   h. $29 + 88 = 117$;   i. $116 + 321 = 437$;
   j. $89 + 100 = 189$
7. a. $9 \times 6 = 54$;   b. $3 \times 9 = 27$;   c. $9 \times 4 = 36$;
   d. $8 \times 9 = 72$;   e. $54 \times 62 = 3348$;   f. $28 \times 15 = 420$;
   g. $14 \times 81 = 1134$;   h. $69 \times 12 = 828$
9. a. 2;   b. 16;   c. 150;   d. 3;   e. 7;   f. 6;
   g. 112;   h. 32
11. a. 4;   b. 6;   c. 15;   d. 26;   e. 12;   f. 16;
    g. 22;   h. 24

CHAPTER 4

Exercise 4.1

1. a. 24;   b. 218;   c. 11,203;   d. 2,130,069
3. a. 1510;   b. 1956;   c. 1969;   d. 1919;   e. 1776;
   f. 1488;   g. 1909;   h. 2764
5. a. MDCCLXXVI: b. MCMLXXXIX: c. MCMLII:
   d. MCMXIX: e. MCMXCIX: f. MMI: g. MMDIX:
   h. MLXVI
9. Addition, subtraction, multiplication, and repetition
11. a. MCDXXIX: b. MCDXCII: c. MDCCLXXV:
    d. MDCCCIII: e. MCMIII: f. MCMXIV

Exercise 4.2

1. a. 2^4;   b. 3^5;   c. 5^3;   d. 7^7;   e. 4^8;   f. 9^{10}
3. a. 10^3;   b. 10^5;   c. 10^6;   d. 10^8
5. a. 1;   b. 1;   c. 1;   d. 1;   e. 1;   f. 1
7. a. 4×10^0 or 4;   b. 4×10^3 or 4000;   c. 4×10^2 or 400;
   d. 4×10^1 or 40;   e. 4×10^6 or 4,000,000;
   f. 4×10^5 or 400,000
9. a. $(4 \times 10^2) + (8 \times 10^1) + (6 \times 10^0)$;
   b. $(1 \times 10^3) + (3 \times 10^2) + (9 \times 10^1) + (2 \times 10^0)$;
   c. $(8 \times 10^4) + (4 \times 10^3)$;   d. $(1 \times 10^4) + (2 \times 10^3) + (6 \times 10^0)$;
   e. 4×10^6;
   f. $(4 \times 10^6) + (3 \times 10^5) + (6 \times 10^4) + (9 \times 10^3) + (2 \times 10^2) +$
   $(1 \times 10^1) + (4 \times 10^0)$;

g. $(8 \times 10^9) + (2 \times 10^8) + (6 \times 10^7) + (1 \times 10^6) + (4 \times 10^3) +$
$(3 \times 10^2) + (1 \times 10^1) + (1 \times 10^0)$;

h. $(9 \times 10^9) + (1 \times 10^8) + (1 \times 10^6) + (1 \times 10^4) + (6 \times 10^3) +$
$(7 \times 10^2) + (7 \times 10^1) + (7 \times 10^0)$

11. a. 8; b. 81; c. 125; d. 144; e. 32; f. 225

Exercise 4.3

1. All of the following numerals are base-eight numerals. 1, 2, 3, 4, 5, 6, 7, 10, 11, 12, 13, 14, 15, 16, 17, 20, 21, 22, 23, 24, 25, 26, 27, 30, 31

3. $15_{\text{(eight)}}$

5. a. $31_{\text{(eight)}}$; b. $114_{\text{(eight)}}$; c. $105_{\text{(eight)}}$; d. $131_{\text{(eight)}}$;
   e. $176_{\text{(eight)}}$; f. $1233_{\text{(eight)}}$; g. $2012_{\text{(eight)}}$; h. $4477_{\text{(eight)}}$;
   i. $7661_{\text{(eight)}}$

7. a. 109; b. 131; c. 1451; d. 450; e. 151; f. 1583

9. a. ten; b. eight; c. twelve; d. five; e. seven;
   f. nine

11. All of the following are base-five numerals. 100, 101, 102, 103, 104, 110, 111, 112, 113, 114, 120, 121, 122, 123, 124, 130.

13. All of the following are base-five numerals. a. 12; b. 14; c. 31; d. 43; e. 100; f. 244; g. 1021; h. 2020; i. 10,312

15. a. $7777_{\text{(eight)}}$; b. $6666_{\text{(seven)}}$; c. $EEEE_{\text{(twelve)}}$

17. a. <; b. =; c. >; d. >; e. <; f. <

19. a. ten^1; b. seven^2; c. eight^3; d. twelve^1; e. five^3;
    g. nine^4

21. a. $3222_{\text{(four)}}$; b. $453_{\text{(seven)}}$; c. $1030_{\text{(six)}}$; d. $22,200_{\text{(three)}}$;
    e. $280_{\text{(nine)}}$; f. $176_{\text{(twelve)}}$

Exercise 4.4

1. All numerals are binary numerals. 1, 10, 11, 100, 101, 110, 111, 1000, 1001, 1010, 1011, 1100, 1101, 1110, 1111, 10,000, 10,001, 10,010, 10,011, 10,100.

3. a. 7; b. 15; c. 47; d. 375; e. 878; f. 3069

5. All numerals are octal numerals. a. 54; b. 411; c. 416; d. 777; e. 2570; f. 3153; g. 2461; h. 10,077

7. a. =; b. <; c. >; d. <; e. =

9. $10_{\text{(two)}}$; $10_{\text{(eight)}}$

11. $110_{\text{(two)}}$

13. a. $-, \wedge, \triangle, -\ominus, --, -\wedge, -\triangle, \wedge\ominus, \wedge -, \wedge\wedge, \wedge\triangle,$
    $\triangle\ominus, \triangle -, \triangle\wedge, \triangle\triangle, -\ominus\ominus$ b. 58; c. $-\ominus\ominus\ominus$;
    d. $\wedge - \ominus$

15. a. (1) 74T; (2) 2052; (3) 13,231; b. (1) T45; (2) 2725; (3) 21,433;
    c. (1) 1073; (2) 3427; (3) 24,230; d. (1) 10E3; (2) 3507; (3) 24,423;
    e. (1) 1139; (2) 3575; (3) 30,132; f. (1) 1149; (2) 3611; (3) 30,204

CHAPTER 5

Exercise 5.1

1. a. 30;   b. 600;   c. 40;   d. 600;   e. 300
3. All numerals are base-eight.   a. 140;   b. 66;   c. 1031

5.

+	0	1	2	3	4	5	6	7	8	9	T	E
0	0	1	2	3	4	5	6	7	8	9	T	E
1	1	2	3	4	5	6	7	8	9	T	E	10
2	2	3	4	5	6	7	8	9	T	E	10	11
3	3	4	5	6	7	8	9	T	E	10	11	12
4	4	5	6	7	8	9	T	E	10	11	12	13
5	5	6	7	8	9	T	E	10	11	12	13	14
6	6	7	8	9	T	E	10	11	12	13	14	15
7	7	8	9	T	E	10	11	12	13	14	15	16
8	8	9	T	E	10	11	12	13	14	15	16	17
9	9	T	E	10	11	12	13	14	15	16	17	18
T	T	E	10	11	12	13	14	15	16	17	18	19
E	E	10	11	12	13	14	15	16	17	18	19	1T

7. All numerals are in base-twelve.   a. 1092;   b. 1033;   c. 18E0
9. All numerals are in base-twelve.   a. 1146;   b. 1T87;   c. 2184
11. All numerals are in base-twelve.   a. 13E6;   b. 1868;   c. 2173
13. All numerals are in base-two.   a. 10,000;   b. 11,000;   c. 11,110
15. All numerals are in base-eight.   a. 121;   b. 267;   c. 246
17. All numerals are in base-two.   a. 1;   b. 11;   c. 11
19. All numerals are in base-eight.   a. 5544;   b. 1124;   c. 3402
21. All numerals are in base-two.   a. 1;   b. 1001;   c. 1011

Exercise 5.2

1. a. Multiplicative identity property;   b. distributive property;
   c. basic sums;   d. commutative property of multiplication.
3. a. $q = 5; r = 1$;   b. $q = 19; r = 5$;   c. $q = 14; r = 0$;
   d. $q = 30; r = 4$;   e. $q = 9; r = 34$;   f. $q = 46; r = 1$
5. All numerals are in base-eight.   a. 32,110;   b. 42,125;
   c. 64,070
7. All numerals are in base-eight.   a. 2, R-60;   b. 20, R-16;
   c. 12, R-46
9. All numerals are in base-five.   a. 202;   b. 141;   c. 242
11. All numerals are in base-five.   a. 44, R-1;   b. 124, R-1   c. 204

13.

×	0	1	2	3	4	5	6	7	8	9	T	E
0	0	0	0	0	0	0	0	0	0	0	0	0
1	0	1	2	3	4	5	6	7	8	9	T	E
2	0	2	4	6	8	T	10	12	14	16	18	1T
3	0	3	6	9	10	13	16	19	20	23	26	29
4	0	4	8	10	14	18	20	24	28	30	34	38
5	0	5	T	13	18	21	26	2E	34	39	42	47
6	0	6	10	16	20	26	30	36	40	46	50	56
7	0	7	12	19	24	2E	36	41	48	53	5T	65
8	0	8	14	20	28	34	40	48	54	60	68	74
9	0	9	16	23	30	39	46	53	60	69	76	83
T	0	T	18	26	34	42	50	5T	68	76	84	92
E	0	E	1T	29	38	47	56	65	74	83	92	T1

15. All numerals are in base-twelve.   a.   35,056;   b.   46,433;
    c.   87,936
17. All numerals are in base-twelve.   a.   2T, R-16;   b.   113, R-47
19. All numerals are in base-two.   a.   1010;   b.   10,000,111
21. $60 \times 64 = 3840$
23. All numerals are in base-eight. $16 \times 24 = 430$
25. All numerals are in base-twelve. $42 \times 26 = T50$

CHAPTER 6

Exercise 6.1

1. a; c; e; f; g; h
3. a. $\{91\}$;   b. $\{4, 5, 6, 7, \ldots\}$;   c. $\{0, 1, 2, \ldots, 13\}$;   d. $\{4\}$;
   e. $\{0, 1, 2, \ldots, 10\}$;   f. $\{0, 1, 2, \ldots, 17\}$;   g. $\{0, 1, 2, 3, 4\}$;
   h. $\{13, 14, 15, 16, \ldots\}$
5. ϕ
7. a.   Twelve plus six equals eighteen;
   b.   Seventeen is less than three plus some number;
   c.   Three times some number is eighteen;
   d.   The sum of five and some number equals twenty-six;
   e.   Seven plus some number is greater than nine;
   f.   Three less than some number is less than or equal to eight;
   g.   Three times some number plus six is greater than or equal to eleven;
   h.   The sum of seven times some number and five equals twenty-six.
9. $\{0, 1, 2, 3, 4\}$
11. $\{11, 12, 13, \ldots\}$
13. $\{36, 37, 38, \ldots\}$

15. a. {3};   b. φ;   c. {3};   d. {3}
17. a. {0, 1, 2};   b. φ;   c. {0, 1, 2};   d. {0, 1, 2}
19. {12$_{(eight)}$}
21. Less than 120 pounds.
23. 92
25. 26

Exercise 6.2

1. {10, 11, 12, . . . , 16}

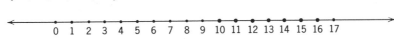

3. {8, 9, 10, 11, 12}

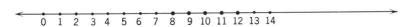

5. {0, 1, 2, 3, 10, 11, 12, . . .}

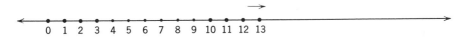

7. {0, 1, 2, 3, 4, 5, 12, 13, 14, . . .}

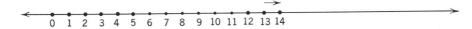

9. {0, 1, 2, . . . , 9}

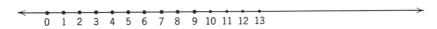

11. {2, 3, . . . , 10}

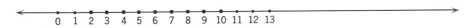

13. {16, 17, 18, . . .}

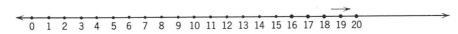

15. {2, 3, 4, 5, . . .}

1.

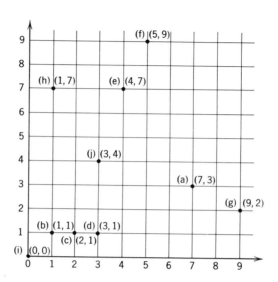

3. $\{(1, 2), (2, 1)\}$

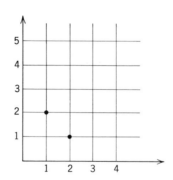

5. $\{(4, 4), (4, 5), (5, 2), (5, 3), (5, 4), (5, 5)\}$

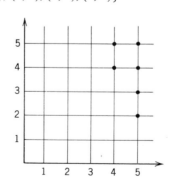

7. $\{(0, 8), (1, 7), (2, 6), (3, 5), (4, 4), (5, 3), (6, 2), (7, 1), (8, 0)\}$

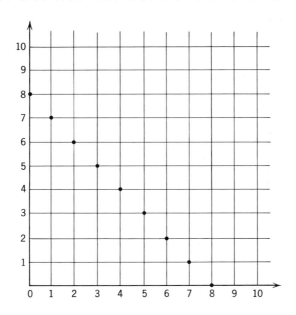

9. $\{(0, 0), (0, 1), (0, 2), \ldots, (0, 9), (1, 0), (1, 1), (1, 2), \ldots,$
$(1, 8), (2, 0), (2, 1), (2, 2), \ldots, (2, 7), (3, 0), (3, 1), (3, 2), \ldots,$
$(3, 6), (4, 0), (4, 1), \ldots, (4, 5), (5, 0), (5, 1), \ldots, (5, 4), (6, 0), (6, 1), (6, 2),$
$(6, 3), (7, 0), (7, 1), (7, 2), (8, 0), (8, 1), (9, 0)\}$

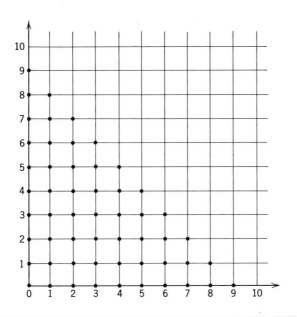

11. $\{(0, 0), (0, 1), (0, 2), \ldots, (0, 8), (1, 0), (1, 1), \ldots,$
    $(1, 6), (2, 0), (2, 1), (2, 2), (2, 3), (3, 0), (3, 1)\}$

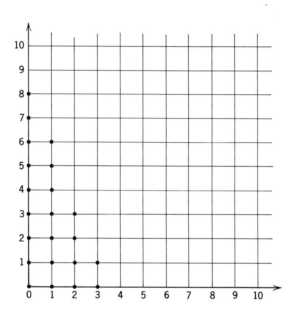

13. $\{(0, 0), (0, 1), (0, 2), (0, 3), (1, 0), (1, 1), (1, 2), (1, 3), (2, 0), (2, 1), (2, 2),$
    $(3, 0), (3, 1), (3, 2), (4, 0), (4, 1), (5, 0), (5, 1), (6, 0), (7, 0)\}$

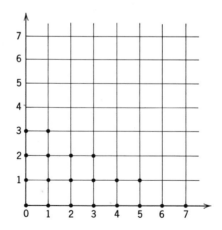

15. $\{(0, 0), (0, 1), (0, 2), (0, 3), (0, 4), (1, 0), (1, 1), (1, 2), (1, 3), (2, 0), (2, 1),$
    $(3, 0)\}$

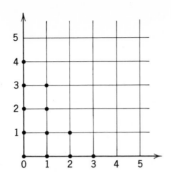

17. $\{(0, 8), (1, 6), (2, 4), (3, 2), (4, 0)\}$

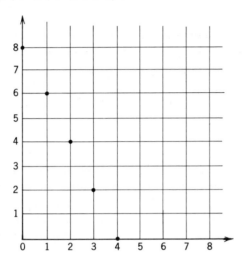

19. $\{(0, 9), (2, 6), (4, 3), (6, 0)\}$

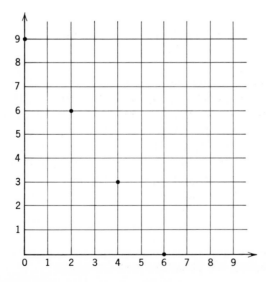

CHAPTER 7

Exercise 7.1

3. a. 16;   b. 49;   c. 144;   d. 225;   e. 625;   f. 10,000
5. a. 2×9; $n = 9$;   b. 2×28; $n = 28$;   c. 2×57; $n = 57$;
   d. 2×128; $n = 128$;   e. 2×250; $n = 250$;
   f. 2×926; $n = 926$;   g. 2×4000; $n = 4000$;
   h. 2×7161; $n = 7161$;   i. $2 \times 384,199$; $n = 384,199$
7. $\{14, 16, 18, 20, 22, 24, 26, 28, 30, 32, 34, 36, 38, 40, 42, 44, 46, 48\}$
9. a. E;   b. E;   c. O;   d. E
11. Odd
13. 1
15. a. Addition and multiplication;   b. multiplication
17. Let $2p + 1$ and $2s + 1$ represent any two odd numbers. Then

$$
\begin{aligned}
(2p + 1)(2s + 1) &= (2p + 1)(2s) + (2p + 1)(1) && \text{Distributive property} \\
&= 4ps + 2s + 2p + 1 && \text{Distributive property} \\
&= 2(2ps + s + p) + 1 && \text{Distributive property} \\
&= 2n + 1 && 2ps + s + p \text{ is a whole} \\
& && \text{number } n \text{ by the} \\
& && \text{closure properties}
\end{aligned}
$$

19. Let $2p$ be any even number and $(2s + 1)$ be any odd number. Then

$$
\begin{aligned}
2p + (2s + 1) &= (2p + 2s) + 1 && \text{associative property of addition} \\
&= 2(p + s) + 1 && \text{distributive property} \\
&= 2n + 1 && p + s \text{ is a whole number } n \text{ by the} \\
& && \text{closure property of addition}
\end{aligned}
$$

Exercise 7.2

1. A natural number greater than 1 is a prime number if it has exactly two divisors.
3. $\{2, 3, 4, 5, 6, \ldots\}$
5. The primes less than 200 are: 2, 3, 5, 7, 11, 13, 17, 19, 23, 29, 31, 37, 41, 43, 47, 53, 59, 61, 67, 71, 73, 79, 83, 89, 97, 101, 103, 107, 109, 113, 127, 131, 137, 139, 149, 151, 157, 163, 167, 173, 179, 181, 191, 193, 197, 199
7. a. $2^3 \times 3^2$;   b. $2^2 \times 17$;   c. $2^2 \times 47$;   d. $2^2 \times 109$
9. a. 0; 3; 6; 9;   b. 1; 4; 7; 10;   c. 2; 5; 8; 11; 14;   d. disjoint
11. a. 1; 2; 3; 4; 6; 12;   b. 1; 17;   c. 1; 2; 3; 6; 9; 18;
   d. 1; 29;   e. 1; 2; 3; 4; 6; 9; 12; 18; 36;   f. 1; 3; 5; 9; 15; 45
13. a. $1^2 + 4^2$;   b. $4^2 + 5^2$;   c. $2^2 + 3^2$;   d. $4^2 + 9^2$;
   e. $5^2 + 8^2$;   f. $5^2 + 6^2$
15. 1; 3; 7; 9

Exercise 7.3

1. a. 2 and 3;   b. 5 and 73;   c. 2 and 3;   d. 2 and 101
3. a. $2^2 \times 3^2$;   b. $2^3 \times 7$;   c. $2^2 \times 3 \times 7$;   d. $2^2 \times 5^2$;
      e. $2^4 \times 7$;   f. 5^3;   g. $7 \times 2^2 \times 5^2$;   h. $2^6 \times 13$;
      i. $3^3 \times 37$;   j. 2×443
5. b; c; d; f; g; h; j
7. b; h
9. a. 421;   b. 23;   c. 11;   d. 11;   e. 61;   f. 337
11. The answers are given in the form of ordered pairs (x, y). The first component gives the value of x and the second component gives the value of y.
    a. $(2, 0)$, $(5, 0)$, $(8, 0)$, $(1, 1)$, $(4, 1)$, $(7, 1)$, $(0, 2)$, $(3, 2)$, $(6, 2)$, $(9, 2)$, $(2, 3)$, $(5, 3)$, $(8, 3)$, $(1, 4)$, $(4, 4)$, $(7, 4)$, $(0, 5)$, $(3, 5)$, $(6, 5)$, $(9, 5)$, $(2, 6)$, $(5, 6)$, $(8, 6)$, $(1, 7)$, $(4, 7)$, $(7, 7)$, $(0, 8)$, $(3, 8)$, $(6, 8)$, $(9, 8)$, $(2, 9)$, $(5, 9)$, $(8, 9)$;   b. $(0, 0)$, $(0, 2)$, $(0, 4)$, $(0, 6)$, $(0, 8)$, $(1, 0)$, $(1, 2)$, $(1, 4)$, $(1, 6)$, $(1, 8)$, $(2, 0)$, $(2, 4)$, $(2, 6)$, $(2, 8)$, $(3, 0)$, $(3, 2)$, $(3, 4)$, $(3, 6)$, $(3, 8)$, $(4, 0)$, $(4, 2)$, $(4, 4)$, $(4, 6)$, $(4, 8)$, $(5, 0)$, $(5, 2)$, $(5, 4)$, $(5, 6)$, $(5, 8)$, $(6, 0)$, $(6, 2)$, $(6, 4)$, $(6, 6)$, $(6, 8)$, $(7, 0)$, $(7, 2)$, $(7, 4)$, $(7, 6)$, $(7, 8)$, $(8, 0)$, $(8, 2)$, $(8, 4)$, $(8, 6)$, $(8, 8)$, $(9, 0)$, $(9, 2)$, $(9, 4)$, $(9, 6)$, $(9, 8)$;   c. $(2, 0)$, $(5, 0)$, $(8, 0)$, $(0, 2)$, $(3, 2)$, $(6, 2)$, $(9, 2)$, $(1, 4)$, $(4, 4)$, $(7, 4)$, $(2, 6)$, $(5, 6)$, $(8, 6)$, $(0, 8)$, $(3, 8)$, $(6, 8)$, $(9, 8)$;   d. $(5, 0)$, $(3, 2)$, $(2, 3)$, $(1, 4)$, $(0, 5)$, $(9, 5)$, $(8, 6)$, $(7, 7)$, $(6, 8)$, $(5, 9)$, $(4, 1)$
13. 7; 7^2; 7^3; . . .
15. a. 1;   b. 1;   c. 1;   d. 1;   e. 1;   f. 1.
17. When the sum of the digits of its numeral is divisible by 9. When the sum of the digits of its duodecimal numeral is divisible by 11.
19. A number is divisible by 7 when the sum of the digits in its octal numeral is divisible by 7.

Exercise 7.4

1. a. 8;   b. 5;   c. 33;   d. 6
3. 1; 2; 3; 4; 6; 12
5. a. 1;   b. 2;   c. 2;   d. 3
7. a. 6;   b. 9;   c. 18;   d. n
9. 1
11. a. 36;   b. 78;   c. 252;   d. 108;   e. 375;   f. 3333
13. pq

CHAPTER 8

Exercise 8.1

1. a. Y;   b. L;   c. U;   d. R;   e. D;   f. J;
      g. C;   h. X;   i. E

3. a. True; b. false; c. false; d. false
5. Six
7. A point

9. a.
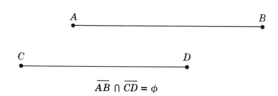

$$\overleftrightarrow{AB} \cap \overline{CD} = \phi$$

b.
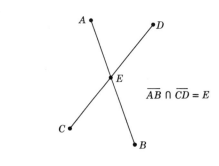

$$\overleftrightarrow{AB} \cap \overline{CD} = E$$

$$\overleftrightarrow{AB} \cap \overline{CD} = \overline{DB}$$

c.

11. Three
13. b; c; f; g

Exercise 8.2

1. a. P; b. R; c. $\overleftrightarrow{PT}$; d. $\overline{PR}$; e. $\overrightarrow{QS}$; f. $\overline{QR}$
3. $\overrightarrow{PR} \cup \overrightarrow{RQ}$ is $\overrightarrow{PQ}$

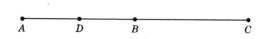

5. a. $\angle APC, \angle DPB, \angle APD, \angle CPB$; b. E, F, K;
   c. $\angle CPA, \angle CPB$; d. $\overrightarrow{PD}; \overrightarrow{PB}$; e. P
7. a. True; b. false; c. true; d. true; e. true;
   f. true
9. An angle is the union of two rays with a common endpoint.
11. a. $\angle BAD$; b. $\angle JBL$; c. $\angle JAM$; d. $\angle PAD$;
    e. B, I, J; f. J, K, L, M; g. $\overrightarrow{BI}, \overrightarrow{BJ}$; h. $\overrightarrow{CK}, \overrightarrow{CL}$

Exercise 8.3

1. a; e; g; h
3. 0
5. a. R, S, T; b. $\overline{RS}$; $\overline{ST}$; $\overline{RT}$; $\angle SRT$; $\angle RTS$; $\angle TSR$
7. a. P; Z; Y; R; b. P; S; T; Q; c. Q; F; S; L; T; H;
   d. B; R; Y; J; Z; E; e. $\{A, D, G, I, K, C\}$

9. a.

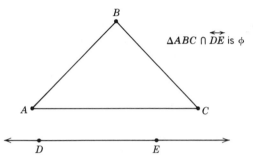

$\triangle ABC \cap \overleftrightarrow{DE}$ is ϕ

b.

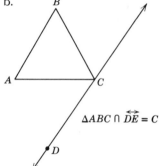

$\triangle ABC \cap \overleftrightarrow{DE} = C$

c.

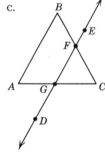

$\triangle ABC \cap \overleftrightarrow{DE}$ is the set consisting of two points G and F

11. a; b; e; h; i
13. A triangle is a polygon with three sides.
15. A quadrilateral is a polygon with four sides.
17. a. $\{G, C\}$; b. E; D; F; c. J; I

Exercise 8.4

9. a. $\overline{CD} \cong \overline{PQ}$; b. $\overline{AB} \cong \overline{EF}$

11.

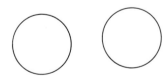

Answers to Odd-Numbered Problems   455

13.

15. a; e; f

Exercise 8.5

1. a. Cylinder;   b. cone;   c. triangular prism;
      d. rectangular prism (cube);   e. pyramid;   f. prism;
      g. pyramid;   h. sphere;   i. cone

CHAPTER 9

Exercise 9.1

1. a. $(2, 8)$; $\frac{2}{8}$;   b. $(11, 12)$; $\frac{11}{12}$;   c. $(3, 6)$; $\frac{3}{6}$;   d. $(6, 16)$; $\frac{6}{16}$;
    e. $(4, 5)$; $\frac{4}{5}$;   f. $(1, 2)$; $\frac{1}{2}$;   g. $(6, 9)$; $\frac{6}{9}$;   h. $(2, 3)$; $\frac{2}{3}$;
    i. $(9, 14)$; $\frac{9}{14}$
3. In the following the numerator is given first.   a. 1; 2;   b. 3; 4;
    c. 3; 5;   d. 5; 8;   e. 1; 3;   f. 9; 16;   g. 12; 79;
    h. 18; 18;   i. 42; 100
5. $\frac{7}{12}$
7. $\frac{30}{46}$
9. a. $\frac{3}{9}$;   b. $\frac{5}{12}$;   c. 3;   d. 100;   e. 16;   f. 72;
    g. 0;   h. $\frac{7}{7}$

Exercise 9.2

1. a. $\frac{4}{20}, \frac{5}{25}, \frac{6}{30}$;   b. $\frac{20}{24}, \frac{25}{30}, \frac{30}{36}$;   c. $\frac{4}{16}, \frac{5}{20}, \frac{6}{24}$;   d. $\frac{4}{64}, \frac{5}{80}, \frac{6}{96}$;
    e. $\frac{20}{28}, \frac{25}{35}, \frac{30}{42}$
3. a. $\{\frac{5}{6}, \frac{10}{12}, \frac{15}{18}, \frac{20}{24}, \ldots\}$;   b. $\{\frac{7}{8}, \frac{14}{16}, \frac{21}{24}, \frac{28}{32}, \ldots\}$;
    c. $\{\frac{6}{7}, \frac{12}{14}, \frac{18}{21}, \frac{24}{28}, \ldots\}$;   d. $\{\frac{1}{4}, \frac{2}{8}, \frac{3}{12}, \frac{4}{16}, \ldots\}$;
    e. $\{\frac{5}{16}, \frac{10}{32}, \frac{15}{48}, \frac{20}{64}, \ldots\}$;   f. $\{\frac{3}{4}, \frac{6}{8}, \frac{9}{12}, \frac{12}{16}, \ldots\}$;
    g. $\{\frac{2}{3}, \frac{4}{6}, \frac{6}{9}, \frac{8}{12}, \ldots\}$;   h. $\{\frac{5}{1}, \frac{10}{2}, \frac{15}{3}, \frac{20}{4}, \ldots\}$
5. a. $\frac{5}{7}$;   b. $\frac{11}{12}$;   c. $\frac{9}{20}$;   d. $\frac{2}{3}$;   e. $\frac{7}{11}$;   g. $\frac{4}{15}$
7. a. 75;   b. 16;   c. 192;   d. 120;   e. 1152;   f. 195
9. a. $\frac{3}{5}$;   b. $\frac{1}{10}$;   c. $\frac{8}{3}$;   d. $\frac{8}{9}$
11. a. $\{\frac{2}{3}, \frac{4}{6}, \frac{6}{9}, \frac{8}{12}, \ldots\}$;   b. $\{\frac{1}{5}, \frac{2}{10}, \frac{3}{15}, \frac{4}{20}, \ldots\}$;
    c. $\{\frac{9}{4}, \frac{18}{8}, \frac{27}{12}, \frac{36}{16}, \ldots\}$;   d. $\{\frac{7}{4}, \frac{14}{8}, \frac{21}{12}, \frac{28}{16}, \ldots\}$;
    e. $\{\frac{3}{100}, \frac{6}{200}, \frac{9}{300}, \frac{12}{400}, \ldots\}$;   f. $\{\frac{2}{15}, \frac{4}{30}, \frac{6}{45}, \frac{8}{60}, \ldots\}$
13. a. Twelfths;   b. half;   c. two;   d. nine
15. a. 18;   b. 12;   c. 15;   d. 7;   e. 4;   f. 25
17. a. $\frac{96}{144}$;   b. $\frac{45}{144}$;   c. $\frac{14}{144}$;   d. $\frac{81}{144}$;   e. $\frac{132}{144}$;   f. $\frac{4}{144}$
19. Yes

CHAPTER 10

Exercise 10.1

1. a. $>$;   b. $<$;   c. $>$;   d. $=$;   e. $<$;   f. $>$;
   g. $>$;   h. $=$;   i. $<$;   j. $>$

3.

5. a. $\{1, 2, 3, 4, 5, 6\}$;   b. $\{1, 2\}$;   c. $\{3, 4, 5, \ldots\}$;
   d. $\{0, 1, 2\}$

7. a. $\frac{1}{2}$;   b. $\frac{5}{6}$;   c. $\frac{5}{3}$;   d. $\frac{51}{100}$;   e. same;   f. $\frac{75}{100}$;
   g. $\frac{5}{17}$;   h. same

9. Phil Jones

Exercise 10.2

1. a. $\frac{3}{4}$;   b. $\frac{3}{4}$;   c. $\frac{7}{5}$;   d. $\frac{3}{5}$;   e. $\frac{5}{8}$;   f. $\frac{5}{6}$;   g. $\frac{1}{2}$;
   h. $\frac{1}{4}$

3. a. $\frac{7}{8}$;   b. $\frac{29}{24}$;   c. $\frac{77}{48}$;   d. $\frac{47}{24}$

5. a. $\frac{1}{2}$;   b. $\frac{1}{2}$;   c. $\frac{1}{5}$;   d. $\frac{1}{4}$

7. a. $\{\frac{5}{3}\}$;   b. $\{\frac{1}{8}\}$;   c. $\{\frac{1}{3}\}$;   d. $\{\frac{7}{16}\}$

9. a. $\frac{7}{12}$;   b. $\frac{13}{24}$;   c. $\frac{15}{16}$;   d. $\frac{5}{1}$

11. a. Commutative property of addition;   b. additive identity property;
   c. associative property of addition;
   d. commutative property of addition;
   e. commutative property of addition.

13. a. 4;   b. 6;   c. 8;   d. 24;   e. 120;   f. 240;
   g. 54;   h. 300

15. $\frac{91}{60}$ hours or 1 hour 31 minutes

17. Gain; $\frac{17}{40}$ of a point.

19. $\frac{19}{60}$

Exercise 10.3

1. a. $\frac{1}{6}$;   b. $\frac{1}{15}$;   c. $\frac{2}{5}$;   d. $\frac{9}{4}$;   e. $\frac{19}{28}$;   f. $\frac{2}{5}$

3. a. $\frac{6}{5}$;   b. $\frac{4}{3}$;   c. $\frac{8}{3}$;   d. $\frac{27}{19}$;   e. $\frac{7}{8}$;   f. $\frac{1}{1}$;   g. $\frac{7}{12}$;
   h. $\frac{5}{2}$

5. a. $\frac{1}{2}$;   b. $\frac{3}{2}$;   c. $\frac{5}{8}$;   d. $\frac{3}{4}$;   e. $\frac{125}{32}$;   f. $\frac{7}{4}$;   g. $\frac{4}{3}$;
   h. $\frac{3}{4}$

7. a. $\{\frac{1}{2}\}$;   b. $\{\frac{4}{5}\}$;   c. $\{\frac{6}{1}\}$;   d. $\{\frac{2}{1}\}$;   e. $\{\frac{12}{1}\}$;
   f. $\{\frac{21}{8}\}$

9. a. $\frac{3}{5} \cdot \frac{2}{3}$;   b. $\frac{1}{3}$;   c. $\frac{1}{3} \cdot \frac{2}{3} + \frac{1}{3} \cdot \frac{5}{6}$;   d. $(\frac{3}{2} \cdot \frac{1}{3}) \cdot \frac{2}{5}$

11. An example is: $(\frac{2}{3} \div \frac{5}{6}) \div \frac{1}{4} = \frac{16}{5}$ but $\frac{2}{3} \div (\frac{5}{6} \div \frac{1}{4}) = \frac{1}{5}$

---

13. An example is: $(\frac{2}{3} - \frac{1}{2}) \div \frac{1}{4} = \frac{2}{3}$ and $(\frac{2}{3} \div \frac{1}{4}) - (\frac{1}{2} \div \frac{1}{4}) = \frac{2}{3}$

15. a. $\frac{5}{24}$; b. $\frac{13}{24}$

17. $\frac{100}{37}$

19. a. $\{\frac{1}{1}\}$; b. $\{\frac{69}{50}\}$; c. $\{\frac{9}{2}\}$; d. $\{\frac{25}{12}\}$

Exercise 10.4

1. a. $\frac{7}{8}$; b. $\frac{1}{2}$; c. $\frac{3}{9}$; d. $\frac{17}{16}$; e. $\frac{242}{15}$; f. $\frac{115}{8}$;
   g. $\frac{267}{12}$; h. $\frac{481}{25}$

3. a. $\frac{\frac{2}{1}}{\frac{3}{1}}$; b. $\frac{\frac{5}{1}}{\frac{8}{1}}$; c. $\frac{\frac{3}{1}}{\frac{16}{1}}$; d. $\frac{\frac{27}{1}}{\frac{28}{1}}$

5. a. $7\frac{1}{4}$; b. $5\frac{5}{8}$; c. $3\frac{1}{4}$; d. $5\frac{3}{5}$; e. $1\frac{19}{50}$; f. $5\frac{17}{24}$;
   g. $25\frac{1}{3}$; h. $19\frac{3}{5}$

7. a; d; e; f

9. $\frac{1}{10}$

11. $\frac{1}{2}$

13. a. $\frac{1}{3}$; b. $\frac{5}{6}$; c. $\frac{8}{9}$; d. $\frac{a}{b}$

Exercise 10.5

1. a; d; e; f

3. a. $\frac{21}{500}$; b. $\frac{1}{4}$; c. $\frac{11}{8}$; d. $\frac{946}{25}$; e. $\frac{29}{8}$; f. $\frac{801}{200}$;
   g. $\frac{143}{125}$; h. $\frac{7}{8}$; i. $\frac{41}{25}$; j. $\frac{1111}{125}$

5. a. $\frac{25}{100}$; 0.25; b. $\frac{5}{10}$; 0.5; c. $\frac{625}{1000}$; 0.625; d. $\frac{48}{100}$; 0.48;
   e. $\frac{6}{10}$; 0.6; f. $\frac{4375}{10000}$; 0.4375; g. $\frac{85}{1000}$; 0.085; h. $\frac{984}{1000}$; 0.984

7. a. 0.17; b. 0.11; c. 0.83; d. 0.27; e. 0.42;
   f. 0.67; g. 0.14; h. 0.89

9. a. One hundred fifteen and three tenths;
   b. one hundred thirty-four thousandths;
   c. one thousand three hundred forty-five and one hundred forty-five hundred thousandths;
   d. fourteen and four hundred sixty-five thousandths;
   e. forty-five and sixty-seven ten thousandths;
   f. seventy and three hundred sixteen ten thousandths

CHAPTER 11

Exercise 11.1

1. a. $^-8$; b. $^+5$; c. $^+7$; d. $^-11$; e. 0; f. $^-100$;
   g. $^+16$; h. $^+139$

3. a. $^+56$; b. $^+73$; c. 0; d. $^+23$; e. $^+56$; f. $^+102$;
   g. $^+14$; h. $^+788$

5. a. $=$; b. $<$; c. $>$; d. $=$; e. $>$; f. $<$;
   g. $<$; h. $>$

7. a; b; d; e; f

9. d
11. φ

Exercise 11.2

1. a. $^+3$;   b. $^+3$;   c. $^-7$;   d. $^-23$;   e. $^-11$;   f. $^-58$;
   g. $^-48$;   h. $^-11$
3. a. $\{^+6\}$;   b. $\{^-12\}$;   c. $\{^-2\}$;   d. $\{^+5\}$;   e. $\{^+6\}$;
   f. $\{^+1\}$;   g. $\{^-12\}$;   h. $\{^+11\}$
5. Positive
7. Negative
9. a. $^+7$;   b. $^-1$;   c. $^-21$;   d. $^-21$;   e. $^-29$;   f. $^-4$
11. a. $^+19$;   b. $^-86$;   c. $^-69$;   d. $^+86$;   e. $^-89$;   f. $^+91$
13. $^-120$
15. \$603
17. $24°$

Exercise 11.3

1. a. $^+21$;   b. $^-72$;   c. $^-45$;   d. $^+66$;   e. $^-608$;
   f. $^-1088$;   g. $^+832$;   h. $^+1296$;   i. $^+437$;   j. $^-2592$
3. a. $^+6$;   b. $^-9$;   c. $^-3$;   d. $^+21$;   e. $^+13$;   f. $^-64$;
   g. $^+35$;   h. $^-39$
5. a. $^+3$; commutative property of multiplication;
   b. $^+6$; distributive property;
   c. $^-8$; associative property of multiplication;
   d. $^+3$; identity element;   e. $^+7$; distributive property
7. a. Positive;   b. negative;   c. negative;   d. positive;
   e. positive
9. a. $=$;   b. $=$;   c. $=$;   d. $=$;   e. $=$
11. Positive

CHAPTER 12

Exercise 12.1

1. b; c; d; e; f; h
3. a. $\frac{4}{-8}$;   b. $\frac{3}{-5}$;   c. $\frac{7}{11}$;   d. $\frac{-16}{-17}$;   e. $\frac{-1}{100}$;   f. $\frac{4}{-7}$;
   g. $\frac{-5}{-9}$;   h. $\frac{12}{16}$
5. a. $(^-6, 8)$;   b. $(^-13, ^-12)$;   c. $(8, 24)$;   d. $(^-35, 8)$;
   e. $(^-20, 4)$;   f. $(^-29, 40)$
7. a. $(6, 4)$;   b. $(^-4, ^-24)$;   c. $(^-2, ^-8)$;   d. $(^-21, 20)$;
   e. $(15, 32)$;   f. $(4, 10)$
9. $(13, ^-6)$; $(13, ^-6)$

Exercise 12.2

1. a. $(3, 4)$;   b. $(^-2, ^-2)$;   c. $(4, ^-3)$;   d. $(5, ^-2)$;
   e. $(^-2, 3)$;   f. $(2, 9)$

3. a. $(^-4, ^-3)$;   b. $(2, ^-3)$;   c. $(^-8, 5)$;   d. $(2, 1)$;
   e. $(^-12, ^-7)$;   f. $(^-3, ^-4)$

5. $\frac{-2}{7}; \frac{2}{7}; -\frac{2}{7}$

7. Yes

9. $(1, 1) \cdot (a, b) = (1 \cdot a, 1 \cdot b)$ Definition 12.4

        $= (a, b)$ Multiplicative identity property for integers

11. $(a, b) + (c, d) = (ad + bc, bd)$ Definition 12.3

            $= (bc + ad, bd)$ Commutative property of addition
                                   of integers

            $= (cb + da, db)$ Commutative property of multiplication
                                     of integers

            $= (c, d) + (a, b)$ Definition 12.3

13. $[(a, b) \cdot (c, d)] \cdot (e, f) = (ac, bd) \cdot (e, f)$ Definition 12.4

                  $= ([ac]e, [bd]f)$ Definition 12.4

                  $= (a[ce], b[df])$ Associative property of
                                         multiplication of integers

                  $= (a, b) \cdot (ce, df)$ Definition 12.4

                  $= (a, b) \cdot [(c, d) \cdot (e \cdot f)]$ Definition 12.4

15. $(a, b) \cdot [(c, d) + (e, f)] = (a, b) \cdot (cf + de, df)$ Definition 12.3

                         $= (a[cf + de], b[df])$ Definition 12.4

$(a, b) \cdot (c, d) + (a, b) \cdot (e, f) = (ac, bd) + (ae, bf)$ Definition 12.4

                    $= ([ac][bf] + [bd][ae], [bd][bf])$

                                               Definition 12.3

                  $= (acbf + bdae, b^2df)$ Commutative
                                               and associative
                                               properties of
                                               multiplication
                                               of integers

                  $= (acf + dae, bdf)$ Definition 12.2

                  $= (a[cf + de], b[df])$ Associative and
                                               commutative
                                               properties of
                                               multiplication
                                               of integers and
                                               distributive
                                               property

Hence $(a, b) \cdot [(c, d) + (e, f)] = (a, b) \cdot (c, d) + (a, b) \cdot (e, f)$

Exercise 12.3

1. a. $<$; b. $>$; c. $=$; d. $>$; e. $>$; f. $>$; g. $>$;
   h. $>$
3. a. $\frac{-1}{7}$, b. $\frac{3}{8}$; c. $\frac{3}{4}$; d. $\frac{10}{9}$; e $\frac{4}{-35}$; f. $\frac{-3}{4}$
5. $\frac{4}{30}$ or $\frac{2}{15}$
7. $ab > 0$
9. $ac < bc$

Exercise 12.4

1. a. No inverses; no additive identity; b. no
3. a. No additive inverses; b. no
5. a. No additive identity; no multiplicative inverses; no closure for addition;
   b. no

CHAPTER 13

Exercise 13.1

1. a; b; e; f
3. a. R; b. ϕ; c. R; d. R; e. Q; f. S; g. Q;
   h. S
5. No; $(\sqrt{2})^2 = 2$
7. a; c; e
9. a. True; b. If r is a real number, then it is an irrational number;
   c. false

CHAPTER 14

Exercise 14.1

1. a. 0; b. 2; c. 2; d. 1; e. 2; f. 0

3.

+	0	1	2	3	4	5	6
0	0	1	2	3	4	5	6
1	1	2	3	4	5	6	0
2	2	3	4	5	6	0	1
3	3	4	5	6	0	1	2
4	4	5	6	0	1	2	3
5	5	6	0	1	2	3	4
6	6	0	1	2	3	4	5

5. All properties are satisfied.
7. a. {5}; b. {5}: c. {4}; d. {4}; e. {5}; f. {4};
   g. {2}; h. {3}.
9. a. 6; b. 7; c. 2; d. 8; e. 0; f. 4; g. 0;
   h. 11
11. M-10 is not satisfied; 2 does not have a multiplicative inverse.

Exercise 14.2

1. a. Satisfies all properties; b. satisfies all properties; c. does not
   satisfy the multiplicative inverse property; d. does not satisfy the
   multiplicative inverse property.
3. 2; 3; 4; 6; 8; 9; and 10
5. one

Exercise 14.3

1. a. A group; b. not a group; set not closed under the operation and
   there is no identity element; c. not a group; every element does not
   have an inverse; d. not a group; zero does not have an inverse;
   e. a group; f. not a group; 0, 2, 3, 4, 6, 8, 9, and 10 do not have inverses; g. a group.

3. Second term

First term

*	R_0	R_1	R_2	V_1	V_2	V_3
R_0	R_0	R_1	R_2	V_1	V_2	V_3
R_1	R_1	R_2	R_0	V_2	V_3	V_1
R_2	R_2	R_0	R_1	V_3	V_1	V_2
V_1	V_1	V_3	V_2	R_0	R_2	R_1
V_2	V_2	V_1	V_3	R_1	R_0	R_2
V_3	V_3	V_2	V_1	R_2	R_1	R_0

R_0, R_1, R_2, V_1, V_2, and V_3 are defined as shown below:

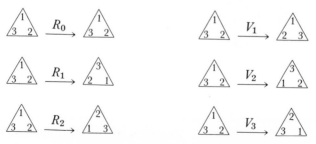

This system is a group, but not a commutative group.

5. a. R_3;   b. R_0;   c. H;   d. R_3;   e. R_2;   f. H;
    g. V;   h. R_1

7. a.

$*$	I	R	S	T
I	I	R	S	T
R	R	S	T	I
S	S	T	I	R
T	T	I	R	S

b. group; c. yes

9. $(x * y) * y^{-1} = x * (y * y^{-1})$ Associative property
    $= x * e$ Inverse property
    $= x$ Identity property

11. If $y^{-1} * x^{-1}$ is the inverse of $x * y$ we must show that $(x * y) * (y^{-1} * x^{-1}) = e$
    and $(y^{-1} * x^{-1}) * (x * y) = e$.
    $(x * y) * (y^{-1} * x^{-1}) = [(x * y) * y^{-1}] * x^{-1}$ associative property
    $= [x * (y * y^{-1})] * x^{-1}$ associative property
    $= (x * e) * x^{-1}$ inverse property
    $= x * x^{-1}$ identity property
    $= e$ inverse property
    That $(y^{-1} * x^{-1}) * (x * y) = e$ is shown in the same manner.

CHAPTER 15

Exercise 15.1

1. 3 units
3. 2 in.
5. 3ft; 1 yd.
7. Acute: a; e; right: c; f; obtuse: d; straight: b
9. a. $30°$;   b. $180°$;   c. $90°$;   d. $150°$;   e. $10°$;   f. $90°$
11. Trapezoids: a; b; c; d; e; f; g; h; parallelograms: a; b; d; f; g; h; rectangles:
    a; b; f; g; squares: a; g
13. a. $110°$;   b. $90°$;   c. $75°$;   d. $20°$;   e. $30°$;   f. $25°$
15. a. $80°$;   b. $135°$;   c. $90°$;   d. $40°$;   e. $105°$;
    f. $60°$;   g. $50°$;   h. $30°$;   i. $92°$
17. a; c; d; f
19. a. one;   b. one

Exercise 15.2

1. a. 12 in.;   b. 240 ft;   c. 68 yd;   d. 29 ft;   e. 28 ft;
    f. 50 yd
3. a. 19 ft;   b. 18 ft;   c. $7\frac{1}{2}$ ft;   d. 96 in.

5. 240 ft; 80 yd
7. 170 in.
9. $5\frac{1}{4}$ ft
11. 14 in.
13. 15 in.; 15 in.; 20 in.
15. 1.4 in.

Exercise 15.3

1. a. base: $\overline{AC}$; altitude: $\overline{BD}$;   b. base: $\overline{RS}$; altitude: $\overline{PQ}$;
   c. base: $\overline{XY}$; altitude: $\overline{WZ}$
3. 7 ft
5. 72 sq in.
7. 56
9. 678.24 sq ft

Exercise 15.4

1. 45 cu in.
3. 576 cu ft
5. 200.96 cu in.
7. $11,498\frac{2}{3}$ cu ft
9. The volume is multiplied by 27

CHAPTER 16

Exercise 16.1

1. 12
3. 16
5. 48
7. 3,628,800 or 10!
9. 6
11. a. 720;   b. 6720;   c. 72;   d. $\frac{1}{30}$;   e. 30;   f. 12
13. 40,320
15. 372

Exercise 16.2

1. a. $\{1, 2, 3, 4, 5, 6\}$;   b. $\{1, 2\}$;   c. $\{1, 3, 5\}$;   d. $\{2, 3, 5\}$;
   e. $\{3, 6\}$
3. a. $\{(1, 1), (1, 2), (1, 3), (1, 4), (1, 5), (1, 6), (2, 1), (2, 2), (2, 3), (2, 4),$
   $(2, 5), (2, 6), (3, 1), (3, 2), (3, 3), (3, 4), (3, 5), (3, 6), (4, 1), (4, 2), (4, 3),$
   $(4, 4), (4, 5), (4, 6), (5, 1), (5, 2), (5, 3), (5, 4), (5, 5), (5, 6), (6, 1), (6, 2),$
   $(6, 3), (6, 4), (6, 5), (6, 6)\}$;   b. $\{(1, 6), (2, 5), (3, 4), (4, 3), (5, 2),$
   $(6, 1)\}$;   c. $\{(6, 5), (5, 6)\}$;   d. $\{(1, 5), (1, 6), (2, 5), (2, 6), (3, 5),$

$(3, 6), (4, 5), (4, 6), (5, 1), (5, 2), (5, 3), (5, 4), (5, 5), (5, 6), (6, 1), (6, 2),$
$(6, 3), (6, 4), (6, 5), (6, 6)\}$

5. a. $\{ABC, ABD, ABE, ABF, ACD, ACE, ACF, ADE, ADF, AEF, BCD,$
   $BCE, BCF, BDE, BDF, BEF, CDE, CDF, CEF, DEF\};$ b. $\{ABC,$
   $ABD, ABE, ABF, ACD, ACE, ACF, ADE, ADF, AEF\};$ c. $\{ABC,$
   $ABD, ABE, ABF\}$

7. a. 8;   b. $\{HHH, HHT, HTH, HTT, THH, THT, TTH, TTT\};$
   c. $\{HHT, HTH, THH\};$ d. $\{HHH, HHT, HTH, HTT\}$

9. a. $\{2\} \cup \{3\} \cup \{4\};$ b. $\{HT\} \cup \{TH\};$
   c. $\{(6, 4)\} \cup \{(4, 6)\} \cup \{(5, 5)\}$

Exercise 16.3

1. a. $\frac{1}{6};$ b. $\frac{1}{2};$ c. $\frac{1}{2};$ d. $\frac{1}{2}$

3. a. $\frac{1}{6};$ b. $\frac{1}{18};$ c. $\frac{11}{36};$ d. $\frac{1}{12};$ e. $\frac{5}{36};$ f. $\frac{1}{9}$

5. a. $\frac{1}{13};$ b. $\frac{1}{2};$ c. $\frac{1}{13};$ d. $\frac{1}{26};$ e. $\frac{1}{4};$ f. $\frac{1}{52}$

7. $\frac{4}{11}$

9. $\frac{3}{5}$

11. a. $\frac{6}{25};$ b. $\frac{2}{25};$ c. $\frac{7}{25};$ d. $\frac{2}{5}$

13. a. $\frac{1}{100};$ b. $\frac{1}{2};$ c. $\frac{1}{4}$

15. a. $\frac{8}{15};$ b. $\frac{2}{5}$

17. No; $0.5 + 0.7 \neq 1$

19. $\frac{5}{6}$

Exercise 16.4

1. a. $\frac{5}{12};$ b. $\frac{1}{18};$ c. $\frac{1}{6};$ d. $\frac{5}{36};$ e. $\frac{5}{18}$

3. a. $\frac{1}{2};$ b. $\frac{1}{2}$

5. a. $\frac{16}{25};$ b. $\frac{18}{25};$ c. $\frac{9}{25};$ d. 0

7. a. $\frac{3}{5};$ b. $\frac{2}{5}$

9. a. $\frac{3}{10};$ b. $\frac{9}{10};$ c. $\frac{3}{5}$

Index

Circle, diameter of, 244
   radius of, 244
Circular reasoning, 218
Circumference, 393–395
Circumscribed, 393
Closure property, of addition, 79, 285, 324–325,
     340, 359, 362, 366
   of multiplication, 84, 291, 330, 332, 340, 361,
     362, 366
Collinear points, 222
Common denominator, 282
Common factor, 210
Common fraction, 261
Common multiple, 214
Commutative group, 370
Commutative operation, 26
Commutative property, of addition, 79, 285,
    324–325, 340, 360, 362, 366
   of multiplication, 85, 291, 331, 332, 361, 362,
    366
Complement, of an angle, 386
   of a set, 31
Complementary angles, 386
Complementary events, 423
Complete factorization, 200, 206–207
Complex fraction, 303–304
Composite numbers, 194–195
Compound statements, 45–51
Concentric circles, 252
Conclusion, 47
Cone, 253, 256
   base of, 255
   lateral surface of, 256
   vertex of, 256
   volume of, 410
Congruence, 242–243
Congruent, 243
   angles, 245
   triangles, 246–249
Conjunction, 45–46, 172
Consecutive primes method, 206–207
Constructing congruent angles, 245
   triangles, 247–249
Contrapositive, 57
Converse, 57–58
Coordinate, 73, 276, 318
Coplanar lines, 223
Corresponding angles, 247
   sides, 247
Counting numbers, 3, 69
Cube, 253
Curve, 219
   simple closed, 235–240

Cylinder, 255–256
   base of, 256
   element of, 256
   volume of, 409

Decagon, 237–238
Decimal, fraction(s), 309–314
   non-terminating, 313
   rational numbers, 307–309
   system of numeration, 109–110
   terminating, 312
Deduction, 61
Deductive system, 61
Degree, 384
Denominator, 263
   common, 282
   least common, 287–288
Dense, 300, 347
Density property of rational numbers, 299–302,
    347–348
Diameter, of circle, 244
   of sphere, 256
Difference, 94–95, 283, 284, 326, 346
Digit, 109, 110
   value of, 110
Direct proof, 62–64
Disjoint sets, 16, 29
Disjunction, 46–47, 173
Distance, 381
Distributive property, 291, 332, 340, 362, 366
   of division over addition, 100
   of division over subtraction, 100
   of intersection over union, 40
   of multiplication over addition, 87–89, 291,
    332, 340, 362, 363
   of multiplication over subtraction, 99
   of union over intersection, 40
Dividend, 155
Divisibility property of a sum, 202
Divisibility rules, 202–206
   for 2, 203–204
   for 3, 204–205
   for 5, 204
   for 9, 205
   for 10, 204
Division, 96–98
   algorithm, 154–162, 211
   and zero, 97–98
   fractions as symbols for, 302–304
   modulo 5, 363–364
   of integers, 332–333
   of nonnegative rational numbers, 293–295

Identity element, 369
   of addition, 80, 285, 324–325, 340, 360, 362, 366
   of multiplication, 87, 92, 291, 331, 332, 340, 362, 366
Implication(s), 47–49
   derived, 57–59
Improper fractions, 266, 304
Improper subset, 10
Inclusive "or", 46
Indirect proof, 64–66
Inequality, 164
Infinite set, 22–23
Inscribed, 393
Integer(s), 316–336
   absolute value of, 318–319
   addition of, 321–325
   and the number line, 317–318
   and whole numbers, 333–334
   division of, 332–333
   multiplication of, 328–332
   negative, 317
   nonnegative, 317
   opposite of, 318
   ordering, 319
   positive, 317
   subtraction of, 325–326
Interior, 232, 236
Intersection of sets, 28–30
Inverse, 57–58
Inverse operations, 93
Irrational numbers, 355, 356
   points, 355
Isosceles triangle, 386

Lattice, 36, 179
   plane, 179–180
   points, 36, 180
Law of contradiction, 54
Law of detachment, 60
Law of excluded middle, 54
Law of substitution, 60
Law of syllogism, 55
Least common denominator, 287–288
Least common multiple, 213–215
Length, 381
Line(s), 219–220
   coplanar, 223
   intersecting, 223
   parallel, 223
   segment, 219
      measurement of, 380–383
   skew, 224

Logic, 43–68
Logical reasoning, 44
Logical system, 61
Logically equivalent, 59
Lowest terms, see Simplest form

Mathematical sentence, 163
Mathematical system(s), 358–379
Measure, 380, 382, 384
Measurement, 382, 383–386
   of angles, 383–386
   of geometric figures, 380–412
   of line segments, 380–383
   standard units of, 383
Member of an equation, 72
Member of a set, 1
Mixed numerals, 304–305
Modulo-five system, 358–364
Modulus, 362
Modus Ponens, 60
Multiple, 13, 156, 213
   common, 214
   least common, 213–215
Multiplication, algorithm, 147–154
   identity element of, 87, 291
   inverse operation of, 93
   modulo 5, 358, 361–363
   of integers, 328–332
   of nonnegative rational numbers, 289–293
   of rational numbers, 338
   of whole numbers, 82–90
   properties of, 84–91, 291–293, 330–332, 361–362
   property of equality, 71
   property of zero, 91–92
   sentence, 98
Multiplicative identity, 87, 92, 290, 291, 331, 332, 340, 362, 366
Multiplicative inverse, 292, 340, 362, 366
Mutually exclusive, 427

n!, 416
Natural number(s), 3, 69
Negation, 49
Noncollinear points, 222
Nonnegative rational numbers, 275–315
   addition of, 281–283
   density property of, 299–302
   division of, 293–295
   equality of, 277–278
   graphs of, 276
   multiplication of, 289–293
   ordering, 278–281

null, *see* Empty set
number property, 22
operations, 25–30, 38–41
overlapping, 17
product, 34–37
replacement, 3, 165
solution, 165–170
subset of, 9–10
truth, see Solution set
union of, 25–28
universal, 15–16
well-defined, 1
Set-builder notation, 3
Set operations, 38–41
Sieve of Eratosthenes, 195–197
Simple closed curve, 235–240, 254
Simplest form, 269–271
Solid geometric figures, 252–258
Solution, 165
Solution sets, 165–169
    graphing, 175–177
Space, 219
    separation of, 226–228
Sphere, 253, 256–257
    center of, 256
    diameter of, 256
    radius of, 256
    volume of, 410
Square, 387
Square numbers, 185–186
Square root, 353
Standard units of measure, 382
Statements, 44
    compound, 45–51
    equivalent, 59
    logically equivalent, 59
Straight angle, 386
Subset(s), 9–12
    improper, 10
    number of, 11–12
    proper, 10
Subtraction, 94–96
    alogrithm, 140–145
    modulo 5, 363–364
    of integers, 325–326
    of nonnegative rational numbers, 283–284
    of rational numbers, 346–347
    properties of, 98–100
    property of equality, 71
    sentence, 96
Subtraction property, 71, 141
Sucessor, 70
Sum, 78, 281, 338

Supplement, 386
Supplementary angles, 386
Symmetric property, 52, 70
Symmetry, 370
Systems of numeration, 103–134
    accidentals of, 104
    ancient, 103–104
    base of, 104
    binary, 127–130
    duodecimal, 122–125
    Egyptian, 103–107
    Hindu-Arabic, 109–110, 116
    octal, 116–122
    place-value with bases other than ten,
        115–134
    principles of, 103
    Roman, 103, 105–107

Tautology, 54–56
Theorem(s), 44, 62, 217
    about whole numbers, 91–93
Transformation group, 373
Transitive property, 52, 70, 72
Transversal, 391
Trapezoid, 387
    area of, 404–405
Triangle(s), 238–239
    acute, 386
    angles of, 238
    area of, 402–403
    classification of, 386
    congruent, 246–249
    constructing congruent, 248–249
    equiangular, 386
    equilateral, 386
    isosceles, 386
    obtuse, 386
    right, 386
    scalene, 386
Triangular numbers, 185–189
Trichotomy principle, 72
Truth set, see Solution set
Truth table, 45
Truth value, 44

Undefined terms, 218
Undefined words, 43–44
Union of sets, 25–28
Unit of measurement, 380, 382, 383
Universal set, 15–16
Universe, *see* Universal set